Advanced Research on Marine Geology and Sedimentology

Advanced Research on Marine Geology and Sedimentology

Guest Editors

Zhuangcai Tian
Shaotong Zhang

MDPI

Basel • Beijing • Wuhan • Barcelona • Belgrade • Novi Sad • Cluj • Manchester

Guest Editors
Zhuangcai Tian
China University of Mining and Technology
Xuzhou
China

Shaotong Zhang
Ocean University of China
Qingdao
China

Editorial Office
MDPI AG
Grosspeteranlage 5
4052 Basel, Switzerland

This is a reprint of the Special Issue, published open access by the journal *Water* (ISSN 2073-4441), freely accessible at: https://www.mdpi.com/journal/water/special_issues/8PV9BJMGP2.

For citation purposes, cite each article independently as indicated on the article page online and as indicated below:

Lastname, A.A.; Lastname, B.B. Article Title. *Journal Name* **Year**, *Volume Number*, Page Range.

ISBN 978-3-7258-5907-8 (Hbk)
ISBN 978-3-7258-5908-5 (PDF)
https://doi.org/10.3390/books978-3-7258-5908-5

Cover image courtesy of Zhuangcai Tian

Contents

Preface

This Special Issue, "Advanced Research on Marine Geology and Sedimentology", explores the dynamic processes shaping our ocean floors, from coastal areas to the deep sea. Its scope spans wave–seabed interactions, shoreline changes, ocean circulation, and sediment properties, showcasing a blend of observational, modeling, and technological studies.

Aiming to advance our fundamental understanding of marine systems and provide insights for sustainable resource management and coastal disaster mitigation, this collection is a response to the growing need to decipher the ocean's role in Earth's history and climate, propelled by new technologies and environmental challenges.

This Special Issue is directed to researchers, students, and professionals in the fields of marine geosciences, oceanography, and environmental policy, offering valuable findings for both scientific advancement and practical application.

Zhuangcai Tian and Shaotong Zhang
Guest Editors

water

Editorial

Advanced Research on Marine Geology and Sedimentology

Zhuangcai Tian [1,*] and Shaotong Zhang [2]

1 State Key Laboratory of Intelligent Construction and Healthy Operation and Maintenance of Deep Underground Engineering, China University of Mining and Technology, Xuzhou 221116, China
2 Frontiers Science Center for Deep Ocean Multispheres and Earth System, Key Lab of Submarine Geosciences and Prospecting Techniques, MOE, College of Marine Geosciences, Ocean University of China, Qingdao 266100, China; shaotong@ouc.edu.cn
* Correspondence: zhuangcaitian@cumt.edu.cn

1. Introduction

The ocean floor is a vast, uncharted territory, rich with geological and sedimentological secrets waiting to be uncovered. Advanced research in marine geology and sedimentology is pivotal for understanding the Earth's history, climate change, and the formation of natural resources [1].

Utilizing cutting-edge technology, such as deep-sea submersibles, autonomous underwater vehicles, and advanced sonar systems, scientists delve into the abyss to map the seafloor, study tectonic plate movements, and collect samples of sediment and rock [2,3]. These studies provide insights into the geological processes that shape our planet, including the formation of underwater volcanoes, the shifting of oceanic trenches, and the creation of abyssal plains [4–6].

Sedimentology, the study of sediments, complements this research by examining the layers of sediment on the ocean floor. These layers act as a historical archive, recording changes in sea level, ocean currents, and past climates [7]. By analyzing the composition and structure of these sediments, researchers can reconstruct ancient environments and track the evolution of marine ecosystems [2,8,9].

2. Overview of the Special Issue

The papers collected in the Special Issue "Advanced Research on Marine Geology and Sedimentology" are diverse, including nine research papers.

- Effect of Relative Wavelength on Excess Pore Water Pressure in Silty Seabeds with Different Initial Consolidation Degrees (Contribution #1).
- Distribution of Excess Pore Water Pressure in Layered Seabed Induced by Internal Solitary Waves (Contribution #2).
- Seasonal Circulation Characteristics of Oceanic System in the Beibu Gulf Based on Observations and Numerical Simulations (Contribution #3).
- Assessment of Storm Surge Disaster Response Capacity in Chinese Coastal Cities Using Urban-Scale Survey Data (Contribution #4).
- Analysis of Shoreline Change in Huizhou-Shanwei Region (China) from 1990 to 2023 (Contribution #5).
- Adaptive Penetration Unit for Deep-Sea Sediment Cone Penetration Testing Rigs: Dynamic Modeling and Case Study (Contribution #6).

 https://doi.org/10.3390/w17213095

- Investigating the Element Geochemical Behavior and Provenance of Surface Sediments in the Offshore Area of Sierra Leone, Africa: Insights from Major and Trace Elements (Contribution #7).
- Turbidity Currents Carrying Shallow Heat Invading Stable Deep-Water Areas May Be an Unrecognized Source of "Pollution" in the Ocean (Contribution #8).
- Features and Constitutive Model of Hydrate-Bearing Sandy Sediment's Triaxial Creep Failure (Contribution #9).

The contributions span from shallow sea to deep sea, including wave-seabed interactions, ocean dynamics, shoreline changes, sediment properties, deep-sea turbidity currents, etc.

Li et al. (Contribution #1) found that the excess pore pressure (EPP) magnitude monotonically increases with wavelength in an initially liquefied seabed, while in an initially consolidated seabed, there is a maximal response wavelength that is inversely related to the consolidation degree. Furthermore, they found two opposite EPP responses to cyclic surface wave loading under varying seabed conditions in initially liquefied and consolidated seabeds. That is, under the same waves, the EPP magnitude is inversely related to the consolidation degree in initially liquefied seabed, while the EPP magnitude is positively related to the consolidation degree in initially consolidated seabed.

Tian et al. (Contribution #2) proposed that increases in saturation and permeability coefficient lead to deeper penetration of excess pore water pressure into the seabed by internal solitary waves (ISWs). Conversely, the effects of shear modulus and porosity are relatively minor and inversely related to the depth of influence of excess pore water pressure. When stratification occurs in the permeability coefficient and saturation of the seabed, significant alterations are observed in the downward propagation of excess pore water pressure. Saturation stratification exhibits similar effects, with soil layers exhibiting higher saturation levels being more conducive to the transmission of excess pore water pressure by ISWs.

Liu et al. (Contribution #3) integrated one-year current observations from four in situ current observation stations (B1−B4) with simulations using the Regional Ocean Modeling System (ROMS) to characterize circulation dynamics in the gulf. Observations show persistent northward subtidal currents west of Hainan Island year-round, primarily sustained by tidal-induced residual currents. These currents briefly reverse southward during strong northerly wind events, whereas subtidal currents in the northern Beibu Gulf are more wind-dependent, showing pronounced seasonal variations. Numerical results confirm that winter circulation is dominated by a basin-wide cyclonic gyre driven by northeasterly monsoons. In summer, circulation in the northern gulf is cyclonic under southeasterly winds, but turns anticyclonic when southwesterly winds prevail, indicating strong sensitivity to summer monsoon wind direction.

Zhu and Cui (Contribution #4) focus on 52 Chinese coastal cities as the research subject by the Hazard–Exposure–Vulnerability (H-E-V) framework and PPRR (Prevention, Preparedness, Response, and Recovery) crisis management theory. The evaluation system for the disaster response capabilities of Chinese coastal cities was constructed based on three aspects: the stability of the disaster-incubating environment (S), the risk of disaster-causing factors (R), and the vulnerability of disaster-bearing bodies (V). The results indicate that Wenzhou has the best comprehensive disaster response capability, while Yancheng has the worst. Moreover, Tianjin, Ningde, and Shenzhen performed well in the three aspects of vulnerability of disaster-bearing bodies, risk of disaster-causing factors, and stability of disaster-incubating environment separately. On the contrary, Dandong (tied with Qinzhou), Jiaxing, and Chaozhou performed poorly in the above three areas.

Li et al. (Contribution #5) obtained the length and structure data of the shorelines in eight periods by manual visual interpretation of Landsat RS (remote sensing) images from 1990 to 2023. The results show that during 33 years, the length of the shorelines increased 15.83 km, with an average growth rate of 0.48 km/y; the value of the intensity of change in the shorelines was 0.08%. The overall change in the fractal dimension of the shorelines was small, between 1.0395 and 1.0673. As far as the influencing factors are concerned, the influence of the natural environment is a long process, and human activities are more capable of changing the length and shape of the shorelines in a short period of time, with factors such as the degree of economic development having a greater impact on the shorelines.

Zhu et al. (Contribution #6) proposed a load-adaptive sediment rig that minimizes zero-velocity points, ensures data continuity, and contributes to sedimentology research. This paper analyzes the mechanical properties and layering patterns of sediment, along with the interaction mechanisms between sediment and mechanical structures. Subsequently, a mechanical structure–sediment integrated model with adaptive control logic is established. Finally, real sediment data are introduced into the physical model for simulation experiments. The simulation results demonstrate that the load-adaptive rig reduces data breakpoints by 50% and increases the maximum single penetration stroke to 1.8 m. Additionally, the load-adaptive rig provides redundancy between penetration force and stroke, automatically reducing penetration force for greater stroke when encountering low-strength sediments and, conversely, sacrificing part of the stroke for greater force.

Hu et al. (Contribution #7) tested the grain size and major and trace elements of 35 surface sediments in the offshore area of Sierra Leone, and made a comparative study with the sediments in the offshore area of China. The results show that sandy silt is the main sediment type in the research area, and the average sediment mean grain size (Mz) is 4.15Φ. The content of Ca in the samples is the highest among the major elements (except Si), with an average of 5.1%. The content of Sr is the highest among the trace elements (except Ti, P, and Mn), with an average of 378.2 μg/g. The results of correlation analysis and factor analysis show that there are three main sources of sediments in the research area, namely, terrigenous weathering products, ilmenite-dominated ore, and oceanic biochemical substances. Compared with the sediments in China offshore, the sediments in the study area are more affected by marine biochemistry and have special ore input characteristics.

Tian et al. (Contribution #8) conducted systematic experiments on warm turbidity currents to understand how sediment-driven turbidity currents lead to mixing in stable stratification using existing environmental entrainment numbers. The experimental results show that the dimensionless numbers Rs, RT, and $R0$ control the flow process of warm turbid plumes, and corresponding functional relationships are summarized. The frequent occurrence of warm turbidity current events caused by increasingly prominent environmental problems cannot be ignored, as it directly affects the deep-water environment of lakes or coastal oceans, which may be an important contribution to heat transfer that has not been evaluated in previous ocean events.

Sun et al. (Contribution #9) take hydrate-bearing sandy sediment as the research object and conduct triaxial compression creep tests at different saturation degrees (20%, 30%, and 40%). The results show that the hydrate-containing sandy sediments have strong creep characteristics, and an accelerated creep phenomenon will occur under the long-term action of high stress. The longstanding destructive power of the specimen progressively increases with the increase in hydrate saturation, but the difference in the triaxial strength of the specimen progressively increases. This indicates that the damage to the hydrate structure during long-term loading is the main factor causing the strength decrease.

3. Conclusions

Advanced research on marine geology and sedimentology is crucial for expanding our understanding of the Earth's systems and for guiding the sustainable use of our oceanic resources. Understanding the geological structure of the ocean floor can aid in the search for mineral and oil resources, while the knowledge of sedimentary processes can inform strategies for environmental conservation and disaster mitigation.

The contributions in this collection demonstrate how different research fields can be combined to address the rapid development of marine geology and sedimentology. The research scope covers multiple sea areas and includes technological innovation. The editor of this Special Issue would like to thank all the authors who participated in the project and invite further scientific activities in this field.

Author Contributions: Conceptualization, Z.T. and S.Z.; writing—original draft preparation, Z.T.; writing—review and editing, Z.T. and S.Z. All authors have read and agreed to the published version of the manuscript.

Funding: This research received no external funding.

Conflicts of Interest: The author declares no conflicts of interest.

List of Contributions:

1. Li, H.; Zhang, Y.; Ma, A.; Wen, M.; Zhao, Z.; Zhang, S. Effect of Relative Wavelength on Excess Pore Water Pressure in Silty Seabeds with Different Initial Consolidation Degrees. *Water* **2025**, *17*, 2829. https://doi.org/10.3390/w17192829.
2. Tian, H.; Jia, L.; Zhao, J.; Wang, L.; Kan, J.; Wu, F.; Tian, Z. Distribution of Excess Pore Water Pressure in Layered Seabed Induced by Internal Solitary Waves. *Water* **2025**, *17*, 2532. https://doi.org/10.3390/w17172532.
3. Liu, G.; Zhang, N.; Yang, Y.; Wang, C. Seasonal Circulation Characteristics of Oceanic System in the Beibu Gulf Based on Observations and Numerical Simulations. *Water* **2025**, *17*, 2365.
4. Zhu, L.; Cui, S. Assessment of Storm Surge Disaster Response Capacity in Chinese Coastal Cities Using Urban-Scale Survey Data. *Water* **2025**, *17*, 2245. https://doi.org/10.3390/w17152245.
5. Li, S.; Gui, F.; Feng, J.; Wang, Y.; Song, Y.; Wang, W.; Lin, C. Analysis of Shoreline Change in Huizhou-Shanwei Region (China) from 1990 to 2023. *Water* **2025**, *17*, 1460. https://doi.org/10.3390/w17101460.
6. Zhu, Y.; Zhang, Z.; Yang, X.; Fei, Z.; Guo, L.; Xue, G.; Liu, Y. Adaptive Penetration Unit for Deep-Sea Sediment Cone Penetration Testing Rigs: Dynamic Modeling and Case Study. *Water* **2025**, *17*, 1159. https://doi.org/10.3390/w17081159.
7. Hu, M.; Zhu, Y.; Chen, W.; Yu, B.; Zhang, P.; Hu, C.; Jin, R. Investigating the Element Geochemical Behavior and Provenance of Surface Sediments in the Offshore Area of Sierra Leone, Africa: Insights from Major and Trace Elements. *Water* **2024**, *16*, 3540. https://doi.org/10.3390/w16233540.
8. Tian, H.; Xu, G.; Zhao, J.; Ren, Y.; Wu, H. Turbidity Currents Carrying Shallow Heat Invading Stable Deep-Water Areas May Be an Unrecognized Source of "Pollution" in the Ocean. *Water* **2024**, *16*, 3521. https://doi.org/10.3390/w16233521.
9. Sun, S.; Zhang, X.; Zhou, Y. Features and Constitutive Model of Hydrate-Bearing Sandy Sediment's Triaxial Creep Failure. *Water* **2024**, *16*, 2947. https://doi.org/10.3390/w16202947.

References

1. Zeng, L.; Wang, C.; Foster, D.A.; Su, M.; Cui, H.; Jia, J. Controls on sediment transport from rivers to trenches in passive and active continental margins. *Sci. China Earth Sci.* **2024**, *67*, 3868–3880. [CrossRef]
2. Li, H.; van Loon, A.J.; Xu, Y.; He, Y.; Yang, Z.; Zhou, W.; Liu, S.; Li, Y. Reservoir characteristics and potential of a shallow-marine fan in the Miocene Huangliu Fm. within the DF1-1 Gas Field (Yinggehai Basin, S China). *Geoenergy Sci. Eng.* **2023**, *230*, 212202. [CrossRef]

3. Yang, Z.; Wang, K.; Xu, W.; Xu, J.; Liu, M.; Hu, L. Multi-proxy sedimentary records in the northeastern South China Sea: Implication for sea level and East Asian Summer Monsoon evolution since 13 ka BP. *Palaeogeogr. Palaeoclimatol. Palaeoecol.* **2025**, *675*, 113099. [CrossRef]
4. Zhu, B.; Pei, H.; Yang, Q. Reliability analysis of submarine slope considering the spatial variability of the sediment strength using random fields. *Appl. Ocean Res.* **2019**, *86*, 340–350. [CrossRef]
5. Chen, H.; Xie, X.; Mao, K.; He, Y.; Su, M.; Zhang, W. Depositional Characteristics and Formation Mechanisms of Deep-Water Canyon Systems along the Northern South China Sea Margin. *J. Earth Sci.* **2020**, *31*, 808–819. [CrossRef]
6. Fan, C.; Tian, Z.; Cui, K.; Huang, J.; Bian, S.; Yang, L.; Zhang, T. A new discovery of source contribution and transport mechanism of clay minerals in Taiwan Canyon-Manila Trench. *Deep. Sea Res. Part I Oceanogr. Res. Pap.* **2025**, *224*, 104571. [CrossRef]
7. Tian, H.; Ren, Y.; Chen, Z.; Tao, W.; Wu, H.; Xu, G. Numerical Study of the Transport Process of Shallow Heat Carried by Turbidity Currents in Deep-Sea Environments. *J. Geophys. Res.-Ocean.* **2023**, *128*, 29. [CrossRef]
8. Wu, J.; Yu, L.; Guo, X.; Xie, Y.; Ma, Z. Application of resistivity in the characterization of shear failure process of gas bearing marine silt. *Mar. Georesources Geotechnol.* **2023**, *41*, 1001–1012. [CrossRef]
9. Zhang, Y.; Sun, Z.; Lei, B.; Li, M.; Guo, X.; Zheng, T.; Luo, J. Morphology of brine-Seawater interface and spatial distribution of submarine Groundwater dischargewindows in the muddy coast. *Geophys. Res. Lett.* **2024**, *51*, e2024GL111193. [CrossRef]

water

MDPI

Article

Features and Constitutive Model of Hydrate-Bearing Sandy Sediment's Triaxial Creep Failure

Shihui Sun [1,2,*], Xiaohan Zhang [1,3] and Yunjian Zhou [1,3]

1 State Key Laboratory of Offshore Natural Gas Hydrates, Beijing 102209, China
2 Sanya Offshore Oil & Gas Research Institute, Northeast Petroleum University, Sanya 572025, China
3 CNOOC Research Institute Co., Ltd., Beijing 100028, China
* Correspondence: sshsmile@163.com

Abstract: In the longstanding development of hydrate-bearing sediment (HBS) reservoirs, slow and permanent deformation of the formation will occur under the influence of stress, which endangers the safety of hydrate development projects. This paper takes hydrate-bearing sandy sediment (HBSS) as the research object and conducts triaxial compression creep tests at different saturation degrees (20%, 30%, and 40%). The results show that the hydrate-containing sandy sediments have strong creep characteristics, and accelerated creep phenomenon will occur under the long-term action of high stress. The longstanding destructive power of the specimen progressively raises with the increase in hydrate saturation, but the difference in the triaxial strength of the specimen progressively increases. This indicates that the damage to the hydrate structure during long-term loading is the main factor causing the strength decrease. Further, a new nonlinear creep constitutive model was developed by using the nonlinear Burgers model in series with the fractional-order viscoplastic body model, which can well describe the creep properties of HBSS at different saturation levels.

Keywords: sandy hydrate; features; constitutive model; triaxial creep

1. Introduction

Natural gas hydrates (NGH) are cage-like compounds formed by methane gas and water at low temperatures and high pressures, which are widely endowed in shallow layers of the deep ocean and permafrost in the onshore area [1,2]. NGH reserves are abundant and have high energy densities, and they are considered one of the most promising clean energy sources to be developed [3,4]. Currently, several countries, including the United States, Japan, China, Canada, and Russia, have carried out NGH test mining, proving the feasibility of NGH exploitation. Nevertheless, enormous challenges, for instance, insufficient stockpile strength, underdeveloped exploration and development technologies, and viable induced geological dangers pose significant challenges to the safe and effective development of NGHs [5,6].

A large number of engineering practices have shown that reservoir creep is closely associated with the longstanding stability and safe operation of mining activities [7]. It is crucial to thoroughly consider the creep behaviour of the rock mass in order to predict and mitigate the risks associated with geological engineering projects [8–11]. Through in situ observation in the field, it is found that the marine NGH reservoir has strong creep characteristics. For this reason, some academics are investigating the creep characteristics of NGH reservoirs [12–14]. Parameswaran et al. carried out creep experiments on THF-containing hydrate sediments, as well as obtained the deformation curves and deformation rates under different axial stresses. Li et al. [15]. carried out separately loaded creep experiments on frozen samples of hydrate-bearing sediment (HBS) and found that the creep behaviour of HBS has a strong temperature sensitivity. The resulting creep deformation and deformation rate increased significantly with increasing temperature. Miyazaki et al. [16].

established an empirical equation for the relationship between creep rate and triaxial loading rate by comparing triaxial experiments with creep experiments at different loading rates. Hu et al. [17] carried out a joint acoustic and creep experiment on a HBS, and established a correlation between acoustic characteristics and the mechanical behaviour of creep. The above studies have laid a solid foundation for the interpretation of the creep behaviour of HBS.

Establishing a creep constitutive model based on the results of creep tests conducted on hydrate-bearing sediments (HBSs) is a crucial aspect of understanding their behaviour under prolonged stress. This model aims to accurately reflect the actual creep conditions these sediments experience in real-world applications. Additionally, determining the relevant parameters for this model is vital, as these parameters influence the predictive accuracy of the model [18,19].

The creep constitutive model serves as a vital tool for accurately characterizing the creep behaviour of hydrate-bearing sediments (HBSs). By providing a comprehensive framework that captures the essential features of creep under varying stress conditions, this model effectively links experimental creep test data with numerical simulations of reservoir deformation [20,21]. This connection is crucial, as it enables researchers and engineers to translate laboratory findings into predictive models that can simulate the behaviour of HBSs in real-world scenarios. This research holds significant engineering application value [22]. The most commonly used basic models for the creep constitutive model of geotechnical materials are the Nishihara mode and the Burgers mode. When this load does not exceed the long-term strength, the classical constitutive model can describe the decelerated and stable creep of geotechnical materials. However, the model faces difficulties in describing the conduct of accelerated creep, for which scholars have proposed various modification strategies [23]. Specifically, these strategies include: (1) introducing damage elements: in different studies, scholars have adopted diverse damage equations to compensate for the model's shortcomings in describing accelerated creep [24]; (2) handling viscoplastic units: fractional processing of the viscous components in viscoplastic units to enhance the model's descriptive capabilities [25]; and (3) introducing winding elements: based on the principle of energy conservation, introducing winding elements to modify classical models more effectively describes the phenomenon of accelerated creep. In addition, Cheng et al. [26]. and Liu et al. [27]. have proposed improved models. Cheng et al. added damping units and nonlinear viscoplastic units to the classical Nishihara model to simulate the accelerated creep stage of soft rocks. Liu et al., on the other hand, developed a new improved model by connecting the Bingham model with the Nishihara damage model. These improvement methods can effectively describe the phenomenon of accelerated creep, providing strong support for research in related fields. In the above studies, there needs to be more data on the creep behaviour of hydrate, and the current research is mainly focused on the influence of hydrate saturation, temperature, and particle size on the creep behaviours of hydrate [28]. The creep constitutive model studies have focused on conventional geotechnical materials, while fewer studies have been conducted on hydrate-bearing sandy sediment (HBSS) [29,30]. The applicability of the developed model to HBSS needs to be further investigated.

Therefore, in order to expose this mechanical reaction and the deformation mechanism of HBSS under long-term loading conditions, triaxial compression creep tests were conducted in this study on HBSS with different saturation levels to analyse the deformation characteristics of HBSS under long-term loading [31,32]. Further, a new nonlinear creep constitutive model is set up by linking the Burgers model in series with the fractional-order viscoplastic body model [33–35]. The model is further validated by fitting it with the creep test results obtained from hydrate-bearing sandy sediments (HBSSs), showcasing its applicability and reliability in real-world conditions. The alignment between the model predictions and experimental data not only confirms the model's effectiveness but also enhances its credibility for practical use. The findings from this study hold significant reference value for safety evaluations related to development projects involving HBSSs. By providing a robust framework for understanding the creep behaviour of these sediments, the model aids in assessing potential risks associated with reservoir deformation and stability.

2. Materials and Methods

2.1. Experimental Instruments

The creep test apparatus is shown in Figure 1. The apparatus includes an autoclave reactor, an enclosure-pressure system, a back-pressure system for the pore-pressure system, a thermostat-box system, and an exhaust-gas-treatment system. During the experiment, the axial load of the specimen was applied through a precision-designed reverse-frame platform, which ensured stable application and precise measurement of the load. At the same time, the circumferential pressure of the specimen was precisely controlled by a high-precision flow pump, which had high stability and accuracy to meet the strict requirements for circumferential pressure in the experiment. The pore pressure was provided by a high-pressure gas source, which could provide stable and adjustable air pressure to meet the requirements for pore pressure in the experiment. According to the basic reservoir characteristics of the hydrate formation, the temperature and pressure control parameters of the device were preferred. The original formation pressure of the hydrate reservoir in the Shenhu Sea of the South China Sea was basically between 15 MPa and 20 MPa [36]. The natural gas hydrate reservoir in the Nankai Trough of Japan had a raw formation pressure of about 15 MPa [37]. The temperature range was 10 °C to 20 °C. Therefore, the maximum loading pressures of these three systems reached 30 kN, 18 MPa, and 12 MPa, ensuring the safety and reliability of the experiment under high-pressure conditions. The experiments were carried out in a thermostatic chamber, and the experimental temperatures were controlled in the range of −30 °C to 50 °C with an accuracy of ±0.2 °C. The actual hydrate formation conditions were fully simulated, which were difficult to achieve under laboratory conditions. Choosing other temperature and pressure conditions (to ensure that the hydrate does not decompose) to carry out hydrate experiments was allowed in previous studies [4]. Therefore, in this experiment, the experimental conditions were chosen to be a temperature of 1 °C and an effective perimeter pressure of 1 MPa.

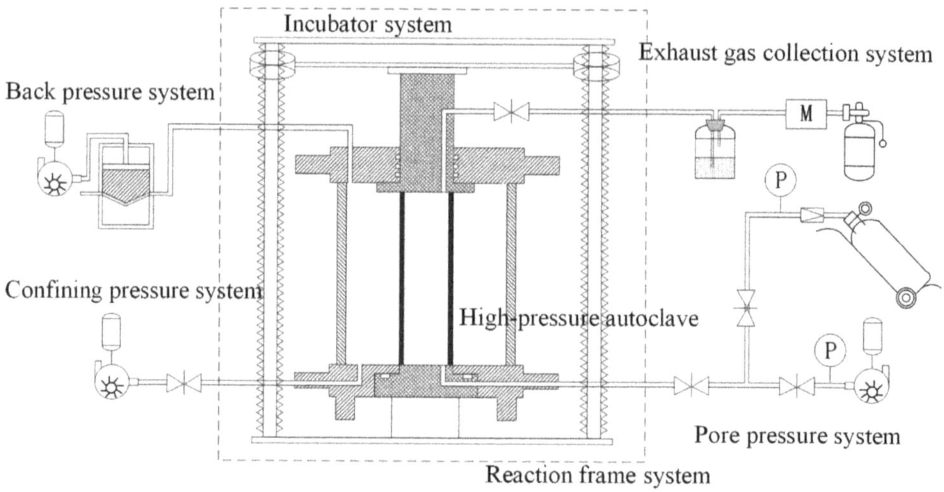

Figure 1. Testing device structural diagram of HBSS (M is the flow meter, P is the pressure gauge).

2.2. Experimental Condition

The matrix of HBSS in the experiment was selected as coarse sand grains with D50 of 0.49 mm and D90 of 0.71 mm. The experimental specimens of HBSS with different degrees of saturation were prepared by the supersaturated gas method. In the generation stage of the experimental hydrate-containing sandy sediments, firstly, the sediment matrix and deionised water were homogeneously mixed and filled into the rubber cylinder of the experimental reactor to fully compact the specimen. Then, the reactor was closed and settled on the reaction frame setup, and the pore pressure, enclosure pressure, and temperature were adjusted to pre-set values. Immediately thereafter, hydrate generation began, methane was continuously consumed, and the pore pressure decreased. When the pore pressure stabilised, it was considered that the deionised water in the specimen was completely converted into hydrate, and the hydrate generation was completed, with a reaction time of about 90 h. Experiments were then carried out.

3. Results and Discussion

3.1. Conventional Triaxial Experiment Results

The bias stress–strain curves of HBSS at different saturation levels are shown in Figure 2. The y-axis represents the bias stress, while the x-axis represents the strain. The test results indicate that under three-dimensional stress, the HBSS specimens exhibited significant axial deformation, with peak strain generally exceeding 0.04. As the saturation degree increased, there was a corresponding rise in the peak strength of the specimens. Additionally, the residual strength of the hydrate-bearing sandy sediment (HBSS) specimens became increasingly pronounced at higher saturation levels. This trend suggests that, as the sediment became more saturated, its ability to withstand shear forces improved significantly. This observation underscores the importance of saturation in determining the mechanical properties of HBSS. Specifically, during the process of resisting sliding at the hydrate-sediment interface under applied confining pressure, the sediments developed a stable residual strength. This stable residual strength is crucial for maintaining structural integrity and stability in environments where these sediments are present.

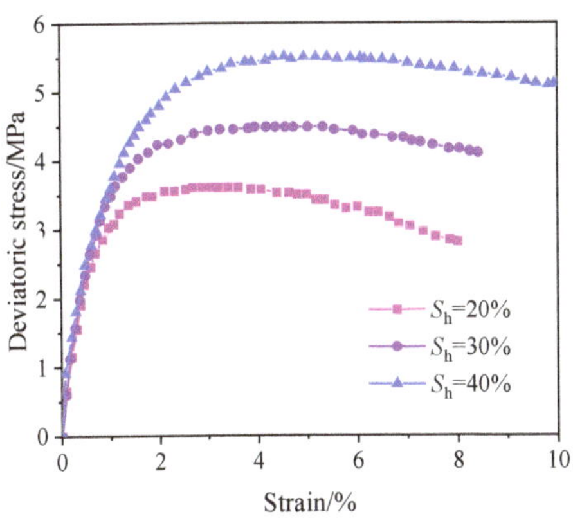

Figure 2. Deviatoric stress–strain curves under different hydrate saturation (20%, 30%, and 40%) of HBSS.

3.2. Triaxial Creep Test Results

Figure 3 presents the creep curves of sandy sediments containing hydrates at various saturation degrees. The curves indicate that, under triaxial-loading conditions, the initial creep strain was relatively small and exhibits a rapid decay, which signifies the presence of a stable creep stage. This stage was characterised by a gradual deformation response to the applied load, suggesting that the sediments were able to maintain a certain level of stability despite the stress. As the axial load was increased, the creep curve transitioned from this stable creep stage to an accelerated creep stage, where the creep rate began to rise significantly. This acceleration indicates that the sediment was experiencing greater strain in response to the increased load, leading to a more pronounced deformation behaviour. Importantly, it has been observed that the amount of deformation corresponding to this accelerated creep phase decreased as the saturation level increased. This finding suggests that higher levels of hydrate saturation played a crucial role in enhancing the brittleness of the methane-containing hydrate sediments. In other words, as the saturation of the hydrate increased, the sediments became less capable of undergoing significant deformation before failure, resulting in a more brittle response.

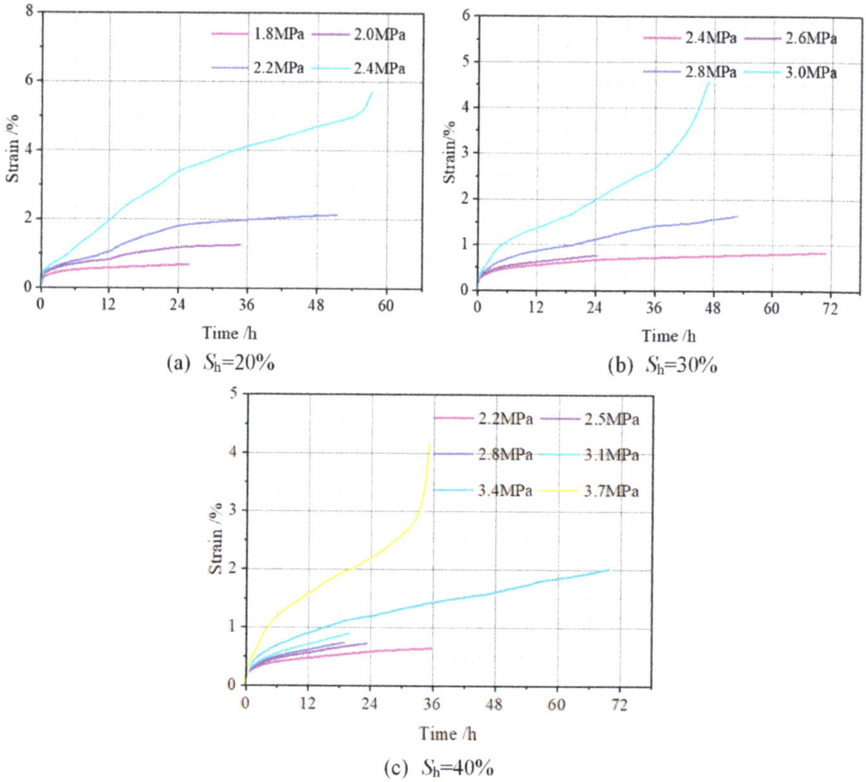

Figure 3. Triaxial creep curves of HBSS (From Ref. [29]).

The creep properties of sandy sediments containing hydrate sediments at different saturation levels were basically the same. Taking the 20% saturation specimen as an example, the axial strain was 0.69% after 26 h of loading at an axial load of 1.8 MPa. At an axial load of 2.0 MPa, the axial strain was 1.26% after 35 h of loading, and at an axial load of 2.2 MPa, the axial strain was 2.16% after 52 h of loading. When the axial load was 2.4 MPa, after 36 h of loading, the specimen creep rate region was stable, the axial strain was 4.13%, and after 54 h of loading, the specimen creep rate began to grow, when the axial strain was 4.95%. The specimen began to creep damage.

3.3. Long-Term Strength

The long-term strength of a hydrate-bearing sandy sediment (HBSS) deposit represents its capacity to withstand damage under prolonged loading conditions. This long-term strength serves as the stratigraphic limit strength, ensuring the safe development of hydrate resources over extended periods. Consequently, accurately calculating the long-term strength is crucial for the sustainable and secure extraction of hydrates. Several methods can be employed to calculate long-term strength, including the excessive creep method, creep rate method, and isochronous stress–strain curve method. In this study, we utilised the excessive creep method to determine the long-term strength of sandy sediments that contain hydrate. The results of these calculations are presented in Table 1. From the data shown in the table, it is evident that the long-term strength of the HBSS increased linearly as hydrate saturation rose. This finding underscores the importance of hydrate saturation in enhancing the stability and strength of the formation. Conversely, it also indicates that the long-term strength of the deposit significantly decreased when hydrate decomposition occurred. In the design of the development programme, the decrease in formation strength due to hydrate decomposition should be fully considered. The long-term strength of sandy sediments containing HBSSs is typically defined by their triaxial strength, which serves as an indicator of their strength class. This is expressed as a parameter that ignores the danger level of the effects of creep for the design of the development programme. A lower strength class also indicates a greater difference between the two and a higher risk of development after neglecting creep. As can be seen in Table 1, the intensity level of sandy sediments with hydrate sediments under different saturation was basically the same, 0.61~0.62, which indicates that the risk of geological hazards was basically the same for different saturated strata after ignoring the creep. However, it is not negligible that the difference between the two increased significantly, which indicates that the damage to the hydrate structure in the long-term loading process was the main factor causing the strength decline.

Table 1. Creep test stress path.

S_h/%	Long-Term Strength/MPa	Triaxial Strength/MPa	Strength Level
20	2.2	3.61	0.61
30	2.8	4.49	0.62
40	3.4	5.52	0.62

4. Description of Full Creep Regions in HBSS: A New Creep Constitutive Model

From the creep curves of the HBSSs, it is obvious that they have accelerated creep. Therefore, we developed a new nonlinear creep constitutive model by concatenating the Burgers model with a fractional-order viscoplastic body model.

4.1. One-Dimensional Creep Constitutive Model

4.1.1. Fractional-Order Viscoplastic Body Models

In linear viscoelastic theory, the mechanical analogue of a purely elastic solid can be represented by a spring, and the mechanical analogue of a purely viscous fluid can be represented by a damper. Various combination models (e.g., Maxwell's model, Kevin's model) can be obtained by connecting the two in series and in parallel to derive the corresponding strain-versus-time relationship. For elastic and viscous units, Hooke's and Newton's laws are obeyed. From the derivative point of view, it can be expressed as follows:

$$\begin{cases} \sigma(t) = E\frac{d^0\varepsilon(t)}{dt^0} \\ \sigma(t) = \eta\frac{d^1\varepsilon(t)}{dt^1} \end{cases} \tag{1}$$

where $\sigma(t)$ is the stress; $\varepsilon(t)$ is the strain; E is the modulus of elasticity of the spring; and η is the coefficient of the viscosity of the viscous element.

For elastic solids, the stress is proportional to the zero-order derivative of the strain, which means the stress is directly proportional to the strain itself. For viscous bodies, the stress is proportional to the first-order derivative of the strain, which means the stress is proportional to the rate of change in the strain, or the strain rate. The stress and strain of a viscoelastic body is intermediate between that of a purely elastic solid and that of a purely viscous fluid, i.e., the stress is proportional to a non-integer of the strain. The intrinsic equation of a viscoelastic body can be described as follows:

$$\sigma(t) = \eta \frac{d^a \varepsilon(t)}{dt^a} \quad (0 \leq a \leq 1) \tag{2}$$

where a is the fractional order and the above equation is known as the Scott–Blair fractional-order component.

Based on the Rieman–Liouville fractional-order calculus theory, the creep constitutive model can be expressed as follows:

$${}_0D_t^{-a} f(t) = \frac{d^{-a}\varepsilon(t)}{dt^{-a}} = \frac{1}{\Gamma(a)} \cdot \int_a^t (t-\tau)^{a-1} f(\tau) d\tau \tag{3}$$

Differential transformations are obtained:

$${}_0D_t^{-a} f(t) = \frac{d^a\varepsilon(t)}{dt^a} = \frac{1}{\Gamma(n-a)} \left(\frac{d}{dt}\right)^n \cdot \int_a^t \frac{f(\tau)}{(t-\tau)a-n+1} d\tau \tag{4}$$

where ${}_0D_t^a$ is the fractional-order derivative operator; t is the time variable; n is the smallest integer greater than a; $f(t)$ is a function of time; and Γ is the Gamma function, defined as follows:

$$\Gamma(z) = \int_0^\infty e^{-x} x^{z-1} dx \quad (Re(z) > 0) \tag{5}$$

where a precondition for the equation to hold, z is a complex number with real part greater than 0. That is, $Re(z) > 0$.

When $\sigma(t)$ is constant, it can be back-calculated from the Rieman–Liouville fractional-order calculus:

$$\varepsilon(t) = \frac{\sigma}{\eta} \cdot \frac{t^a}{\Gamma(1+a)} \quad (0 < a < 1) \tag{6}$$

When σ and η are set to the same value, different fractional-order creep curves are shown in Figure 4. The figure illustrates that as the fractional order (a) gradually increases to one, the creep behaviour of the fractional-order viscoplastic body becomes increasingly consistent with the creep curve of a Newtonian body. Conversely, as the fractional order (a) decreases step by step to zero, the creep curve of the fractional-order viscoplastic body approaches the creep curve of an elastic element. Between zero and one, the strain characteristics expressed by the fractional-order viscoplastic body are not like the strain of the elastic element that is constant with time, nor like the strain of the Newtonian body that shows a linear relationship with time, but show a nonlinear increment. The creep curves at different creep rates and stress levels can be reflected by setting different parameters. When both σ and η are assigned the same value, a variety of fractional-order creep curves emerge, as depicted in Figure 4. This figure demonstrates that as the fractional order (a) gradually approaches one, the creep behaviour of the fractional-order viscoplastic body increasingly aligns with the creep curve typical of a Newtonian fluid. This indicates that, at this stage, the sediment exhibits characteristics similar to those of a classic Newtonian body, where the strain response is linear over time. On the other hand, as the fractional order (a) is systematically reduced to zero, the creep curve of the fractional-order viscoplastic body begins to resemble that of an elastic element. In this scenario, the sediment behaves in a more rigid manner, with strain characteristics that are distinctly different from those of a viscous fluid. Importantly, for values between zero and one, the strain behaviour demonstrated by the fractional-order viscoplastic body is neither consistent with the constant strain behaviour of an elastic element over time nor with the linear strain behaviour of a Newtonian body. Instead, it exhibits a nonlinear increment, reflecting a more complex relationship between stress and strain. This unique response underscores the versatility of the fractional-order model in capturing a wide range of sediment behaviours. Moreover, by adjusting various parameters, the creep curves can be tailored to represent different creep rates and stress levels, thereby allowing for a comprehensive analysis of the sediment's performance under diverse conditions. This flexibility makes the fractional-order viscoplastic model a powerful tool for studying the creep characteristics of sediments across different loading scenarios.

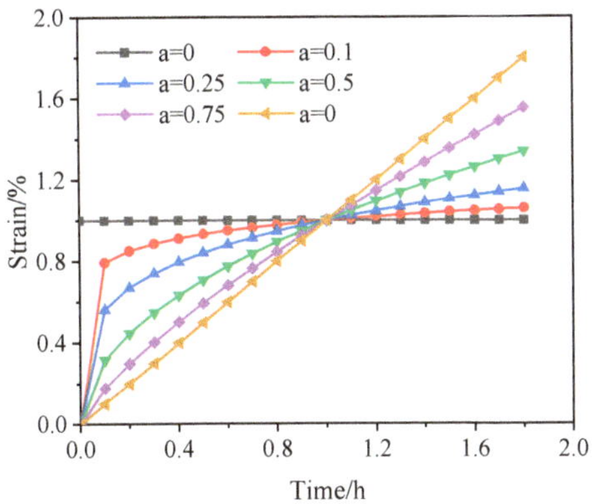

Figure 4. Creep strain variation curves of fractional-order viscoplastic body with different fractional-order parameters (a).

4.1.2. Nonlinearisation of Viscous Elements

In the creep-element model, the creep parameter is constant and cannot qualitatively describe the nonlinear properties of HBSS. During decelerated creep, the viscosity coefficient of HBSS increases with time. Therefore, the viscosity coefficient in the Kelvin model is assumed to be a function of time as shown in Equation (7):

$$\eta(t) = \eta_1 e^{-\lambda t} \tag{7}$$

where η is the viscosity coefficient of the viscous element, in MPa·h, and λ is the creep parameter, in dimensionless.

The derivatives of Equation (7) are executed to obtain the first-order derivative equation of the viscous parameter as shown in Equation (8).

$$\dot{\eta}(t) = -\eta_1 \lambda e^{-\lambda t} \tag{8}$$

The modified viscosity parameters were substituted into the Kelvin model to obtain the modified eigenstructural equations as shown in Equation (9):

$$\sigma = E_1 \varepsilon + \eta(t)\dot{\varepsilon} \tag{9}$$

where σ is the stress, in MPa, E_1 is the elastic model, in MPa, and ε is the creep strain, in dimensionless.

This is obtained by substituting Equation (8) into Equation (9) and solving the differential equation.

$$\varepsilon_1 = \frac{\sigma}{E_1} - \frac{A}{E_1}\left(\exp\left(-\frac{E_1}{\eta\lambda}e^{\lambda t} + BE_1\right)\right) \tag{10}$$

Substituting the boundary conditions $t = 0$ and $\varepsilon = 0$ into Equation (10) yields the modified Kelvin model creep equation as shown in Equation (11).

$$\varepsilon_1 = \frac{\sigma}{E_1}\left(1 - \exp\left(-\frac{E_1\left(e^{\lambda t} - 1\right)}{\eta_1 \lambda}\right)\right) \tag{11}$$

4.1.3. Fractional-Order Nonlinear Creep Modelling

A nonlinear fractional-order creep model (Figure 5) was developed by connecting the Burgers model in series with the fractional-order viscoplastic body model, which can comprehensively describe the nonlinear creep characteristics of HBSS. The model includes a Kelvin body (I), a nonlinear Maxwell body (II), and a fractional-order viscoplastic body (III).

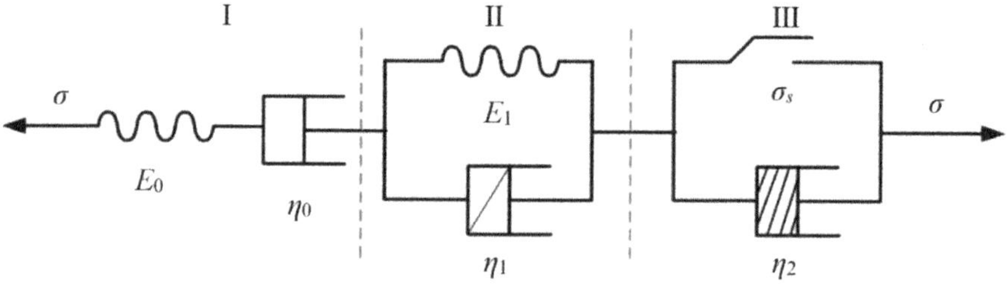

Figure 5. The mechanical model of the HBSS nonlinear creep constitutive model. (**I**) Kelvin body, (**II**) nonlinear Maxwell body, (**III**) fractional-order viscoplastic body.

The Kelvin body model creep equation, as shown in Equation (12).

$$\varepsilon_0 = \frac{\sigma}{E_0} + \frac{\sigma}{\eta_0}t \tag{12}$$

The fractional-order nonlinear creep model is obtained by superimposing Equations (6), (11) and (12) as shown in Equation (13):

$$\begin{cases} \varepsilon(t) = \frac{\sigma}{E_0} + \frac{\sigma}{\eta_0}t + \frac{\sigma}{E_1}\left(1 - \exp\left(-\frac{E_1\left(e^{\lambda t}-1\right)}{\eta_1\lambda}\right)\right), \sigma < \sigma_s \\ \varepsilon(t) = \frac{\sigma}{E_0} + \frac{\sigma}{\eta_0}t + \frac{\sigma}{E_1}\left(1 - \exp\left(-\frac{E_1\left(e^{\lambda t}-1\right)}{\eta_1\lambda}\right)\right) + \frac{\sigma-\sigma_s}{\eta_2}\cdot\frac{t^a}{\Gamma(1+a)}, \sigma \geq \sigma_s \end{cases} \tag{13}$$

where E_0 is the instantaneous modulus of elasticity, in MPa.

4.2. Three-Dimensional Creep Constitutive Model

In the three-dimensional stress state, the stress tensor inside the material can be decomposed into the spherical stress tensor σ_m and the deviatoric stress tensor S_{ij}, which are expressed as follows:

$$\begin{cases} \sigma_m = \frac{1}{3}(\sigma_1 + \sigma_2 + \sigma_3) = \frac{1}{3}\sigma_{kk} \\ S_{ij} = \sigma_{ij} - \delta_{ij}\sigma_m = \sigma_{ij} - \frac{1}{3}\delta_{ij}\sigma_{kk} \end{cases} \tag{14}$$

where δ_{ij} is the Kronecker function. It is generally accepted that the spherical stress tensor can only change the volume of the material but not its shape. And the bias stress tensor S_{ij} can only cause shape change. Similarly, the strain tensor can be decomposed into the ball strain tensor ε_m and the bias strain tensor e_{ij} accordingly, with the expression:

$$\begin{cases} \varepsilon_m = \frac{1}{3}(\varepsilon_1 + \varepsilon_2 + \varepsilon_3) = \frac{1}{3}\varepsilon_{kk} \\ e_{ij} = \varepsilon_{ij} - \delta_{ij}\varepsilon_m = \varepsilon_{ij} - \frac{1}{3}\delta_{ij}\varepsilon_{kk} \end{cases} \tag{15}$$

From the above equation:

$$\varepsilon_{ij} = e_{ij} + \delta_{ij}\varepsilon_m \tag{16}$$

For the three-dimensional stress state the Hook body has

$$\begin{cases} \sigma_m = 3K\varepsilon_m \\ S_{ij} = 2Ge_{ij} \end{cases} \tag{17}$$

where K is the bulk modulus and G is the shear modulus. To satisfy the isotropic conditions of the material, it is assumed that the elastic strain is caused by the spherical stress tensor, while the creep is caused by the deviatoric stress tensor. The creep equation of the material under a three-dimensional stress state is

$$\begin{cases} \varepsilon_m = \frac{\sigma_m}{3K} \\ e_{ij} = \frac{S_{ij}}{2G_1} + \frac{S_{ij}}{2\eta_1}t + \frac{S_{ij}}{2G_2}\left[1 - \exp\left(-\frac{G_2}{\eta_2}t\right)\right] \end{cases} \tag{18}$$

where G_1, G_2 is the shear modulus. K, G_1, G_2, η_1, and η_2 are model parameters, which are determined by fitting the test results.

Maxwell body in a three-dimensional state of stress:

$$e_{ij}^M = \frac{S_{ij}}{2G_0} + \frac{S_{ij}}{2H_0}t \tag{19}$$

Kelvin body in a three-dimensional state of stress:

$$e_{ij}^K = \frac{S_{ij}}{2G_2}\left[1 - \exp\left(\frac{G_2}{H_2}t\right)\right] \tag{20}$$

Fractional-order viscoplastic bodies under three-dimensional stress states:

$$e_{ij}^L = \frac{S_{ij} - \sigma_s}{2H_2} \cdot \frac{t^a}{\Gamma(1+a)} \tag{21}$$

Superimposing Equations (19)–(21), the total strain in three-dimensional stress is obtained and can be expressed as Equation (22).

$$\begin{cases} \varepsilon(t) = \frac{S_{ij}}{2G_0} + \frac{S_{ij}}{2H_0}t + \frac{\sigma_m\delta_{ij}}{3K_0} + \frac{S_{ij}}{2G_1}\left(1 - \exp\left(-\frac{G_1\left(e^{\lambda t}-1\right)}{H_1\lambda}\right)\right), \sigma < \sigma_s \\ \varepsilon(t) = \frac{S_{ij}}{2G_0} + \frac{S_{ij}}{2H_0}t + \frac{\sigma_m\delta_{ij}}{3K_0} + \frac{S_{ij}}{2G_1}\left(1 - \exp\left(-\frac{G_1\left(e^{\lambda t}-1\right)}{H_1\lambda}\right)\right) + \frac{S_{ij}-\sigma_s}{2H_2} \cdot \frac{t^a}{\Gamma(1+a)}, \sigma \geq \sigma_s \end{cases} \tag{22}$$

According to the theory of plastic mechanics, we can obtain the following relationship:

$$\begin{cases} S_{ij} = \sigma_{ij} - \sigma_m\delta_{ij} \\ \sigma_m = \frac{1}{3}(\sigma_1 + \sigma_2 + \sigma_3) \end{cases} \tag{23}$$

where σ_m is the average stress, in MPa, σ_1, σ_2 and σ_3 are the first, second, and third principal stresses, in MPa, respectively.

In the conventional triaxial test, $\sigma_2 = \sigma_3$ and Equation (23) is deformed as Equation (24) (Liu et al.) [38].

$$\begin{cases} S_{11} = \frac{2}{3}(\sigma_1 - \sigma_3) \\ \sigma_m = \frac{1}{3}(\sigma_1 + 2\sigma_3) \end{cases} \tag{24}$$

Bringing Equation (24) into Equation (22), the HBSS three-dimensional creep constitutive equation is obtained as Equation (25).

$$\begin{cases} \varepsilon(t) = \frac{\sigma_1-\sigma_3}{3G_0} + \frac{\sigma_1-\sigma_3}{3H_0} + \frac{\sigma_1+2\sigma_3}{9K_0} + \frac{\sigma_1-\sigma_3}{3G_1}\left(1 - \exp\left(-\frac{G_1\left(e^{\lambda t}-1\right)}{H_1\lambda}\right)\right), \sigma_1 - \sigma_3 < \sigma_s \\ \varepsilon(t) = \frac{\sigma_1-\sigma_3}{3G_0} + \frac{\sigma_1-\sigma_3}{3H_0} + \frac{\sigma_1+2\sigma_3}{9K_0} + \frac{\sigma_1-\sigma_3}{3G_1}\left(1 - \exp\left(-\frac{G_1\left(e^{\lambda t}-1\right)}{H_1\lambda}\right)\right) \\ \quad + \frac{\sigma_1-\sigma_3-\sigma_s}{3H_2}t + \frac{\sigma_1-\sigma_3-\sigma_s}{3G_2} \cdot \frac{t^a}{\Gamma(1+a)}, \sigma_1 - \sigma_3 \geq \sigma_s \end{cases} \tag{25}$$

4.3. Model Parameter Identification

In this study, we utilised a planned solution function for nonlinear regression analysis. By fitting the creep curve, we were able to obtain the parameters for the optimal fractional-order nonlinear creep model specific to hydrate-bearing sandy sediments (HBSSs), as shown in Table 2. The results of the curve fitting for the experimental data are presented in Figure 6. In Figure 6, the solid line illustrates the fitting results, while the scattered points correspond to the experimental data depicted in Figure 3. Remarkably, the average correlation coefficient for the model's fit reaches a high value of 0.98. This strong correlation indicates that the parameters in Table 2 effectively represent the creep behaviour of HBSSs throughout all testing stages.

Table 2. Model parameter fitting results.

S_h/%	σ_1–σ_3 /MPa	G_0 /MPa	K_0 /MPa	H_0 /MPa·h	G_1 /MPa	λ /10^{-2}	H_1 /MPa·h	H_2 /MPa·h	G_2 /MPa	a	R^2
20	1.8	365	425	2415	168	0.01	1235	/	/	/	0.984
	2	456	431	1214	40	−3.59	942	/	/	/	0.982
	2.2	429	426	1240	43	4.61	1159	/	/	/	0.995
	2.4	456	442	2405	75	13.15	4208	429	3.64	0.63	0.999
30	2.4	1434	1664	461	119	−2.82	2183	/	/	/	0.997
	2.6	996	1061	530	127	−4.39	1819	/	/	/	0.994
	2.8	2246	2063	328	39	−1.52	1920	/	/	/	0.995
	3	1270	1060	1022	194	55.59	1232	132	7.93	8.01	0.999
40	2.2	812	799	1375	71	−5.60	2127	/	/	/	0.995
	2.5	941	877	1727	103	−4.35	1572	/	/	/	0.994
	2.8	1071	955	2113	176	0.01	1203	/	/	/	0.984
	3.1	1264	1099	2457	136	0.01	1277	/	/	/	0.990
	3.4	1477	1366	898	62	0.01	1976	/	/	/	0.999
	3.7	638	2400	1914	155	35.24	1754	249	5.66	6.43	0.989

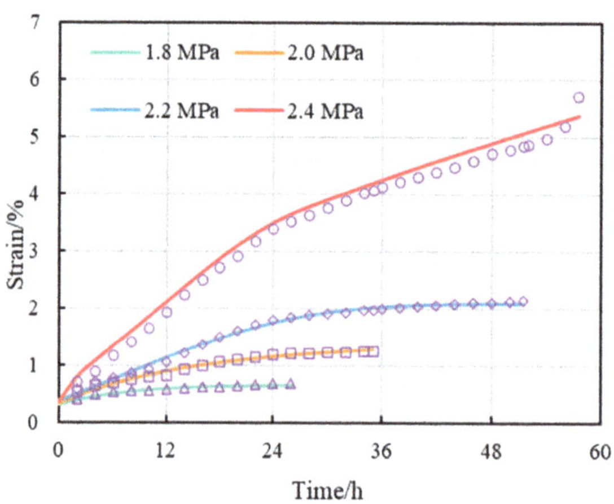

Figure 6. *Cont.*

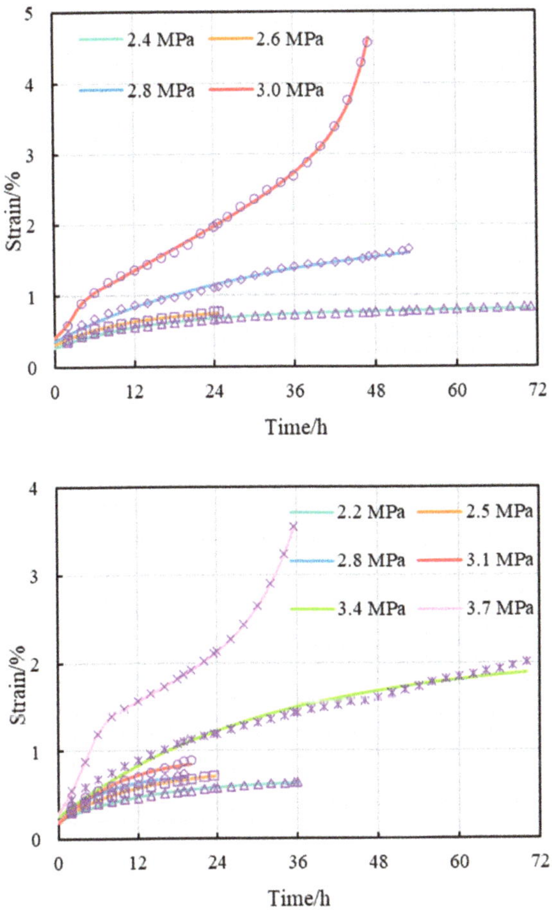

Figure 6. The curve fitting for the experimental data.

4.4. Model Verification

The model comparison work is based on the experimental data of creep strain with a saturation of 40%. We compared the model developed in this study with the Nishihara model and the Burgers model. The calculation results are shown in Figure 7. The fit correlation coefficient of the Nishihara model to the experimental data was 0.971, and the fit correlation coefficient of the Burgess model to the experimental data was 0.971. Both of them were smaller than the fit correlation coefficients of the models developed in this study, which were 0.989. It can be seen that the model developed in this study can better express the nonlinear acceleration behaviour of the HBSS, and is more suitable for conducting the deformation analysis of the HBSS [39].

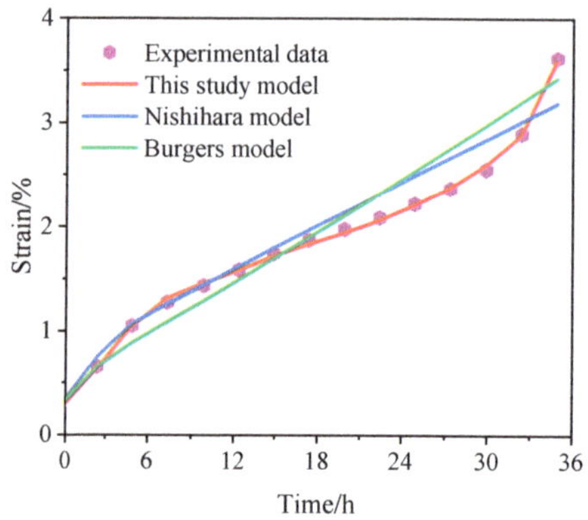

Figure 7. Comparison of this study model with the Nishihara model and Burgers model calculations.

5. Conclusions and Prospects

The long-term strength of HBSS is measured by their triaxial strength, which shows their strength class. A lower strength class means a bigger difference and higher risk if creep is ignored. The strength of sandy sediments with hydrates is similar under different saturation levels, indicating similar geological hazard risks. However, the difference between the two increases significantly, suggesting that damage to the hydrate structure during long-term loading is the main cause of strength decline.

This study creatively introduces the fractional-order nonlinear creep model by utilising the Riemann–Liouville fractional calculus framework and nonlinear damage theory, aiming to accurately determine the damage characteristic structure under three-dimensional stress conditions. To verify the effectiveness and reliability of this model, we apply it to the fracturing stage of the nonlinear creep model and compare it with the triaxial creep experimental results. The results show a high degree of consistency between the two, thus fully demonstrating the accuracy and reliability of this model in practical applications.

This study compares a new model with the traditional Burgers model and the Nishihara model in terms of their ability to replicate the creep behaviour of HBSS. Through comparative analysis, the new model exhibits significant advantages over the other two models in replicating the creep behaviour.

However, there are still deficiencies in this study, such as the lack of correlation equations between model parameters and saturation/stress, etc., and the inability of the model to be extended for applications. These are limited by the size of the experimental sample. In the next step, we will carry out more experiments to enrich the parameter samples, establish the relationship between the saturation/stress and the model parameters, and improve the applicability and dexterity of using the model.

Author Contributions: Methodology, S.S.; validation, Y.Z. and X.Z.; formal analysis, X.Z.; writing—original draft preparation, S.S.; writing—review and editing, Y.Z. All authors have read and agreed to the published version of the manuscript.

Funding: This research was funded by Open-ended Fund of State Key Laboratory of National Gas Hydrates (Grant No. 2022-FKJJ-SHW).

Data Availability Statement: Data is contained within the article.

Conflicts of Interest: Authors Xiaohan Zhang and Yunjian Zhou were employed by the company CNOOC Research Institute Co., Ltd. The remaining authors declare that the research was conducted in the absence of any commercial or financial relationships that could be construed as a potential conflict of interest.

References

1. Li, A.; Deng, H.; Zhang, H.; Jiang, M.; Liu, H.; Xiao, Y.; Wen, J. Developing a Two-Step Improved Damage Creep Constitutive Model Based on Soft Rock Saturation-Loss Cycle Triaxial Creep Test. *Nat. Hazards* **2021**, *108*, 2265–2281. [CrossRef]
2. Li, Y.; Liu, L.; Jin, Y.; Wu, N. Characterization and Development of Marine Natural Gas Hydrate Reservoirs in Clayey-Silt Sediments: A Review and Discussion. *Adv. Geo-Energy Res.* **2021**, *5*, 75–86. [CrossRef]
3. Wu, P.; Li, Y.; Wang, L.; Wang, L.; Sun, X.; Liu, W.; Song, Y. Triaxial Tests on the Overconsolidated Methane Hydrate-Bearing Clayey-Silty Sediments. *J. Pet. Sci. Eng.* **2021**, *206*, 109035. [CrossRef]
4. Dong, L.; Li, Y.; Liao, H.; Liu, C.; Chen, Q.; Hu, G.; Liu, L.; Meng, Q. Strength Estimation for Hydrate-Bearing Sediments Based on Triaxial Shearing Tests. *J. Pet. Sci. Eng.* **2020**, *184*, 106478. [CrossRef]
5. Li, J.; Ye, J.; Qin, X.; Qiu, H.; Wu, N.; Lu, H.; Xie, W.; Lu, J.; Peng, F.; Xu, Z.; et al. The First Offshore Natural Gas Hydrate Production Test in South China Sea. *China Geol.* **2018**, *1*, 5–16. [CrossRef]
6. Priest, J.A.; Hayley, J.L. Strength of Laboratory Synthesized Hydrate-Bearing Sands and Their Relationship to Natural Hydrate-Bearing Sediments. *JGR Solid. Earth* **2019**, *124*, 12556–12575. [CrossRef]
7. Sun, X.; Yao, D.; Qu, J.; Sun, S.; Qin, Z.; Tao, L.; Zhao, Y. A Novel Transient Hole Cleaning Algorithm for Horizontal Wells Based on Drift-Flux Model. *Geoenergy Sci. Eng.* **2024**, *233*, 212517. [CrossRef]
8. Morshedifard, A.; Masoumi, S.; Abdolhosseini Qomi, M.J. Nanoscale Origins of Creep in Calcium Silicate Hydrates. *Nat. Commun.* **2018**, *9*, 1785. [CrossRef]
9. Zhu, Y.; Chen, C.; Luo, T.; Song, Y.; Li, Y. Creep Behaviours of Methane Hydrate-Bearing Sediments. *Environ. Geotech.* **2019**, *9*, 199–209. [CrossRef]
10. Königsberger, M.; Irfan-Ul-Hassan, M.; Pichler, B.; Hellmich, C. Downscaling Based Identification of Nonaging Power-Law Creep of Cement Hydrates. *J. Eng. Mech.* **2016**, *142*, 04016106. [CrossRef]
11. Sun, X.; Luo, T.; Wang, L.; Wang, H.; Song, Y.; Li, Y. Numerical Simulation of Gas Recovery from a Low-Permeability Hydrate Reservoir by Depressurization. *Appl. Energy* **2019**, *250*, 7–18. [CrossRef]
12. Lei, L.; Seol, Y. Pore-Scale Investigation Of Methane Hydrate-Bearing Sediments Under Triaxial Condition. *Geophys. Res. Lett.* **2020**, 47. [CrossRef]
13. Gao, Y.; Chen, Y.; Zhao, X.; Wang, Z.; Li, H.; Sun, B. Risk Analysis on the Blowout in Deepwater Drilling When Encountering Hydrate-Bearing Reservoir. *Ocean. Eng.* **2018**, *170*, 1–5. [CrossRef]
14. Fereidounpour, A.; Vatani, A. An Investigation of Interaction of Drilling Fluids with Gas Hydrates in Drilling Hydrate Bearing Sediments. *J. Nat. Gas. Sci. Eng.* **2014**, *20*, 422–427. [CrossRef]
15. Li, Y.; Liu, C.; Liu, L.; Sun, J.; Liu, H.; Meng, Q. Experimental Study on Evolution Behaviors of Triaxial-Shearing Parameters for Hydrate-Bearing Intermediate Fine Sediment. *Adv. Geo-Energy Res.* **2018**, *2*, 43–52. [CrossRef]
16. Miyazaki, K.; Yamaguchi, T.; Sakamoto, Y.; Aoki, K. Time-Dependent Behaviors of Methane-Hydrate Bearing Sediments in Triaxial Compression Test. *Int. J. JCRM* **2011**, *7*, 43–48. [CrossRef]
17. Hu, Q.; Li, Y.; Sun, X.; Chen, M.; Bu, Q.; Gong, B. Integrating Test Device and Method for Creep Failure and Ultrasonic Response of Methane Hydrate-Bearing Sediments. *Rev. Sci. Instrum.* **2023**, *94*, 025105. [CrossRef]
18. Zhu, Z.; Luo, F.; Zhang, Y.; Zhang, D.; He, J. A Creep Model for Frozen Sand of Qinghai-Tibet Based on Nishihara Model. *Cold Reg. Sci. Technol.* **2019**, *167*, 102843. [CrossRef]
19. Deng, H.; Dai, G.; Azadi, M.R.; Hu, Y. Drained Creep Test and Creep Model Evaluation of Coastal Soft Clay. *Indian Geotech. J.* **2021**, *51*, 1283–1298. [CrossRef]
20. Liu, B.; Zhan, L.; Lu, H.; Zhang, J. Advances in Characterizing Gas Hydrate Formation in Sediments with NMR Transverse Relaxation Time. *Water* **2022**, *14*, 330. [CrossRef]
21. Wu, Q.; Lin, N.; Li, L.; Chen, F.; Zhang, B.; Wu, Q.; Xv, X.; Wang, X. Experimental Study on the Kinetics of the Natural Gas Hydration Process with a NiMnGa Micro-/Nanofluid in a Static Suspension System. *Water* **2022**, *14*, 1–17. [CrossRef]
22. Lin, H.; Zhang, X.; Wang, Y.; Yong, R.; Fan, X.; Du, S.; Zhao, Y. Improved Nonlinear Nishihara Shear Creep Model with Variable Parameters for Rock-Like Materials. *Adv. Civ. Eng.* **2020**, *2020*, 7302141. [CrossRef]
23. Li, G.; Wang, Y.; Hu, Y.; Wang, D.; Yang, X.; Li, Y.; Zhou, Z.; Zhang, S. Shear Creep Mechanical Properties and Damage Model of Mudstone in Open-Pit Coal Mine. *Sci. Rep.* **2022**, *12*, 5148. [CrossRef]
24. Zhou, H.W.; Wang, C.P.; Han, B.B.; Duan, Z.Q. A Creep Constitutive Model for Salt Rock Based on Fractional Derivatives. *Int. J. Rock Mech. Min. Sci.* **2011**, *48*, 116–121. [CrossRef]
25. Yan, B.; Guo, Q.; Ren, F.; Cai, M. Modified Nishihara Model and Experimental Verification of Deep Rock Mass under the Water-Rock Interaction. *Int. J. Rock Mech. Min. Sci.* **2020**, *128*, 104250. [CrossRef]
26. Cheng, H.; Zhang, Y.; Zhou, X. Nonlinear Creep Model for Rocks Considering Damage Evolution Based on the Modified Nishihara Model. *Int. J. Geomech.* **2021**, *21*, 04021137. [CrossRef]
27. Liu, J.; Wu, F.; Zou, Q.; Chen, J.; Ren, S.; Zhang, C. A Variable-Order Fractional Derivative Creep Constitutive Model of Salt Rock Based on the Damage Effect. *Geomech. Geophys. Geo-Energ. Geo-Resour.* **2021**, *7*, 46. [CrossRef]
28. Chen, M.; Li, Y.; Zhang, Y.; Qi, M.; Wu, N. Recent Advances in Creep Behaviors Characterization for Hydrate-Bearing Sediment. *Renew. Sustain. Energy Rev.* **2023**, *183*, 113434. [CrossRef]
29. Li, Y.; Hu, Q.; Wu, N.; Wang, H.; Sun, X.; Hu, G.; Sun, Z.; Jiang, Y. Acoustic Characterization for Creep Behaviors of Marine Sandy Hydrate-Bearing Sediment. *Sci. Rep.* **2023**, *13*, 22199. [CrossRef]

30. Liu, X.; Liu, L.; Li, Z.; Yao, Z. Experimental Analysis on Creep Properties of Frozen Silty Mudstone Considering Conservation of Energy. *J. Test. Eval.* **2021**, *49*, 417–434. [CrossRef]
31. Liu, W.; Zhou, H.; Zhang, S.; Jiang, S.; Yang, L. A Nonlinear Creep Model for Surrounding Rocks of Tunnels Based on Kinetic Energy Theorem. *J. Rock Mech. Geotech. Eng.* **2023**, *15*, 363–374. [CrossRef]
32. Chen, W.; Wan, W.; Xie, S.; Kuang, W.; Peng, W.; Wu, Q.; Tong, S.; Wang, X.; Tang, X. Features and Constitutive Model of Gypsum's Uniaxial Creep Damage Considering Acidization. *Geofluids* **2020**, *2020*, 8874403. [CrossRef]
33. Nguyen, N.N.; Nguyen, A.V.; Dang, L.X. The Inhibition of Methane Hydrate Formation by Water Alignment underneath Surface Adsorption of Surfactants. *Fuel* **2017**, *197*, 488–496. [CrossRef]
34. Shan, R.; Bai, Y.; Ju, Y.; Han, T.; Dou, H.; Li, Z. Study on the Triaxial Unloading Creep Mechanical Properties and Damage Constitutive Model of Red Sandstone Containing a Single Ice-Filled Flaw. *Rock Mech. Rock Eng.* **2021**, *54*, 833–855. [CrossRef]
35. Shukla, A.; Joshi, Y.M. Boltzmann Superposition Principle for a Time-Dependent Soft Material: Assessment under Creep Flow Field. *Rheol. Acta* **2017**, *56*, 927–940. [CrossRef]
36. Qin, X.; Liang, Q.; Ye, J.; Yang, L.; Qiu, H.; Xie, W.; Liang, J.; Lu, J.; Lu, C.; Lu, H.; et al. The Response of Temperature and Pressure of Hydrate Reservoirs in the First Gas Hydrate Production Test in South China Sea. *Appl. Energy* **2020**, *278*, 115649. [CrossRef]
37. Yamamoto, K.; Wang, X.-X.; Tamaki, M.; Suzuki, K. The Second Offshore Production of Methane Hydrate in the Nankai Trough and Gas Production Behavior from a Heterogeneous Methane Hydrate Reservoir. *RSC Adv.* **2019**, *9*, 25987–26013. [CrossRef]
38. Liu, H.Z.; Xie, H.Q.; He, J.D.; Xiao, M.L.; Zhuo, L. Nonlinear Creep Damage Constitutive Model for Soft Rocks. *Mech. Time-Depend. Mater.* **2017**, *21*, 73–96. [CrossRef]
39. Tian, Z.; Liu, Y.; Zhang, X.; Zhang, Y.; Zhang, M. Formation Mechanisms and Characteristics of the Marine Nepheloid Layer: A Review. *Water* **2022**, *14*, 678. [CrossRef]

water

MDPI

Article

Turbidity Currents Carrying Shallow Heat Invading Stable Deep-Water Areas May Be an Unrecognized Source of "Pollution" in the Ocean

Hao Tian [1,2], Guohui Xu [3,*], Jingtao Zhao [1,2,*], Yupeng Ren [4] and Hanru Wu [3]

1 Qingdao Institute of Marine Geology, China Geological Survey, Qingdao 266237, China; tianhao199207@163.com
2 Laboratory for Marine Mineral Resources, Qingdao Marine Science and Technology Center, Qingdao 266237, China
3 Key Laboratory of Marine Environmental Geological Engineering, Ocean University of China, Qingdao 266100, China; wuhr0825@126.com
4 Qingdao National Laboratory of Marine Science and Technology, Qingdao 266061, China; renyupeng@ouc.edu.cn
* Correspondence: xuguohui@ouc.edu.cn (G.X.); zhaojingtao113@163.com (J.Z.)

Abstract: When turbidity currents carrying shallow heat enter stable stratified lakes or oceans, they can trigger changes in temperature, dissolved chemicals, oxygen concentrations, and nutrient mixing through the stable stratified environmental water. Although it is common for warm turbidity currents to invade stable regions, the impact of turbidity current characteristics on environmental entrainment and the impact of temperature changes caused by the mixing of warm turbidity currents with the environment remains poorly understood. In this study, systematic experiments on warm turbidity currents were conducted to understand how sediment-driven turbidity currents lead to mixing in stable stratification using existing environmental entrainment numbers. The experimental results show that the dimensionless numbers R_s (the ratio of the change in environmental water concentration caused by salinity to sediment load), R_T (the ratio of the change in environmental water concentration caused by temperature difference to sediment load), and R_0 (non-dimensional density ratio) control the flow process of warm turbid plumes, and corresponding functional relationships are summarized. The frequent occurrence of warm turbidity currents events caused by increasingly prominent environmental problems cannot be ignored, as it directly affects the deep-water environment of lakes or coastal oceans, which may be an important contribution to heat transfer that has not been evaluated in previous ocean events.

Keywords: turbidity current; carrying heat; temperature structure; environmental entrainment; marine environment

1. Introduction

Turbidity currents are sediment-rich gravity currents that tend to be generated in relatively shallow, warm shelf environments, moving down slopes and spreading into deep-water lakes or deep oceans [1–5], as shown in Figure 1. The intrusion of external heat that breaks the stable and persistent energy stratification of pristine lakes or oceans is often referred to as "thermal pollution", and the steady state of lakes and coastal oceans is greatly affected by them. They will destabilize the stratification of heat, oxygen concentration, nutrients, and dissolved chemicals in the ambient water column, directly affecting the abundance and community structure of organisms in the marine environment [6–10]. In addition, the stable deep-sea environment makes deep-sea organisms highly sensitive to small temperature changes. Nematodes are a major component of marine sediments, accounting for more than 90% of the abundance of all benthic organisms in the deep sea [11,12]. Small species such as nematodes, which typically reproduce between days and months [13], respond strongly and rapidly to changes in temperature. The relationship between the magnitude of temperature changes and changes in species richness provides evidence that small temperature changes of 0.1 C or less are sufficient to cause significant changes in the biodiversity and community structure of deep-sea nematode assemblages [14]. Therefore, whether turbidity currents from the shallower and warmer parts of the continental shelf have a "thermal pollution" effect during their migration from the seabed to deeper waters, thus affecting the temperature of the ambient water column, has not previously been considered in the field of global heat transport systems.

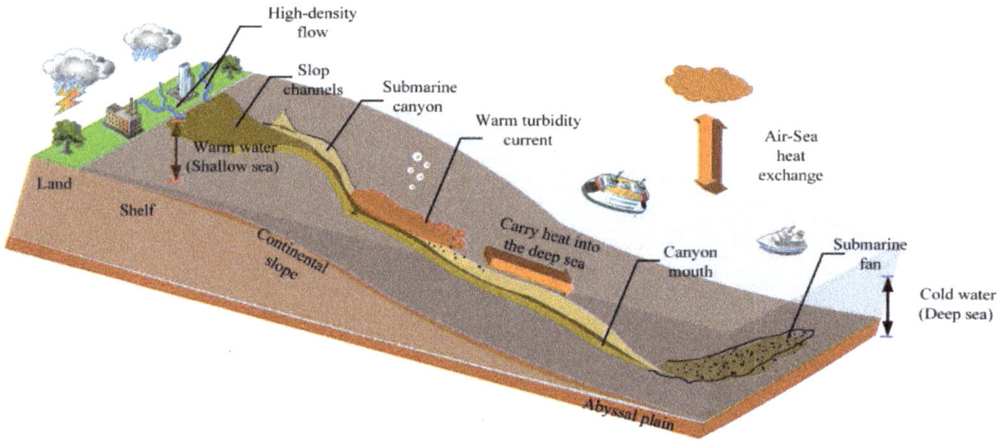

Figure 1. A schematic of the energy and matter changes affected by turbidity current transmission into ocean systems. (Note: Different triggers cause turbidity currents to enter the deep ocean environment along the ocean floor and carry heat from upper source areas further from the continental shelf into the deep ocean domain. Turbidity currents can act as a mechanism for transporting upper-layer heat, thus producing processes that alter the energy and material of the ocean system).

For high-density rivers flowing into the ocean, turbidity currents transport most of the sediment that is transported from the continental shelf or shallow delta environments to the deep-sea plain [15,16]. Sediments that move during these events are typically suspended not in cold, salty water that is characteristic of ocean basins but in fresher or warmer surface sea water or river water that has a lower density. The fate of the suspended sediment depends on the initial volume density of the turbidity current [17–19]. The volume density of the turbidity current depends on the initial density of river water (caused by the initial freshness caused by temperature) and sediment concentration (salinity that can be dissolved in water and concentration of undissolved sediment). The potential density of the ocean depends on the salinity, temperature, and depth of the sea water: the typical density of sea water is about $\rho_a \approx 1031\ \mathrm{kg/m^3}$ at a depth of 1 km and $\rho_a \approx 1040\ \mathrm{kg/m^3}$ at a depth of 2 km [20]. Due to the stable stratification of density, energy (heat), oxygen concentration, nutrients, and dissolved chemicals in the ocean, the warm turbidity currents generated in the continental shelf environment entering the deep sea are likely to contain light environmental water bodies and shallow heat, which may break the stable stratification of environmental water bodies along the turbidity current movement path. Table 1 summarizes the mean water temperature in the turbidity trigger areas and the ambient water temperature of the deeper seabed of the turbidity current path under different trigger conditions, indicating that there is a clear temperature difference between the ambient water temperature in the turbidity source areas and the migration areas. Most lakes and oceans exhibit significant density stratification or material stratification for most of the year, and it is important to understand the process of turbidity currents carrying heat and how to change this stratification.

Table 1. A comparison of ambient water temperature in the trigger source area and along the movement of turbidity currents.

Location of Turbidity Current	Time	Trigger Mechanism	Trigger Source Temperature	Ambient Temperature of Water Body Along the Movement of Turbidity Currents	Source
Gaoping Canyon	7–10 August 2009	Storms cause high-density floods	Gaoping River temperature: above 28 °C	Sea area: 0–200 m: 30–15 °C 200–500 m: 15–8 °C 500–1000 m: 8–5 °C 1000–4000 m: 5–3 °C	[21,22]
Gaoping Canyon	14–17 September 2016	Storms cause high-density floods	Gaoping River temperature: above 28 °C	Sea area: 0–200 m: 30–15 °C 200–500 m: 15–8 °C 500–1000 m: 8–5 °C 1000–4000 m: 5–3 °C	[23,24]
Eel Bay Canyon	12 January–3 April 2000	Storms, not directly related to river floods	0–100 m: 14–10 °C	Sea area: 100–200 m: 10–9 °C 200–400 m: 9–7 °C 400–1500 m: 7–3 °C	[25,26]
Monterey Canyon	17–19 December 2002	Highly consistent with the storm activity	0–200 m: 14–9 °C	Sea area: 200–500 m: 9–5 °C 500–1500 m: 5–3 °C	[26]
Monterey Canyon	Winter	Failure of canyon head or slope due to sediment accumulation	0–200 m: 14–9 °C	Sea area: 200–500 m: 9–5 °C 500–1500 m: 5–3 °C	[26,27]
Congo Canyon	December 2019–March 2010	Increased flow of high-density rivers	Congo River water temperature: above 26 °C	Sea area: 0–200 m: 30–15 °C, 200–500 m: 15–8 °C Under 500 m: 8–3 °C	[28,29]
Val Canyon	17–19 December 2008 5–8 February 2009	High-density flow during river flooding and some local storms	Annual average temperature of Val River: 17–25 °C	Sea area: 0–510 m: 25–13 °C 510–1280 m: 13–3 °C	[30,31]
Lion Bay Canyon	12 November 2003	Rivers flooded during the storm	0–100 m: 17–16 °C	Sea area: 100–200 m: 16–15 °C 200–300 m: 15–13 °C Under 300 m: 13–3 °C	[32]

Note: There is a negative correlation between the ambient temperature of the water body and an increase in depth along the turbidity current path. The temperature data are supplemented from https://argo.ucsd.edu/data/data-visualizations Ocean data system (accessed on 2 December 2024).

Heat-carrying turbidity currents can transport shallow warm water masses to deep-water areas, generating intense convective mixing, overturning the water column, and releasing dissolved gases into the atmosphere. This can have devastating impacts on fish populations and potentially lead to habitat destruction in the surrounding area, such as degassing events in other so-called "killer lakes" such as Nyos and Monoun [33]. However, there is currently no research on turbulent mixing processes originating from shallow warm turbidity currents, and most previous studies on reverse buoyant flows have been driven by the sedimentation behavior of density flows. For example, Sparks (1993) first investigated sedimentation as the main mechanism leading to a decrease in stacking density during turbulent mixing through experiments [34]. Cantelli (2008) studied turbidity currents generated by underwater volcanic eruptions and investigated the deposition and lofting of two different density and particle size sand suspensions in relatively cold environments, indicating that the lofting process affects downstream mixing and sediment thickness [35]. Gladstone (2010) conducted a sedimentary model study of turbidity current experiments with reverse buoyancy, and an important feature of the results is the non-uniqueness of the deposit structure: different initial current compositions can generate deposits with very similar bed profiles and grading characteristics, highlighting the difficulty of reconstructing mother flow properties from field data [36]. Steel et al. (2016, 2017) conducted field observations and laboratory experiments, indicating that the initial characteristics of turbidity currents can change convective processes, affecting velocity, sediment thickness, and range [37–40]. Lu (2022) conducted an experiment on sediment-laden rivers into stratified water bodies, dividing turbidity currents into overflows, interflows, or underflows, depending on density contrast, and using dimensionless parameters to describe the velocity characteristics of turbidity convection [41]. Although it is known that the particle size and sediment concentration of turbidity currents significantly affect convective processes (velocity structure, concentration structure, sediment thickness, and range), there is little theoretical research on the mechanism of temperature's impact on turbidity current movement characteristics, and whether shallow heat carried by turbidity currents can cause "pollution" to the stable hydrological environment in deep-water areas. As a result, there is little understanding of this knowledge gap. This knowledge gap reflects a lack of detailed information on the mechanism of turbidity currents carrying shallow heat to deep-water areas and how temperature conditions affect turbidity current convective mixing changes.

This research seeks to explore sediment and thermal transport in turbidity currents, analyzing how the initial characteristics of warm currents affect the stable stratification of aquatic environments and their primary role in turbidity current motion, excluding complex elements such as wind stress and wave action. In particular, we study the effects of dimensionless parameters R_s (the ratio of changes in ambient water concentration due to salinity to sediment load) and R_T (the ratio of changes in ambient water concentration due to temperature difference to sediment load) on the turbulent mixing of turbidity currents. Based on descriptions of warm turbidity current experiments, we explore how the dynamics of turbidity currents depend on the relative contributions of temperature gradient, sediment load, and turbidity current volume. Non-dimensional density ratios R_0 are used to investigate the relative importance of buoyancy-driven and convective motions. By understanding the behavior of warm turbidity currents in laboratory experiments, we use existing environmental entrainment functions E to understand how sediment-driven turbidity currents can lead to stable mixing. A simplified application is to estimate how much heat induced by warm turbidity currents triggered by extreme hydrological events or earthquakes can carry to deeper layers, potentially disrupting stable stratification in lakes or deep oceans. This encourages us to consider the role of turbidity events in thermal transport in bottom channels, which may be an important mechanism for energy exchange in lake and deep-sea environments.

2. Theory

In multiphase gravity flows, the buoyancy of the flow induced by the relative density difference caused by temperature cannot be ignored. This will inevitably affect the movement and mixing of reverse buoyancy turbidity currents. We introduce the density ratio theory proposed by Lu et al. (2022) and consider the contribution of temperature gradients to turbulent mixing and convective processes [40]. Establishing a simplified system helps to understand the dynamic process of rivers entering the receiving environment with sediment and heat, and this model ignores waves' influence on bottom currents' heat transport. The initial density of the turbidity current is ρ_M, the density of ambient water within the turbidity current is ρ_0, and the density of ambient water in the ocean or lake is ρ_E. In addition, the relative contributions of temperature, sediment, and salinity in each layer can be specified according to the following formula:

$$\begin{aligned} \rho_M &= \rho_0 + \Delta\rho_c \\ \rho_E &= \rho_0 + \Delta\rho_s + \Delta\rho_T \end{aligned} \tag{1}$$

Generally, ρ_0 is the initial density of the river, $\Delta\rho_c$ is the increase in sediment density difference, and $\Delta\rho_T$ can increase or decrease the density, such as in the case of a river with high temperature or a glacier river. It should be noted that $\Delta\rho_T$ is the density difference caused by the temperature difference between the river and the ambient water. Therefore, if the river is hotter than the ambient water ρ_E, $\Delta\rho_T$ is positive. The negative sign in the equation remains unchanged. Equation (1) ensures that the temperature contribution changes ρ_M. $\Delta\rho_c$ is the density difference caused by sediment, which is always positive, and $\Delta\rho_s$ is the density difference caused by salinity, which is also always positive.

From Equation (1), two dimensionless variables can be derived: the first ratio compares the magnitude of temperature differences with the incoming sediment load, defined as:

$$R_T = \Delta\rho_T / \Delta\rho_c \tag{2}$$

The second ratio compares the difference in "freshness" between the environmental water and the river with the incoming sediment load, defined as:

$$R_S = \Delta\rho_s / \Delta\rho_c \tag{3}$$

These two dimensionless parameters can now be used to describe the formation of overflows, interflows, and underflows, as shown in Figure 2. Overflows occur when $\rho_M < \rho_E$ and $1 - R_T < R_S$; interflows occur when $\rho_M = \rho_E$ and $1 - R_T = R_S$, respectively. If the incoming density is $\rho_M > \rho_E$, underflows occur when $1 - R_T > R_S$.

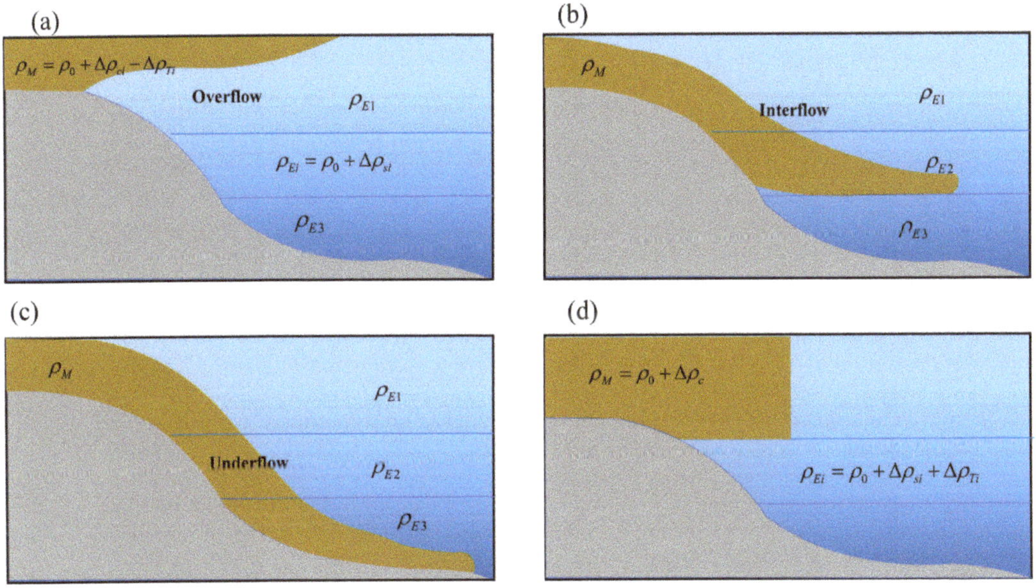

Figure 2. Schematic diagram of three forms of motion of warm turbidity currents entering lakes or oceans. (**a**) Overflow (**b**) Interflow (**c**) Underflow (**d**) Initial state (revised based on research by [40]).

In summary, the new information is first used to estimate when convection may be important in various river plumes. Equations (2) and (3) can be combined to define the following mechanisms:

$$\begin{aligned} &Overflows: R_T + R_S > 1; \\ &Interflows: R_T + R_S = 1; \\ &underflows: R_T + R_S < 1 \end{aligned} \tag{4}$$

Specifically, the convective lofting and environmental mixing activity of turbidity currents are driven by density anomalies generated by sediment load $\Delta\rho_c$ and environmental water density (salinity) $\Delta\rho_s$ as well as water freshness (temperature) $\Delta\rho_T$. When the sediment load disappears, the density of the remaining underflow is now $\rho_M = \rho_0$. Therefore, when ρ_M is less than the density of the lower fluid, i.e., $\rho_0 < \rho_0 + \rho_s + \rho_T$, the rise and mixing of "low-density" substances in the turbid bottom flow and interflow may occur under the following conditions: $0 < R_s + R_T$, which indicates the lower limit of turbulent mixing rise phenomenon. The interval between the turbidity current bottom current and the interflow during the most severe convective mixing is $0 < R_s + R_T < 1$. So, letting $R_0 = R_s + R_T$, by designing experiments with different R_0 values, the fluid dynamic characteristics of turbidity currents can be studied.

Turbulent mixing at the interface of turbidity currents causes ambient water to be entrained into the turbidity current, thereby diluting the flow. Dilution increases the flow thickness, reduces the flow velocity, promotes sediment settling, changes the distribution of heat and sediment concentration within the turbidity current, and causes changes in the stability of environmental water stratification. Therefore, environmental entrainment (E) directly affects the propagation of turbidity currents by changing the density excess, which can quantify the convective activity in laboratory experiments and is an important dimensional parameter for understanding the convective intensity and heat mixing processes of turbidity currents. Morton et al.'s early work provided a well-known theory that the entrainment rate at any given point in a fluid is proportional to the characteristic velocity at that point [42].

$$E = \frac{w_e}{U} \tag{5}$$

For turbidity currents, entrainment is usually quantified as a dimensionless parameter E, which is expressed as the ratio between the ambient entrainment velocity w_e and the characteristic velocity U of turbidity currents. Jacobson (2014) defined w_e for sediment-laden water flow based on the rate of change A of the cross-sectional area of the water flow [43].

$$w_e = \frac{UdA}{x_f dx_f} \tag{6}$$

where E is the environmental entrainment coefficient; ω_e is the environmental entrainment velocity; U is the velocity of the turbidity current; x_f is the distance of turbidity current migration; and dA is the vertical influence area of turbidity flow. In this study, dA is taken as the envelope area where the temperature exceeds 0.1 °C during the turbidity movement process per unit time.

3. Experimental Design

To simulate the process of turbidity current events carrying shallow heat to the deep sea, indoor experiments were designed to study heat transfer characteristics during warm turbidity current movement, as well as the relationship between initial characteristics (temperature, concentration, volume) and convective motion. A series of lock–release experiments were carried out using a similar apparatus as He (2017) and Han (2022) [44,45], as shown in Figure 3. The system included an acrylic rectangular flume, and a control system for the lock–release inflow. The rectangular flume had a length of 12 m, a width of 0.5 m, and a height of 0.7 m. A slope with a height of 0.25 m and an angle of 4 degrees (refer to the average slope of the continental slope) was placed on the right side of the experimental device. The slope material was made of acrylic board. The lock gate divided the rectangular flume into two parts: the right (upstream) part contained the warm heavy current (sediment–water mixture), and the left (downstream) part filled with homogeneous ambient fluid was composed of pure water at room temperature. In turbidity current preparation, the sum of the volume fractions of the two phases is 1, i.e., $V = \int_V \phi_i dV$, where $\phi = \frac{V_s}{V_s+V_w} = \frac{m_s}{\rho_s(V_s+V_w)}$, so the turbidity concentration (ρ_m) can be expressed as $\rho_m = (1-\phi)\rho_w + \phi\rho_s$, where ϕ is the particle volume fraction; V is the volume of turbidity current; V_s is the volume of sediment; V_w is the volume of water; m_s is the sediment mass; and ρ_s is the sediment density.

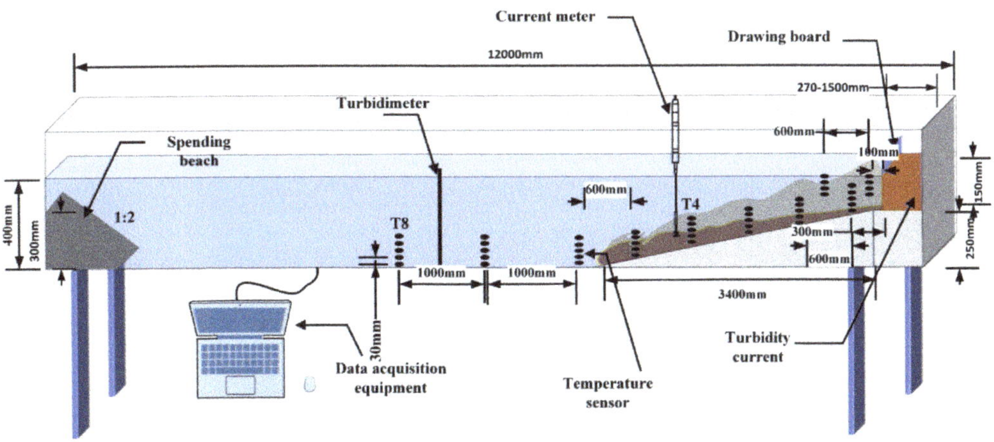

Figure 3. Schematic of experimental device.

Real-time water temperature was collected using temperature sensors (model: PT100) with a frequency of 50 Hz and an accuracy of 0.5%. To monitor the temperature changes caused by convective mixing and lofting during the movement of turbidity currents and their impact on the vertical temperature of ambient water bodies, 10 flow direction sensor groups were established, each consisting of 5 temperature sensors distributed at different positions in the turbulent flow height range (Figure 4a). Previous studies have shown that turbidity flow is mainly composed of silt and clay [15,16,46,47]. Therefore, sedimentary soil from the Yellow River Delta was selected for the experiment, with a median particle size of 0.026 mm and a grading curve shown in Figure 4b.

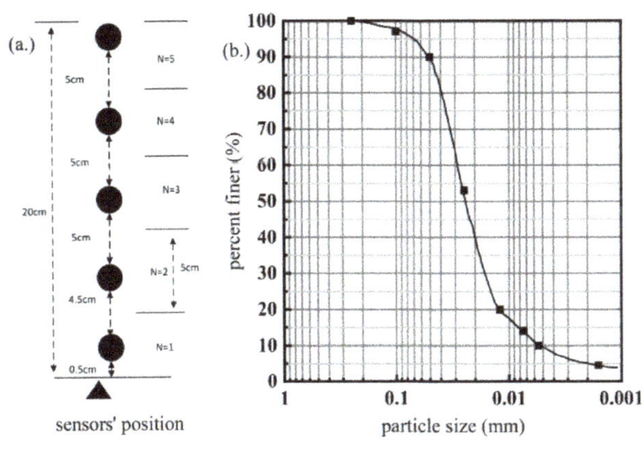

Figure 4. A vertical layout of the sensors and particle grading curve of the experimental soil.

Based on Equation (1), turbidity currents with varying temperatures and volumes were designed and tested, as detailed in Table 2. The purpose of this study is to investigate the factors contributing to the intensity of turbidity currents, including freshwater density, temperature gradient, sediment load, and turbidity current volume. To prepare the warm turbidity current, pure water was uniformly heated to the desired temperature using a heater, which caused a change in $\Delta\rho_T$ [48]. The soil sample was then heated to the same temperature using a constant-temperature electric heating box before being mixed to create the turbidity current. The experiment was conducted in a straight fixed channel, and the channel-slope profile was designed to replicate the actual shape observed in the field. A comprehensive set of temperature and velocity data was collected from turbidity current experiments to analyze the flow dynamics, entrainment characteristics, and flow evolution of the partially constrained flow.

Table 2. Parameters for experiment conditions.

R_0	Sediment Concentration/ a	$\Delta\rho_c$/kg/m^3	V/L	Turbidity Water Temperature/°C	Freshness of Turbid Water Bodies /kg/m^3	Ambient Water Temperature/°C	ρ_E/kg/m^3	ΔT	$\Delta\rho_c$/ kg/m^3
0.500	0.006	10.00	20	40.80	992.20	25.80	997.20	15.0	5.0
0.100	0.030	50.00	20	40.80	992.20	25.80	997.20	15.0	5.0
0.050	0.062	100.00	20	40.80	992.20	25.80	997.20	15.0	5.0
0.025	0.124	200.00	20	40.80	992.20	25.80	997.20	15.0	5.0
0.017	0.186	300.00	20	40.80	992.20	25.80	997.20	15.0	5.0
0.013	0.186	300.00	20	37.80	993.20	25.80	997.20	12.0	4.0
0.010	0.186	300.00	20	34.80	994.20	25.80	997.20	9.0	3.0
0.007	0.186	300.00	20	31.80	995.20	25.80	997.20	6.0	2.0
0.003	0.186	300.00	20	28.80	996.20	25.80	997.20	3.0	1.0
0.010	0.186	300.00	40	34.80	994.20	25.80	997.20	9.0	3.0
0.010	0.186	300.00	60	34.80	994.20	25.80	997.20	9.0	3.0
0.010	0.186	300.00	80	34.80	994.20	25.80	997.20	9.0	3.0
0.010	0.186	300.00	100	34.80	994.20	25.80	997.20	9.0	3.0

4. Experimental Results

4.1. Temperature Monitoring of Vertical Structural Changes

Upon loosening the lockbox door, the mud instantly collapsed, generating a turbidity current that advanced along the bed. As the turbidity current flowed, it presented a distinct interface with the surrounding water. Turbulent mixing occurred, resulting in curling and expansion backwards. The interface ceased to be distinct, and smaller soil particles remained suspended in the surrounding water. The process is a typical turbidity current, as shown in Figure 5.

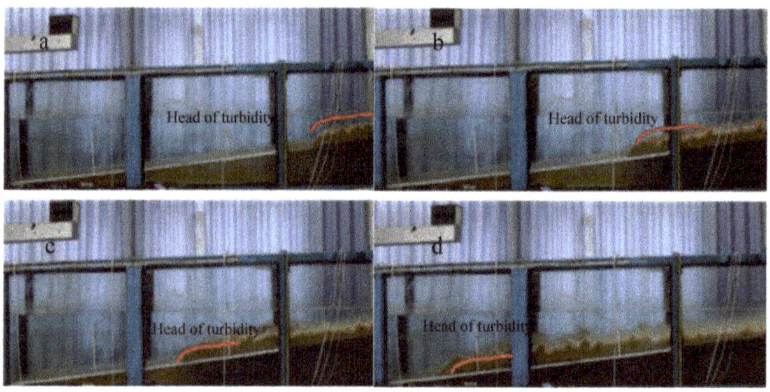

Figure 5. Transport process during the turbidity heat transfer experiment. (**a**–**d**) show the heights of turbidity current flowing through a slope at different times.

The dynamic distribution of the maximum thermal shock temperature through each group of temperature sensors as the turbidity current passes through them is illustrated in Figures 6 and 7. As the turbidity current moved, the temperature value exponentially decreased with height, and the vertical temperature distribution was a nearly empirical power relation, $H = A + B * \Delta T^C$, eventually reaching ambient water temperature. The vertical distribution is like the vertical stratification of sediment concentration in turbidity currents [49]. Some experimental results have been listed, and the remaining experimental groups are presented in Appendix A.

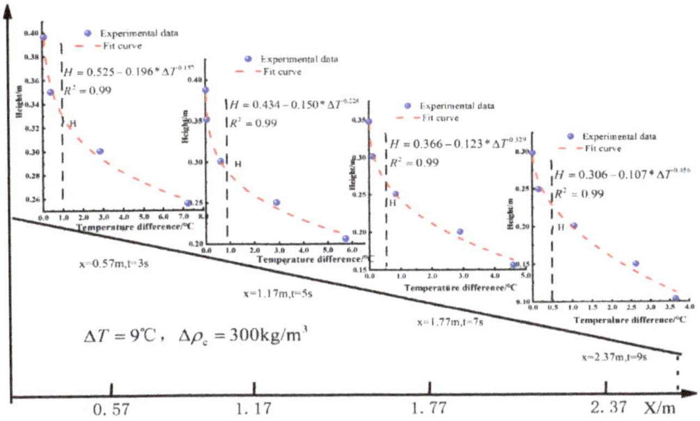

Figure 6. Maximum thermal shock temperature when passing through each group of sensors during turbidity migration (0–2.37 m) under experimental condition $\Delta T = 9$ °C, $\Delta\rho_c = 300\ \mathrm{kg/m^3}$. (Note: The maximum thermal shock temperature when the turbidity current passes through each group of sensors in the migration process of 2.37–5.57 m is shown in the Appendix A. The red dashed line represents the temperature distribution curve when the turbidity current is transmitted).

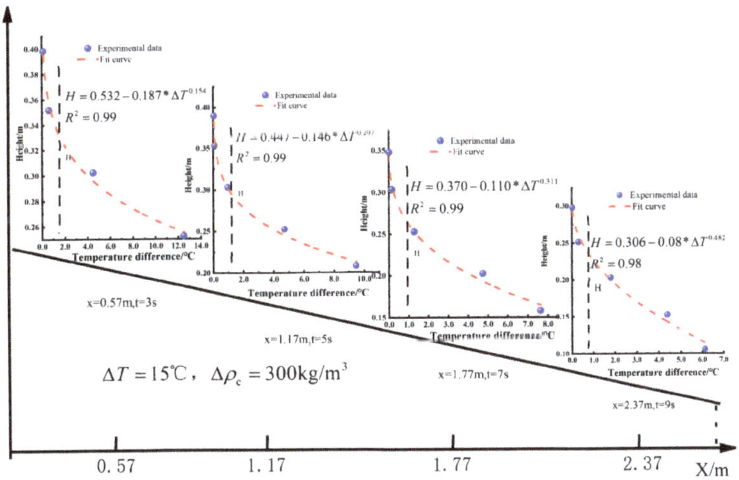

Figure 7. Maximum thermal shock temperature when passing through each group of sensors during turbidity migration (0–2.37 m) under experimental condition $\Delta T = 15$ °C, $\Delta\rho_c = 200\ \mathrm{kg/m^3}$. (Note: Similar experimental groups are concealed, and the remaining figure is in Appendix A).

4.2. Monitoring the Impact of Temperature on Turbidity Current Motion

To monitor temperature changes during turbidity current movement, an array of temperature sensors was used. The central difference method processed sensor data during the turbulent flow to map turbidity current development. We selected the time ($t = 24$ s) when the turbidity current ($\Delta\rho_c = 300\ \mathrm{kg/m^3}, V = 20$ L) head reached the sensor array $T8$ (5.57 m) to monitor the turbidity current impact range A and calculated the number of environmental entrainments E for consistent calculation. Figures 8–10 show the temperature distribution and influence curve along the centerline of the water tank; the horizontal axis is the distance x in the turbulent flow direction, and the vertical axis is the height H of the turbidity current.

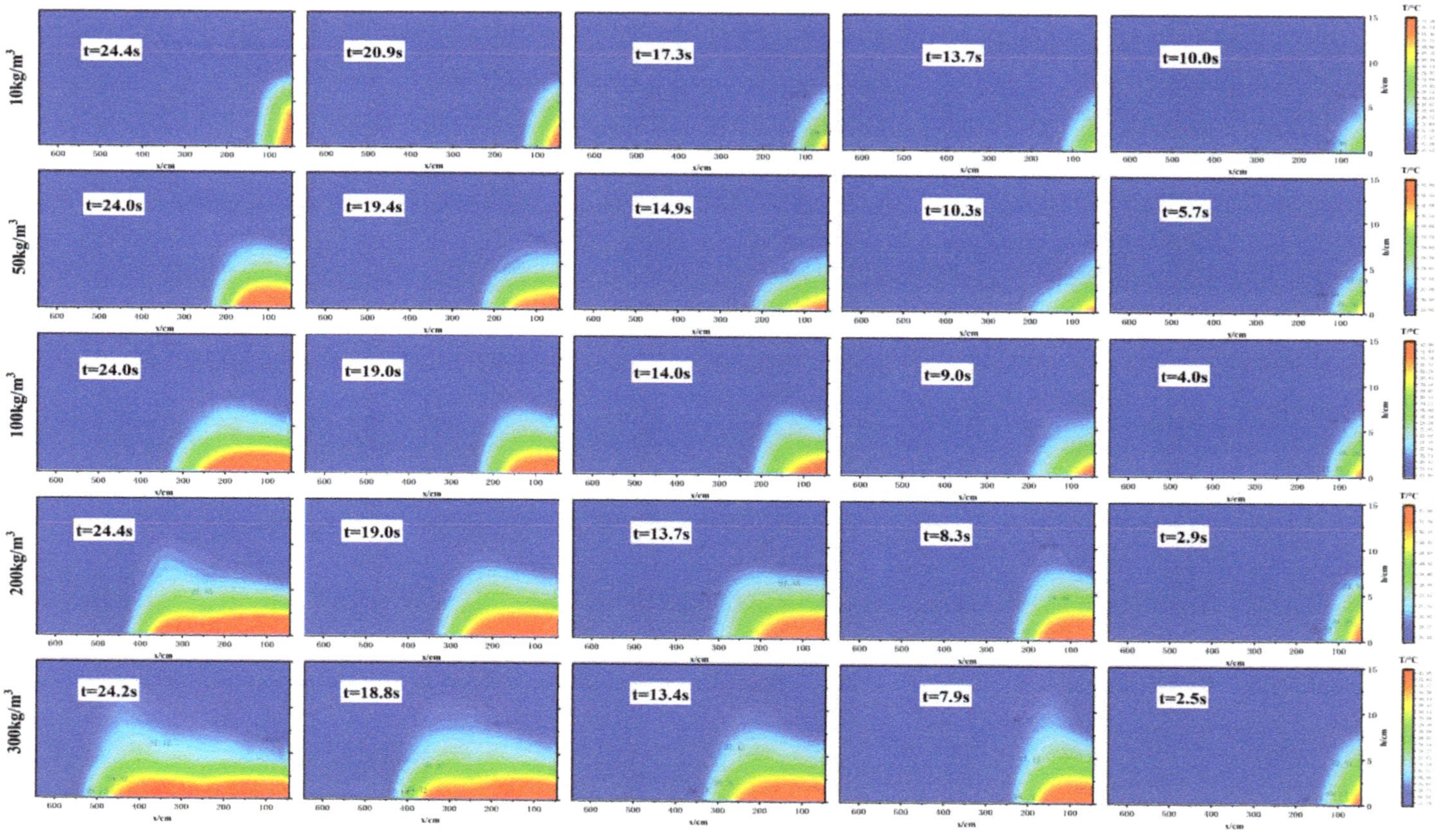

Figure 8. Distribution of temperature profiles and range of influence of turbidity currents of different concentrations 24 s after baffle removal.

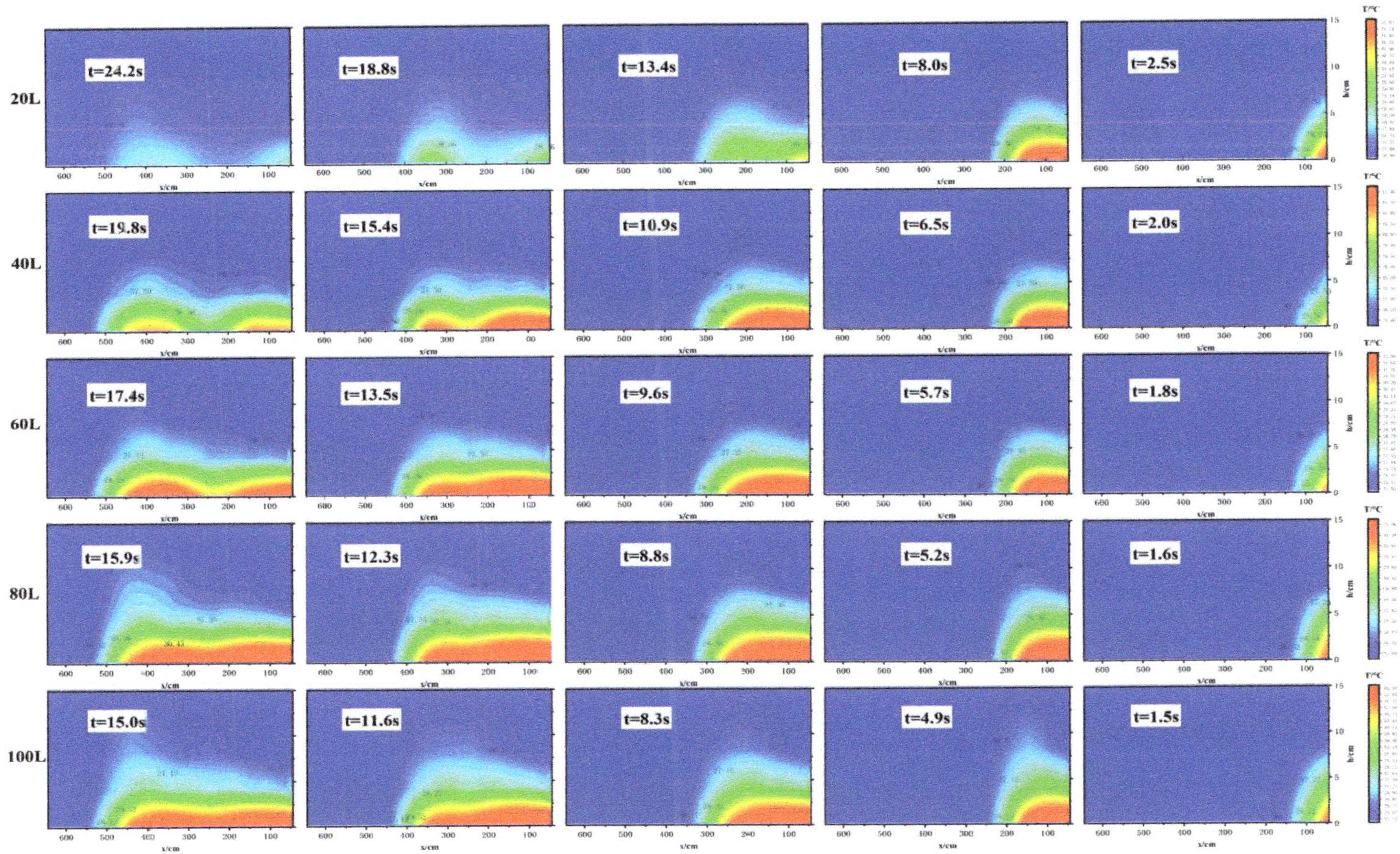

Figure 9. Distribution and influence range of temperature profiles during turbidity flow motion under different temperature differences.

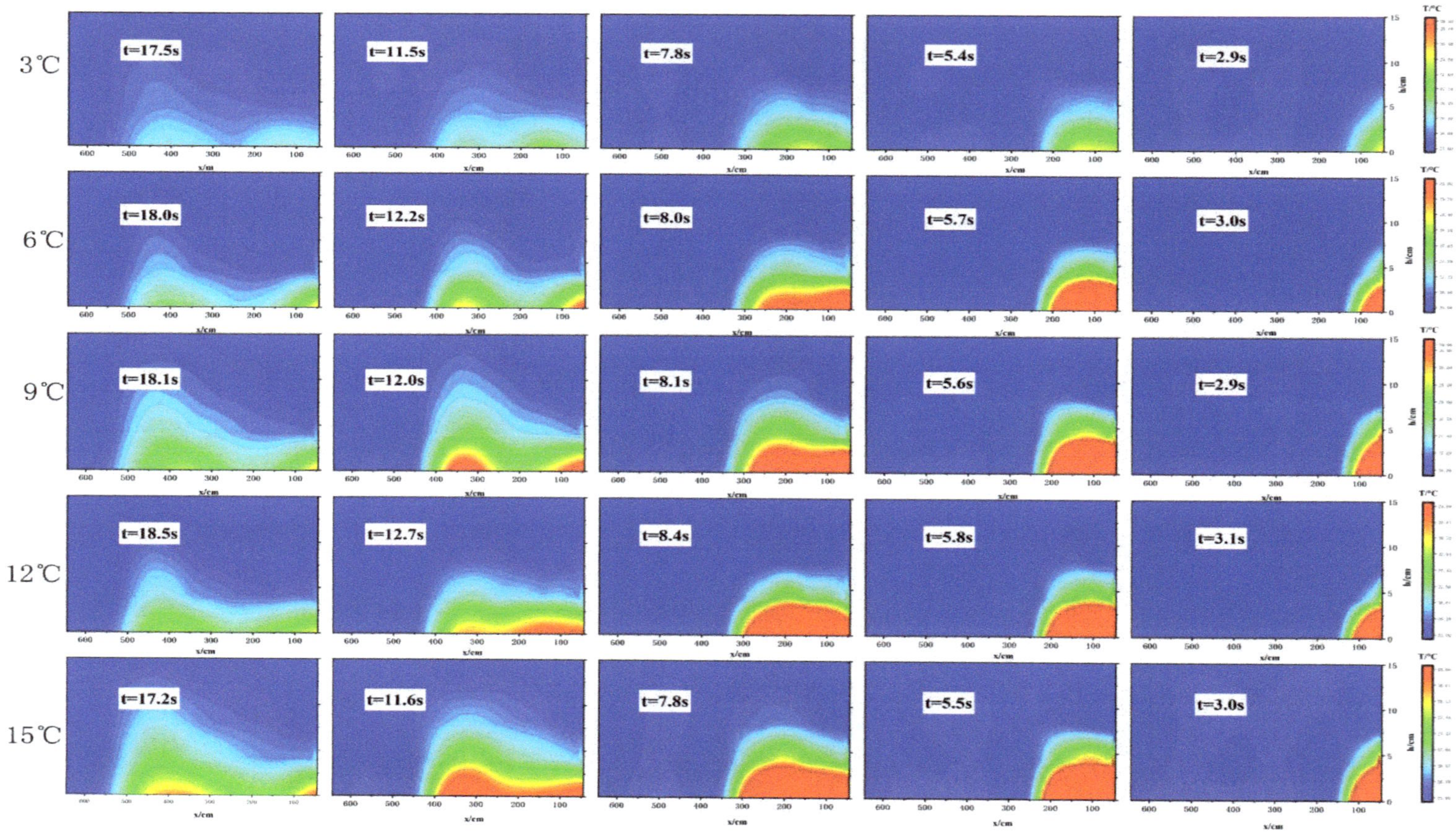

Figure 10. Distribution and influence range of temperature profiles during turbidity flow motion under different temperature differences.

5. Discussion

5.1. Vertical Temperature Distribution in Turbidity Current

Temperature distribution was assessed during all experiments. We selected the temperature variation of the fourth group of sensor arrays ($T4$) in the middle of the flow direction to analyze the vertical structure of temperature during the turbidity current movement process, in order to achieve dimensionless feature calculation of the vertical structure. Normalizing temperature and elevation with depth-averaged temperature and turbidity current thickness, respectively, achieved optimal similarity structure. Mean temperature and mean depth concepts were introduced, and dimensionless experimental data were calculated by normalizing temperature with mean depth using Equation (7). Although there is some variability, the similarity profile represents a good vertical temperature distribution. Figure 8 shows a dimensionless temperature distribution.

$$\Delta T_m = \frac{\int_0^h \Delta T dz}{H} \tag{7}$$

where ΔT_m is depth-averaged temperature, ΔT is the vertical temperature curve, and H is turbidity current height obtained by experimentally fitting the temperature curve, as shown in Figure 11.

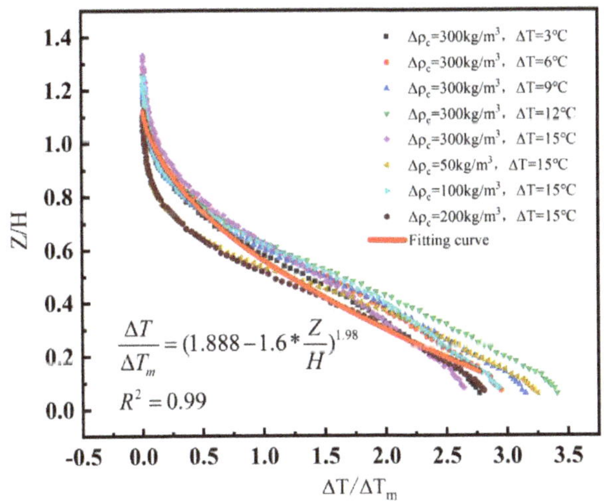

Figure 11. Dimensionless temperature distribution curve during turbidity movement.

To accurately analyze the dynamic heat transfer process of turbidity flows, it is crucial to understand the vertical temperature distribution within them. This study has employed dimensional analysis fitting to determine an empirical power relationship: $\frac{\Delta T}{\Delta T_m} = (1.888 - 1.6 * \frac{Z}{H})^{1.98}$, as shown in Figure 11, which allows us to observe that temperature values decrease exponentially with an increase in height until they equilibrate with their surrounding water bodies. This vertical temperature distribution is like the vertical stratification observed in turbidity currents' sediment concentration [49–51]. Using principles of fluid dynamics and heat transfer, it was possible to determine that vertical dispersion of sediment concentration in a turbidity current was primarily influenced by turbulent mixing and suspension sedimentation, useful insights into entrainment, and effects on the wider oceans. The efficacy of these processes was largely dependent on factors such as the concentration and density distribution of suspended sediments, as well as their particle size. Heat was associated with the sediments and water body of the turbidity system, and the movement of the turbidity current was characterized by the following trend: the sediment concentration was high close to the bed bottom in the vertical direction, and the high concentration hindered sedimentation and inhibited turbulence [3], causing the lower half of the turbidity flow to be relatively stable due to lower amounts of convective heat transfer, resulting in the formation of laminar or weakly turbulent high-concentration near-bed layers containing high heat contents [52,53]. In contrast, the particle size and concentration of the sediments in the upper part of the flow profile were small, which allowed for greater degrees of turbulent mixing and convective heat transfer, which also determined the vertical distribution characteristics for heat (Figure 12).

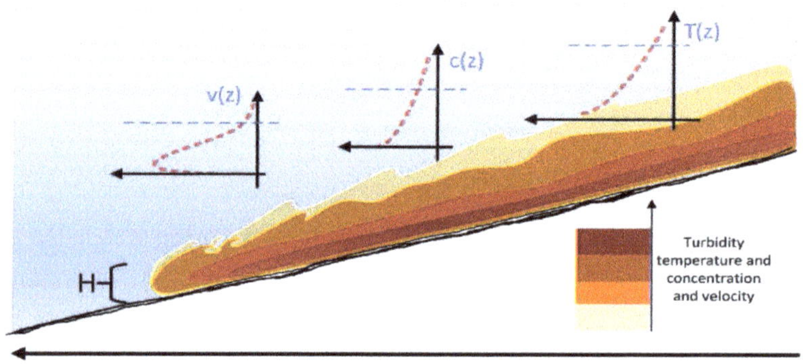

Figure 12. Schematic diagram of heat transport mode and internal temperature distribution of turbidity current in seabed.

A typical heat distribution structure in turbid flow has been summarized, and by performing dimensionless processing through $\frac{\Delta T}{\Delta T_m} = (1.888 - 1.6 * \frac{z}{h})^{1.98}$, it was found that when the turbidity current flowed through each position, the proportion of the heat in the lower half and the total heat flow in that region were always more than 70%, indicating that the heat content of the turbidity current was mainly concentrated in the lower half during the migration process, effectively preserving heat, and areas closer to the bottom bed experienced more heat transfer and greater impact. It can also be theoretically proven that warmed turbidity currents will lead to elevating seafloor temperatures and may trigger the massive release of methane from gas hydrates buried on margins [54,55]. Our experimental results confirmed that the influence of the turbidity current on the bottom bed temperature and heat distribution cannot be ignored.

5.2. The Relationship Between E and Ri

The Richardson gradient number (Ri) is a measure of flow stability that controls the mixing rate between turbulent and ambient fluids. It is determined by the ratio of buoyancy gradient to shear and can be used to predict potential convective behavior and mixing velocity of sediment in a river [18,19]. The Ri formula is given by:

$$Ri = \frac{g'h}{U^2} \tag{8}$$

where g' is reduced gravity, $g' = g\frac{\rho_{turbidity} - \rho_{ambient}}{\rho_{ambient}}$ U is the average velocity of the flow direction, and h is turbidity current height.

However, previous studies have not considered the relationship between initial turbidity current temperature, turbidity current interstitial fluid concentration, sediment content, and ambient fluid density with Ri number [18,49,56]. Based on current indoor experiments, a study was conducted on the relationship between turbidity current stability indicators and environmental entrainment coefficients, considering initial turbidity current temperature, turbidity current interstitial fluid concentration, and sediment content. The calculation results are shown in Table 3.

Table 3. Calculation of *Ri* and environmental entrainment in each group of experiments.

$\Delta T/^{\circ}C$	Migration Distance/m	Time/s	Velocity/m/s	Turbidity Concentration /kg/m^3	Reduced Gravity /g'	Turbidity Current Height/m	*Ri*	*E*
3.0	5.52	24.0	0.23	1297.00	2.996	56.421	4.231	0.004
6.0	5.52	24.0	0.23	1296.00	2.986	56.233	4.217	0.013
9.0	5.52	24.0	0.23	1295.00	2.976	56.044	4.203	0.017
12.0	5.52	24.0	0.23	1294.00	2.966	55.855	4.189	0.019
15.0	5.52	24.0	0.23	1293.00	2.956	55.667	4.175	0.021
15.0	4.80	24.0	0.20	1193.00	1.954	49.275	3.695	0.025
15.0	3.36	24.0	0.14	1093.00	0.952	45.961	3.447	0.036
15.0	2.40	24.0	0.10	1043.00	0.451	44.701	3.352	0.055
15.0	1.44	24.0	0.06	1003.00	0.050	15.682	1.176	0.104
9.0	6.72	24.0	0.28	1295.00	2.976	0.0750	4.203	0.018
9.0	7.68	24.0	0.32	1295.00	2.976	0.0753	4.203	0.017
9.0	8.40	24.0	0.35	1295.00	2.976	0.0751	4.203	0.016
9.0	8.88	24.0	0.37	1295.00	2.976	0.0756	4.203	0.015

The environmental entrainment rate is a key factor in the spatial and temporal development of fluids, which can help explain why turbidity currents can travel thousands of kilometers [57]. We consider temperature conditions within the range of fluid motion stability and analyze whether temperature (carrying shallow heat in turbidity currents) affects the convective intensity of turbidity currents (environmental entrainment coefficient *E*). Figure 13 shows the relationship between the environmental entrainment coefficient *E* and *Ri* considering temperature effects, as shown in $E = \frac{0.129-0.028*Ri}{1-0.07*Ri}$. The turbidity current flow considering the influence of temperature conforms to the traditional Turner (1986) form [58], but the values are different, showing that turbidity currents carrying shallow heat (relatively high temperature) can affect convective flow and environmental mixing during motion. We consider the influence of temperature on the buoyancy reversal points of turbidity currents after a certain distance of movement and the convective mixing characteristics of turbidity currents and explain that the similarity of bed characteristics with "classical" turbidity currents have positive significance. Therefore, in turbidity current events carrying shallow heat to the deep sea, temperature conditions cannot be ignored in the study of fluid dynamics of turbidity currents. Unfortunately, the *Ri* number cannot demonstrate the role and importance of temperature conditions.

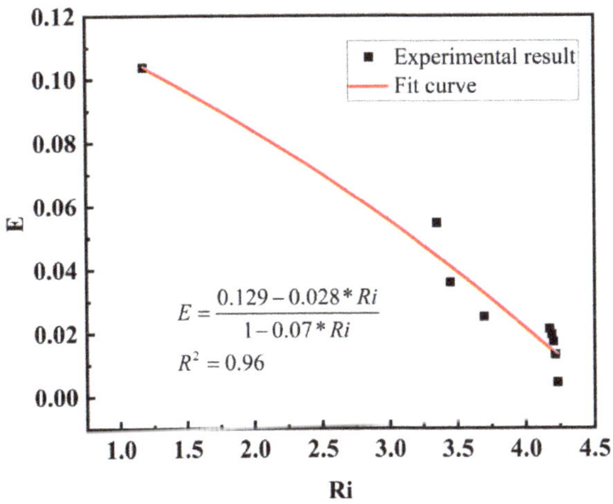

Figure 13. The relationship between environmental entrainment coefficient and Richardson number.

5.3. The Relationship Between E and R_0

The relationship between E and R_i mentioned above cannot characterize the potential convective behavior and convective intensity (entrainment coefficient) of turbidity currents, nor can it reflect the influence of temperature conditions on the characteristics of turbidity current motion. Therefore, using the dimensionless density function R_0 established earlier, the functional relationship between the dimensionless density function R_0 and the environmental entrainment coefficient E is analyzed to intuitively express the role of temperature conditions in turbidity current convective behavior and quantitatively analyze the intrinsic relationship between temperature conditions and convective intensity during turbidity current motion.

When considering the effect of temperature on convective mixing of turbidity currents, as expected, turbidity currents with lower initial temperatures are more conservative in terms of environmental entrainment and convective mixing from the source compared to high-temperature turbidity currents (Table 4). This is because the smaller temperature difference results in a smaller density difference between the interstitial fluid in the initial turbidity current and the surrounding water body, leading to more stable environmental entrainment during the progressive settling process of particles and less likelihood of generating large environmental convective mixing entrainment. Therefore, the relatively stable turbulent motion of the turbidity current is maintained for a longer period of time. Similarly, turbid flow environments with larger temperature differences and lighter interstitial fluids have higher entrainment coefficients (Figure 14) and are closer to the "source" for mixing and diffusion than flows where temperature differences have less impact on interstitial fluid concentration. These results are consistent with previous research on turbidity currents [34,53].

Table 4. Calculation of R_0 and environmental entrainment in each group of experiments.

R_0	x_f	dA	U	ω_e	E
0.0033	5.3000	4.8463	0.2304	0.0010	0.0044
0.0067	5.3000	4.8463	0.2304	0.0030	0.0131
0.0100	5.3000	4.8463	0.2304	0.0039	0.0171
0.0134	5.3000	4.8463	0.2304	0.0045	0.0195
0.0167	5.3000	4.8463	0.2304	0.0049	0.0213
0.0250	4.5700	0.5254	0.1987	0.0050	0.0250
0.0500	3.3100	0.3935	0.1439	0.0052	0.0359
0.1000	2.3100	0.2915	0.1004	0.0055	0.0546
0.5000	1.3000	0.1753	0.057	0.0059	0.1037

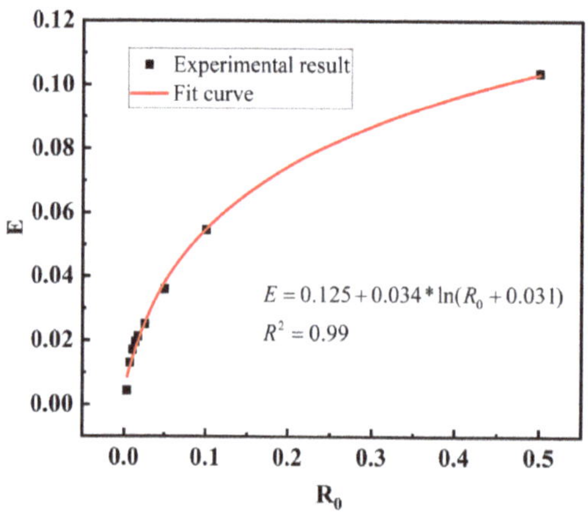

Figure 14. The relationship between environmental entrainment coefficient and R_0.

Two simple dimensionless parameters, R_s and R_T, can fully explain the flow behavior of various turbidity currents and effectively describe the dimensionless functional relationship between turbidity current convection intensity and R in different states. Figure 14 shows that the relative intensity of mixed diffusion and turbulent mixing of turbidity currents depends on R_0, which is the influence of the "temperature and salinity" of turbidity currents, sediment content, and surrounding water density on environmental entrainment. The positive correlation trend between R_0 and convective intensity E is analyzed, and the functional relationship is $E = 0.125 + 0.034\ln(R_0 + 0.031)$. Therefore, the equations of these two key dimensionless parameters can not only quantify the influence of temperature or salinity on the convective layout and turbulent mixing of the source turbidity current but also reflect that the heat carried by the turbidity current will migrate to deeper water areas, changing the stable thermal stratification of the surrounding water.

5.4. Implications and Hypotheses of Turbidity Currents in Marine Environments

To describe the ability of turbidity currents to carry their own heat with different temperature differences and volumes, nine sets of experiments were analyzed. After the turbidity currents flowed over a slope of 5570 mm, the heat carried by the turbidity currents and the initial heat of the turbidity currents were treated as dimensionless. To simplify the calculation, the measurement values of the eighth group of sensors were taken as the average temperature value of each layer in the plane section, and the density of the turbidity current was estimated to be 1000 kg/m^3. The calculation results are shown in Table 5.

$$\phi_i = \frac{Q_i}{Q_0} = \frac{\sum_{1}^{5}\int_{0}^{t} \rho_w v_{TC} d h_{TC} c_w \Delta T d\tau}{\rho_0 V c_0 \Delta T_0} \tag{9}$$

where ϕ_i is the percentage heat content in each layer relative to the total heat content when the turbidity current flows through an interface; Q_i is heat; ρ is fluid density; c is the specific heat of fluid, $c_0 = c_S a - (1-a)c_w$; and V is fluid volume. When the turbid current flows through each layer of an interface, the density and velocity adopt the average value in the vertical direction, where v is the fluid velocity; ΔT is the instantaneous temperature difference between the layer of the turbidity current and the surrounding water body; d is the interface width; and t is the time taken for the turbidity current to pass through the interface.

Table 5. Calculation of heat transfer efficiency of turbidity current.

Turbidity Volume/ L	Turbidity Concentration /kg/m^3	ΔT	c_w/J/(kg·K)	c_s/J/(kg·K)	Initial Heat/kJ	Heat Flux at 5.57 m/kJ	Heat Transport Efficiency/ ϕ
20	1295	3	4182	1182	286.791	38.717	0.135
20	1295	6	4182	1182	573.581	115.8637	0.202
20	1295	9	4182	1182	860.372	212.512	0.247
20	1295	12	4182	1182	1147.163	308.587	0.269
20	1295	15	4182	1182	1433.953	559.242	0.39
40	1295	9	4182	1182	1720.744	578.170	0.336
60	1295	9	4182	1182	2581.116	1022.122	0.396
80	1295	9	4182	1182	3441.488	1369.712	0.398
100	1295	9	4182	1182	4301.860	1772.366	0.412

Indoor experiments allow the visualization of natural turbidity events flowing into lakes or oceans by initiating and developing turbidity currents with similar behavior. In fluid dynamics, flow similarity generally requires geometric similarity, motion similarity (Reynolds number criterion), and dynamic similarity (Froude criterion), but it is difficult to achieve both flow similarity and motion similarity simultaneously due to turbidity currents being gravity-driven flows where gravity acts on suspended sediment particles.

Therefore, the dynamic similarity of turbidity currents is mainly conserved by the Froude similarity [59]. Meanwhile, the Froude (Fr) similarity has a long history in designing laboratory-scale hydrodynamic models, such as rivers, spillways, debris flows, and turbidity currents [60,61].

$$Fr = \frac{U}{\sqrt{g'h}} \tag{10}$$

$$\frac{L_1}{L_2} = \lambda_L, \frac{g'_1}{g'_2} = \lambda_{g'}, \frac{U_1}{U_2} = \sqrt{\lambda_L \lambda_{g'}} \tag{11}$$

Turbidity current events triggered by high-density river floods or earthquakes usually flow along the riverbed, so the influence of waves, wind speed, etc., is not considered. The Froude similarity criterion is used to estimate how much heat turbidity currents can carry into deeper areas, to evaluate the impact of actual turbidity current events on environmental water bodies. In this study, turbidity currents are approximated as conservative flows (e.g., density flows of high-speed turbulence or saltwater flows), and the sediment setting velocity can be set to zero. On the premise of maintaining geometric similarity and dynamic similarity, a turbidity current with a volume of 1.6×10 m^3, an initial height of 50 m, an initial sediment concentration of 80 kg/m^3, and a velocity of 5 m/s can still carry its own 10% heat into deeper waters after moving 2120 m on a 4° slope with a temperature difference of 3 °C from the ambient water (the scaling ratio is $\lambda_L = 400, \lambda_{g'} = 1.23$). Meanwhile, a turbidity current with a volume of 8×10 m^3 and the same flow conditions can carry more than 42% of its own heat into the deep sea with a temperature difference of 9 °C from the ambient water. The scaling of operating conditions based on laboratory conditions belongs to the ideal state, but this reflects that turbidity currents have good heat-carrying efficiency, which is in the same order of magnitude as the heat-carrying efficiency of real turbidity current events calculated by Tian's numerical calculation [9]. These results lead to some predictions and hypotheses about temperature changes and the original energy distribution of the deep-sea system, encouraging us to consider the role of turbidity current events in heat transport in the seabed channel, which may be a mechanism for temperature changes in the deep-sea environment.

It should be noted that the influence of wave action on the heat-carrying characteristics of turbidity currents was ignored in this study. Although changes in wave amplitude have no significant effect on the propulsion of turbidity currents and do not change their velocity, wave motion causes a "piston-like" oscillation in the vertical concentration distribution of turbidity currents [62–64]. Perhaps waves play an indispensable role in the transport of heat carried by turbidity currents in shallow waters [65,66], and future research will take wave factors into account to further study the impact of wave action on environmental entrainment during turbidity current heat transport.

A single turbidity current event (submarine sediment density flow) can transport over 100 km^3 of sediment [67–69], ten times the annual sediment flux from all of the world's rivers [3, 70]. These flows are the longest sediment density flows recognized on Earth and can achieve prodigious run-out distances of more than 1500 km, also inputting large amounts of heat and mass to marine environments over long periods of time via long-distance migration. From limited field data on deep-sea turbidity currents, the temperature of deep-sea environment water rises during turbidity current events, indicating that a large amount of high-temperature heat in the source area moves with the turbidity flow towards the deep-sea environment. The published data show that turbidity currents reaching 1020–1445 m water depth increase the ambient temperature by 2–3 °C [26], even 1–2 °C at 3000 m water depth [71], which changes the original heat distribution of submarine channels, deep-sea canyons and deep-sea plains, significantly affecting changes in the temperature and original heat distribution of marine environments, as shown in Figure 1. Therefore, the increase in the frequency and intensity of turbidity currents driven by future climate change may have a significant impact on the organic matter supply of deep-sea ecosystems and the heat stored in continental margins and ocean basins. As an indispensable part of the marine environmental system, turbidity currents carrying heat into the sea need to be further studied in future research on marine sedimentary geology, marine environmental pollution, and ocean heat distribution.

6. Conclusions

1. We quantitatively describe the convection and environmental mixing dynamics of shallow high-temperature turbidity currents at different sediment concentrations using the environmental entrainment coefficient E and summarize the relationship between the environmental entrainment function E and Ri considering temperature effects: $E = \frac{0.129-0.028Ri}{1-0.07Ri}$. Two simple dimensionless parameters Rs, R_T effectively describe flow factors and flow patterns during turbidity current motion. These two parameters quantify the ratio of temperature, salinity, and sediment content, summarize the relationship between the dimensionless density ratio R_0 reflecting turbidity current convection and environmental mixing and the environmental entrainment number E: $E = 0.125 + 0.034\ln(R + 0.031)$ and describe the impact of upper warm turbidity currents on the stable stratification of lake or coastal marine environments.
2. During the heat transport process of turbidity currents, the temperature distribution in the vertical direction is exponential: $\Delta T/\Delta T_m = (1.89 - 1.6 \times z/H)^{1.98}$, and 70% of the heat is concentrated in the lower half of the turbidity current for rapid transport, making the turbidity current efficiently store transported heat, reducing the vertical convective heat transfer loss, and facilitating the long-range transport of heat carried by turbidity. The continuous input of the turbidity current will be accompanied by a large amount of energy transfer, which will affect the temperature distribution in the regional marine environment.
3. Heat transfer at the sea–air interface has difficulty changing the energy distribution in the deep sea, and the effect of turbidity currents is opposite to the effect of river-to-ocean heat transfer or ocean surface-to-bottom heat transfer, directly affecting the thermal distribution of the internal or underpart of the ocean environment. The introduction of upper-layer heat into the deep sea by turbidity currents is an important component of the global heat transfer system that cannot be ignored, and it is also a complement to the heat transfer process of the global heat transfer system.

Author Contributions: Writing—original draft, H.T.; Writing—review & editing, G.X., J.Z., Y.R. and H.W. All authors have read and agreed to the published version of the manuscript.

Funding: This project was provided by the Qingdao Postdoctoral Funding Project (No. QDBSH20240202079), National Natural Science Foundation of China (No. 41976049), China Marine Geological Survey Project (No. DD20243114) and Natural Science Foundation of Shandong Province (No. ZR2021MD074).

Data Availability Statement: The sea surface temperature and marine water temperature data were obtained from hydrological monitoring platforms (https://argo.ucsd.edu/data/data-visualizations), and the ambient water temperature in the trigger source area data were derived from the following resources available in the public domain [20–31]. Data sets for this research are available at this site (https://doi.org/10.5281/zenodo.8008740).

Acknowledgments: We would like to thank graduate students Zhiyuan Chen, Meng Li, Zihan Zhang, and Cheng Zirui for their help with sampling and sample processing. In addition, we would also like to thank the experts of the "Hanhai inspirational" academic forum of College of Marine Geosciences, Ocean University of China, for their help in revising the paper.

Conflicts of Interest: The authors declare no conflict of interest.

Appendix A

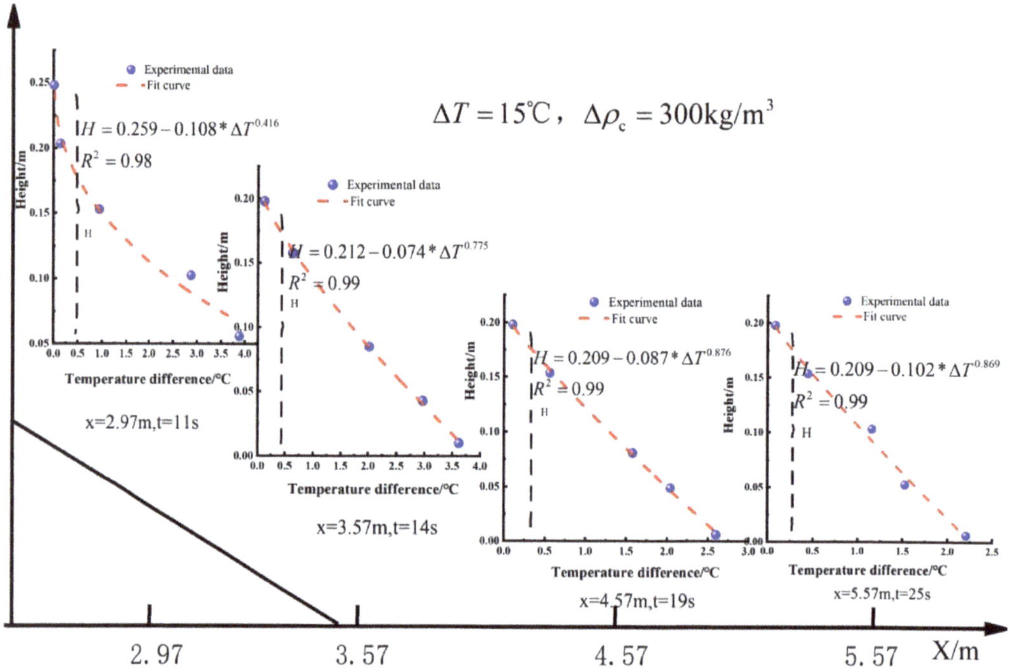

Figure A1. Maximum thermal shock temperature when passing through each group of sensors during turbidity migration (2.37–5.57 m) under experimental condition $\Delta T = 15$ °C, $\Delta\rho_c = 300$ kg/m^3.

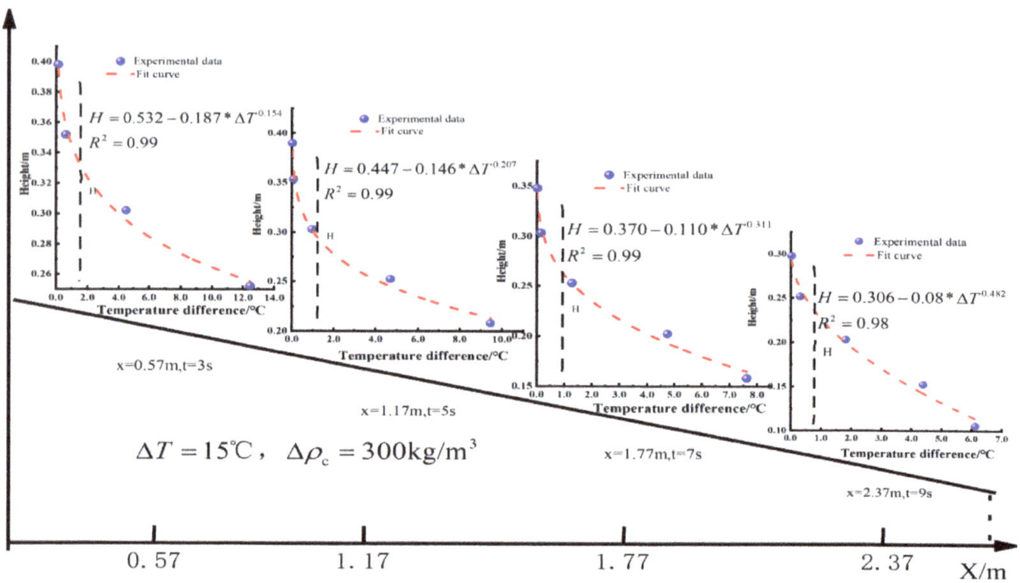

Figure A2. Maximum thermal shock temperature when passing through each group of sensors during turbidity migration (0–2.37 m) under experimental condition $\Delta T = 15$ °C, $\Delta\rho_c = 300$ kg/m^3.

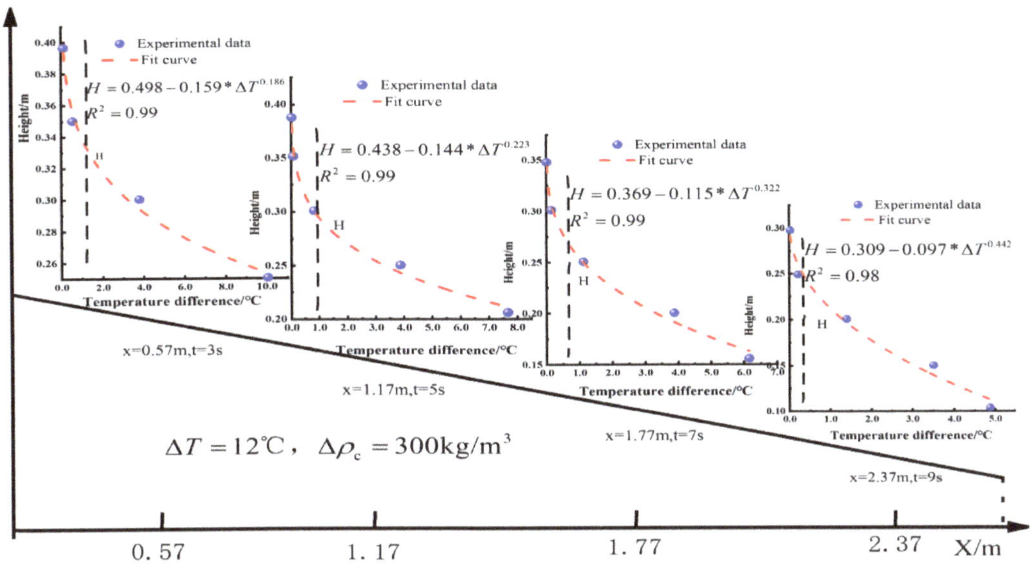

Figure A3. Maximum thermal shock temperature when passing through each group of sensors during turbidity migration (0–2.37 m) under experimental condition $\Delta T = 12$ °C, $\Delta\rho_c = 300$ kg/m^3.

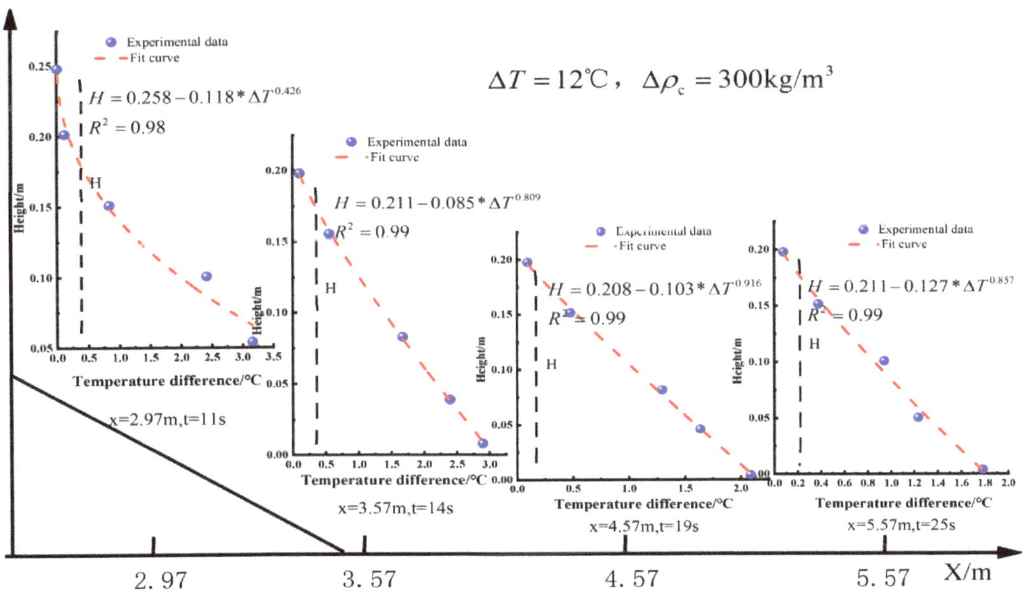

Figure A4. Maximum thermal shock temperature when passing through each group of sensors during turbidity migration (2.37–5.57 m) under experimental condition $\Delta T = 12$ °C, $\Delta\rho_c = 300$ kg/m^3.

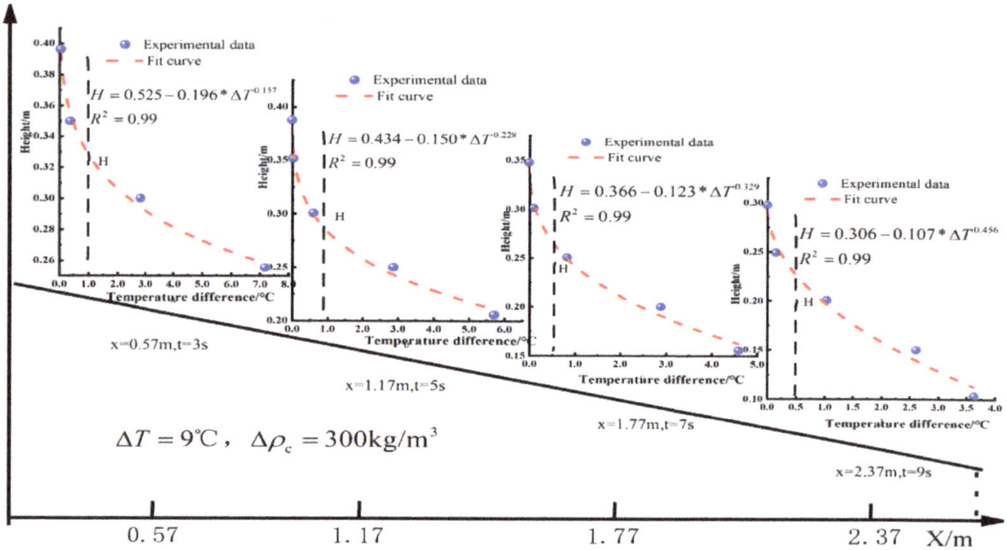

Figure A5. Maximum thermal shock temperature when passing through each group of sensors during turbidity migration (0–2.37 m) under experimental condition $\Delta T = 9$ °C, $\Delta\rho_c = 300$ kg/m^3.

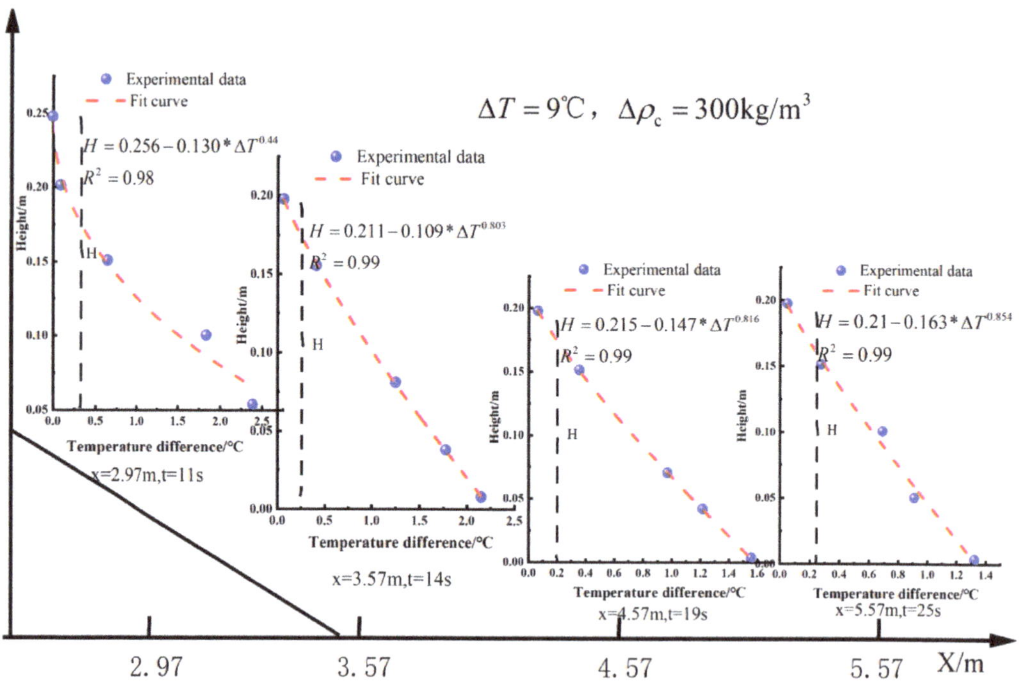

Figure A6. Maximum thermal shock temperature when passing through each group of sensors during turbidity migration (2.37–5.57 m) under experimental condition $\Delta T = 9$ °C, $\Delta\rho_c = 300$ kg/m^3.

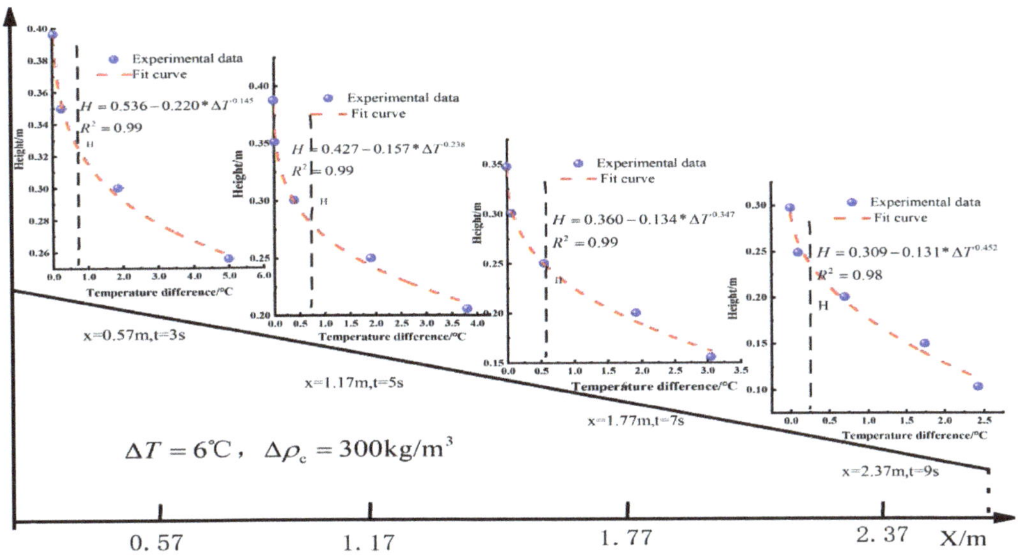

Figure A7. Maximum thermal shock temperature when passing through each group of sensors during turbidity migration (0–2.37 m) under experimental condition $\Delta T = 6\ ^\circ\text{C}$, $\Delta\rho_c = 300\ \text{kg/m}^3$.

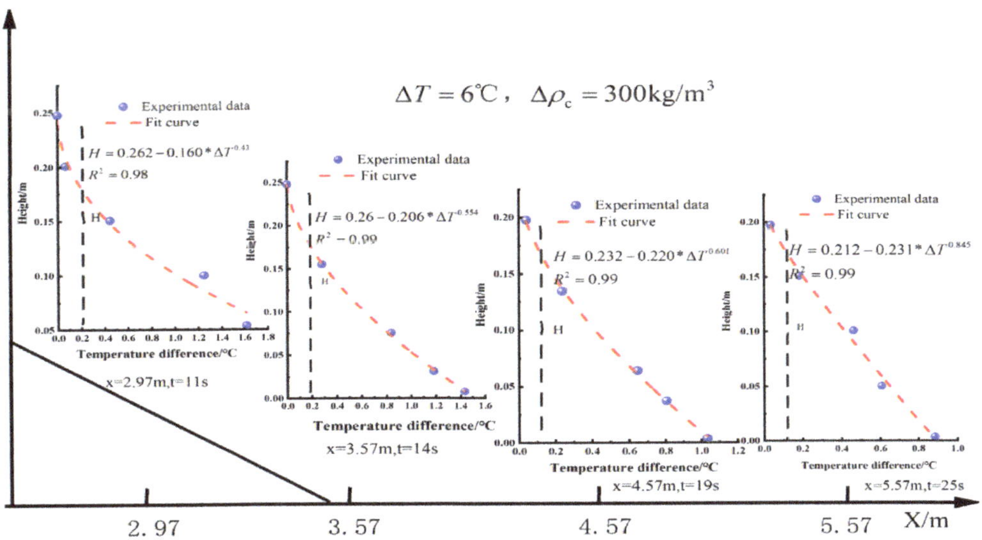

Figure A8. Maximum thermal shock temperature when passing through each group of sensors during turbidity migration (2.37–5.57 m) under experimental condition $\Delta T = 6\ ^\circ\text{C}$, $\Delta\rho_c = 300\ \text{kg/m}^3$.

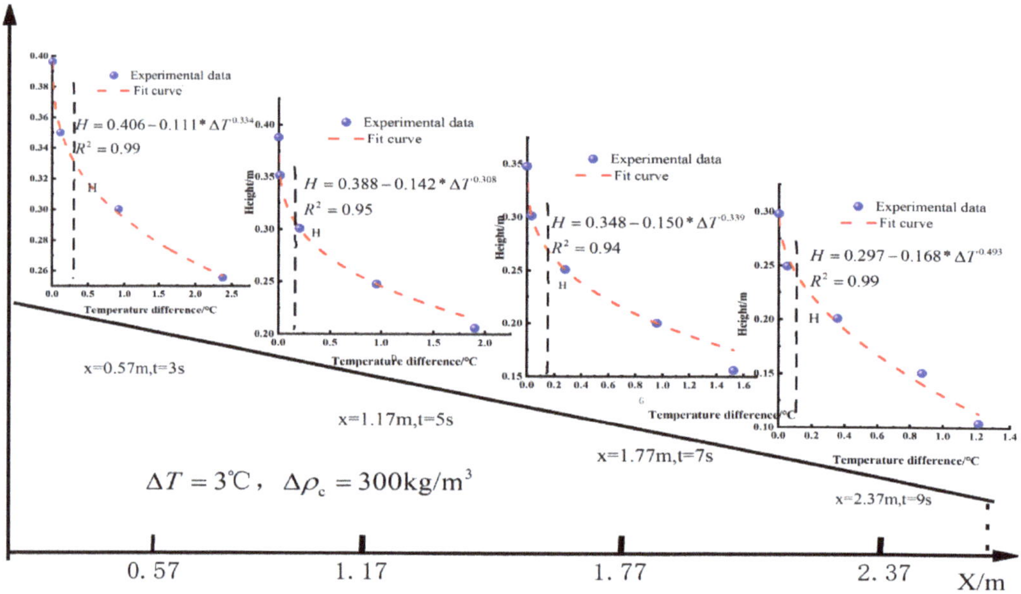

Figure A9. Maximum thermal shock temperature when passing through each group of sensors during turbidity migration (0–2.37 m) under experimental condition $\Delta T = 3$ °C, $\Delta\rho_c = 300$ kg/m^3.

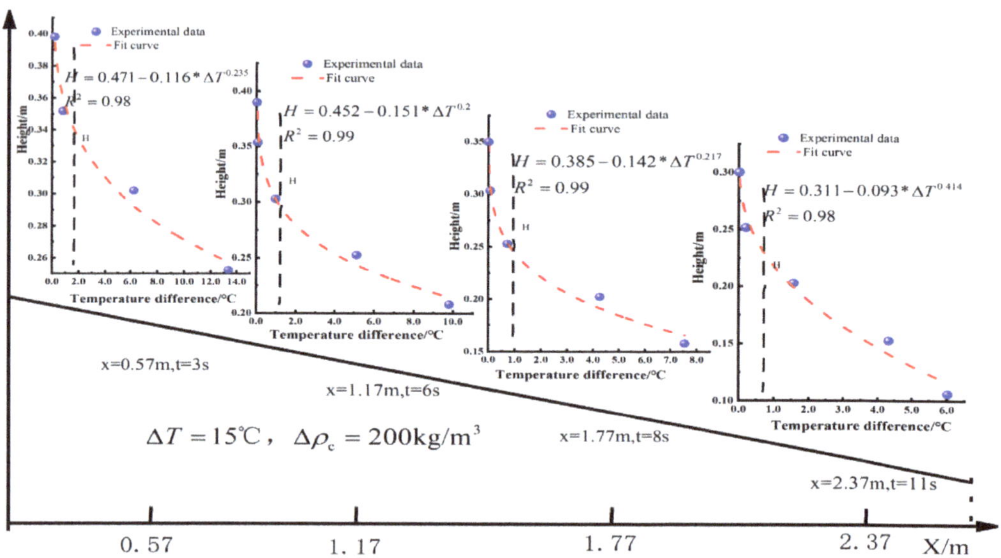

Figure A10. Maximum thermal shock temperature when passing through each group of sensors during turbidity migration (0–2.37 m) under experimental condition $\Delta T = 15$ °C, $\Delta\rho_c = 200$ kg/m^3.

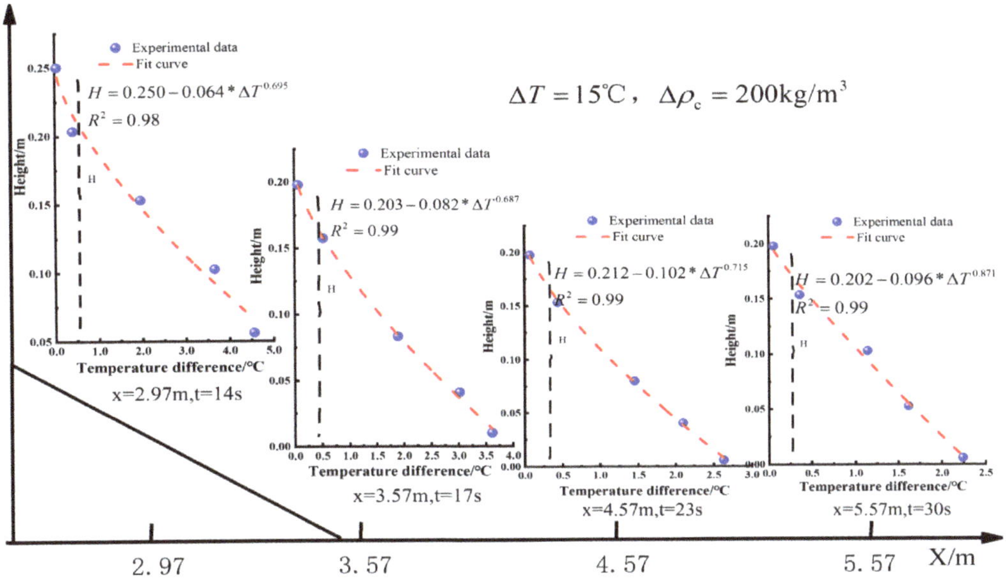

Figure A11. Maximum thermal shock temperature when passing through each group of sensors during turbidity migration (2.37–5.57 m) under experimental condition $\Delta T = 15$ °C, $\Delta\rho_c = 200$ kg/m^3.

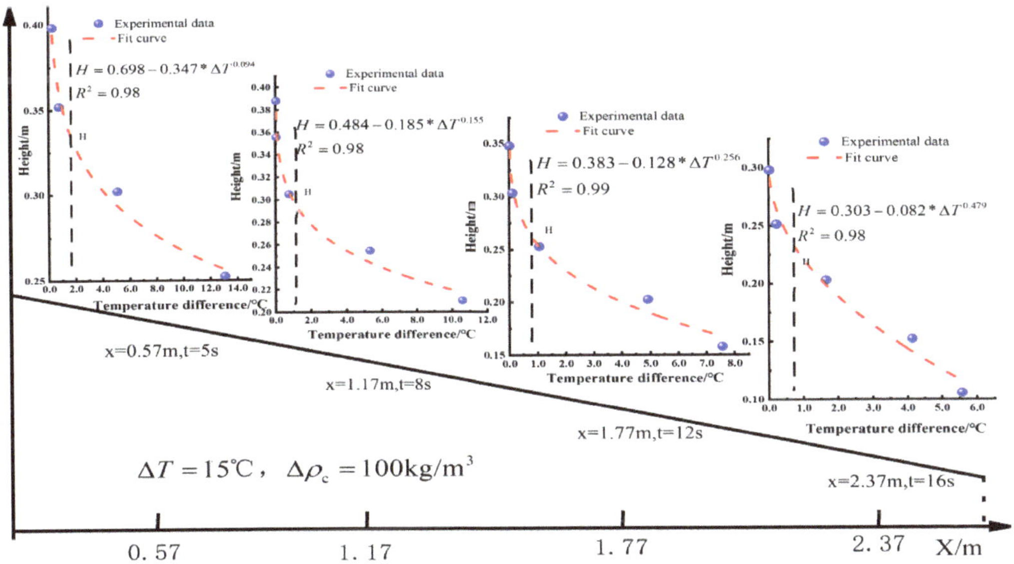

Figure A12. Maximum thermal shock temperature when passing through each group of sensors during turbidity migration (0–2.37 m) under experimental condition $\Delta T = 15$ °C, $\Delta\rho_c = 100$ kg/m^3.

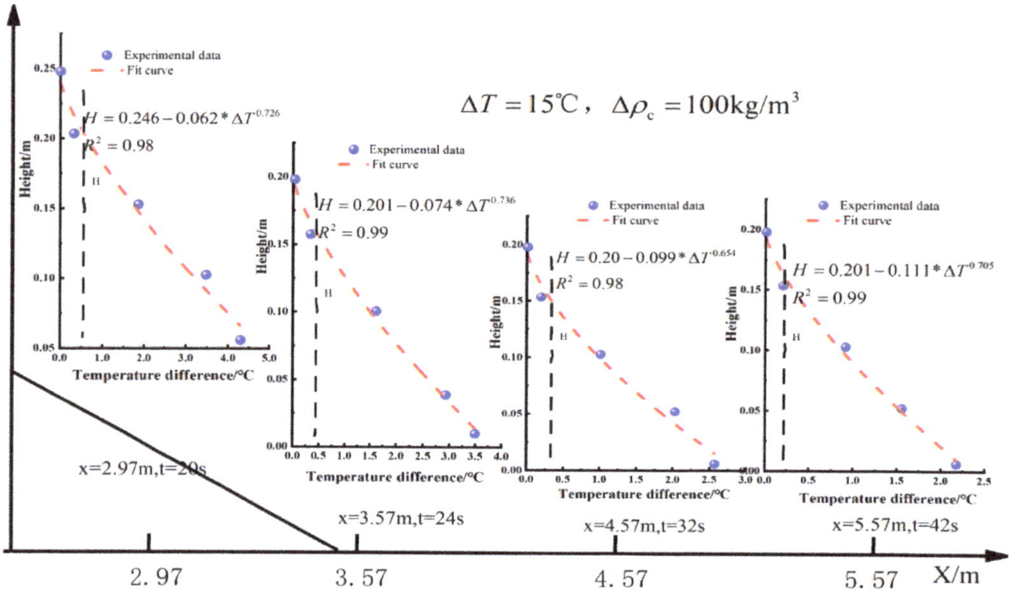

Figure A13. Maximum thermal shock temperature when passing through each group of sensors during turbidity migration (2.37–5.57 m) under experimental condition $\Delta T = 15$ °C, $\Delta\rho_c = 100$ kg/m^3.

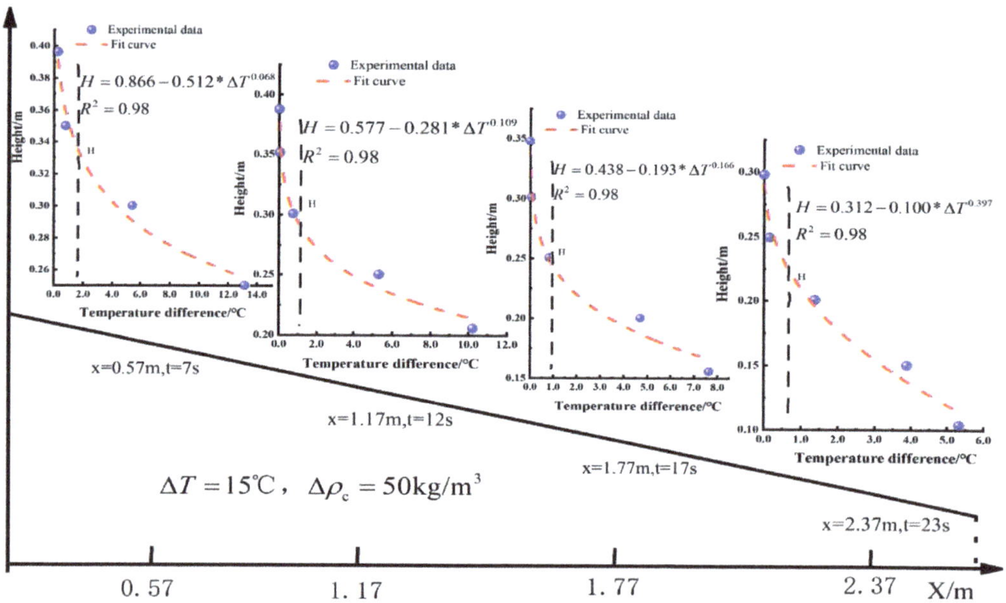

Figure A14. Maximum thermal shock temperature when passing through each group of sensors during turbidity migration (0–2.37 m) under experimental condition $\Delta T = 15$ °C, $\Delta\rho_c = 50$ kg/m^3.

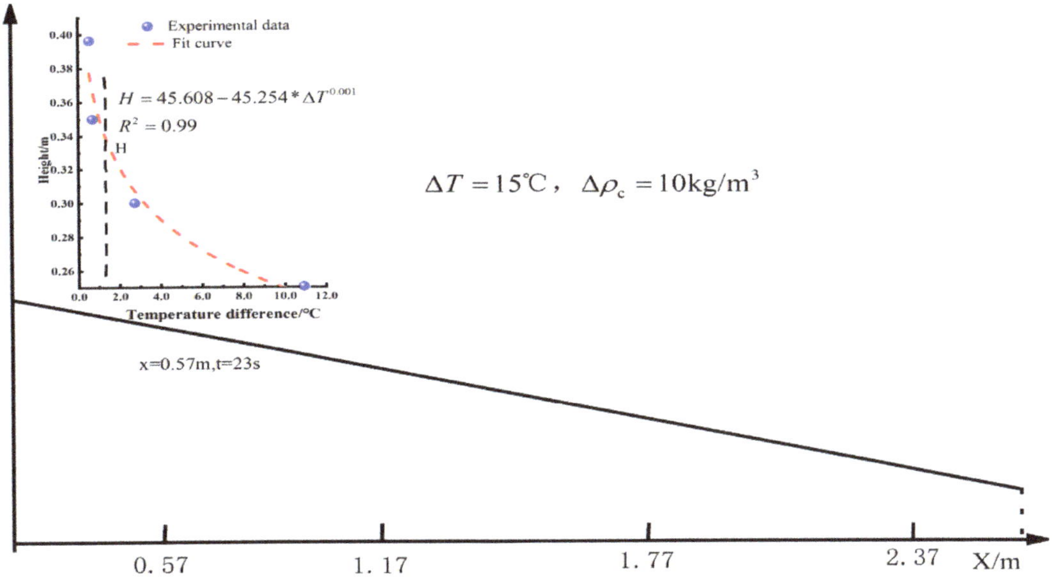

Figure A15. Maximum thermal shock temperature when passing through each group of sensors during turbidity migration (0–2.37 m) under experimental condition $\Delta T = 15$ °C, $\Delta\rho_c = 10$ kg/m^3.

References

1. Khripounoff, A.; Vangriesheim, A.; Babonneau, N.; Crassous, P.; Dennielou, B.; Savoye, B. Direct observation of intense turbidity current activity in the Zaire submarine valley at 4000 m water depth. *Mar. Geol.* **2003**, *194*, 151–158. [CrossRef]
2. Piper, D.J.; Normark, W.R. Processes that initiate turbidity currents and their influence on turbidites: A marine geology perspective. *J. Sediment. Res.* **2009**, *79*, 347–362. [CrossRef]
3. Talling, P.J.; Masson, D.G.; Sumner, E.J.; Malgesini, G. Subaqueous sediment density flows: Depositional processes and deposit types. *Sedim* **2012**, *59*, 1937–2003. [CrossRef]
4. Xu, J.; Noble, M.; Rosenfeld, L.K. In-situ measurements of velocity structure within turbidity currents. *GeoRL* **2004**, *31*. [CrossRef]
5. Tian, Z.; Huang, J.; Xiang, J.; Zhang, S. Suspension and transportation of sediments in submarine canyon induced by internal solitary waves. *PhFl* **2024**, *36*, 022112. [CrossRef]
6. Dalrymple, R.W.; Mackay, D.A.; Ichaso, A.A.; Choi, K.S. Processes, morphodynamics, and facies of tide-dominated estuaries. *Princ. Tidal Sedimentol.* **2012**, 79–107. [CrossRef]
7. Hage, S.; Galy, V.; Cartigny, M.; Acikalin, S.; Clare, M.; Gröcke, D.; Hilton, R.; Hunt, J.; Lintern, D.; McGhee, C. Efficient preservation of young terrestrial organic carbon in sandy turbidity-current deposits. *Geo* **2020**, *48*, 882–887. [CrossRef]
8. Kane, I.A.; Clare, M.A. Dispersion, accumulation, and the ultimate fate of microplastics in deep-marine environments: A review and future directions. *Front. Earth Sci.* **2019**, *7*, 80. [CrossRef]
9. Pohl, F.; Eggenhuisen, J.T.; Kane, I.A.; Clare, M.A. Transport and burial of microplastics in deep-marine sediments by turbidity currents. *Environ. Sci. Technol.* **2020**, *54*, 4180–4189. [CrossRef]
10. Tian, H.; Ren, Y.; Chen, Z.; Tao, W.; Wu, H.; Xu, G. Numerical Study of the Transport Process of Shallow Heat Carried by Turbidity Currents in Deep-Sea Environments. *J. Geophys. Res. Ocean.* **2023**, *128*, e2022JC019478. [CrossRef]
11. Danovaro, R.; Bianchelli, S.; Gambi, C.; Mea, M.; Zeppilli, D. α-, β-, γ-, δ-and ε-diversity of deep-sea nematodes in canyons and open slopes of Northeast Atlantic and Mediterranean margins. *Mar. Ecol. Prog. Ser.* **2009**, *396*, 197–209. [CrossRef]
12. Lambshead, P.J.D. *Marine Nematode Biodiversity*; Volume 1: Nematode morphology, physiology, and ecology; CABI Publishing: Wallingford, UK, 2004; pp. 438–468.
13. Heip, C.; Vincx, M.; Vranken, G. The ecology of marine nematodes. *Oceanogr. Mar. Biol.* **1985**, *23*, 399–489.
14. Danovaro, R.; Dell'Anno, A.; Pusceddu, A. Biodiversity response to climate change in a warm deep sea. *Ecol. Lett.* **2004**, *7*, 821–828. [CrossRef]
15. Paull, C.K.; Talling, P.J.; Maier, K.L.; Parsons, D.; Xu, J.; Caress, D.W.; Gwiazda, R.; Lundsten, E.M.; Anderson, K.; Barry, J.P. Powerful turbidity currents driven by dense basal layers. *Nat. Commun.* **2018**, *9*, 4114. [CrossRef]
16. Simmons, S.; Azpiroz-Zabala, M.; Cartigny, M.; Clare, M.; Cooper, C.; Parsons, D.; Pope, E.; Sumner, E.; Talling, P. Novel acoustic method provides first detailed measurements of sediment concentration structure within submarine turbidity currents. *J. Geophys. Res. Ocean.* **2020**, *125*, e2019JC015904. [CrossRef]

17. Hage, S.; Cartigny, M.J.; Sumner, E.J.; Clare, M.A.; Hughes Clarke, J.E.; Talling, P.J.; Lintern, D.G.; Simmons, S.M.; Silva Jacinto, R.; Vellinga, A.J. Direct monitoring reveals initiation of turbidity currents from extremely dilute river plumes. *Geophys. Res. Lett.* **2019**, *46*, 11310–11320. [CrossRef]
18. Kelly, R.; Dorrell, R.; Burns, A.; McCaffrey, W. The structure and entrainment characteristics of partially confined gravity currents. *J. Geophys. Res. Ocean.* **2019**, *124*, 2110–2125. [CrossRef]
19. Traer, M.; Hilley, G.; Fildani, A.; McHargue, T. The sensitivity of turbidity currents to mass and momentum exchanges between these underflows and their surroundings. *J. Geophys. Res. Earth Surf.* **2012**, *117*, 7. [CrossRef]
20. Gill, A.E. *Atmosphere-Ocean Dynamics*; Academic Press: Cambridge, MA, USA, 1982; Volume 30.
21. Hwang, J.-S.; Souissi, S.; Dahms, H.-U.; Tseng, L.-C.; Schmitt, F.G.; Chen, Q.-C. Rank-abundance allocations as a tool to analyze planktonic copepod assemblages off the Danshuei river estuary (Northern Taiwan). *Zool. Stud.* **2009**, *48*, 49–62.
22. Sparkes, R.B.; Lin, I.-T.; Hovius, N.; Galy, A.; Liu, J.T.; Xu, X.; Yang, R. Redistribution of multi-phase particulate organic carbon in a marine shelf and canyon system during an exceptional river flood: Effects of Typhoon Morakot on the Gaoping River–Canyon system. *Mar. Geol.* **2015**, *363*, 191–201. [CrossRef]
23. Wang, R.-M.; You, C.-F.; Chu, H.-Y.; Hung, J.-J. Seasonal variability of dissolved major and trace elements in the Gaoping (Kaoping) River Estuary, Southwestern Taiwan. *J. Mar. Syst.* **2009**, *76*, 444–456. [CrossRef]
24. Zhang, Y.; Liu, Z.; Zhao, Y.; Colin, C.; Zhang, X.; Wang, M.; Zhao, S.; Kneller, B.J.G. Long-term in situ observations on typhoon-triggered turbidity currents in the deep sea. *Geo* **2018**, *46*, 675–678. [CrossRef]
25. Puig, P.; Ogston, A.S.; Mullenbach, B.; Nittrouer, C.; Parsons, J.; Sternberg, R. Storm-induced sediment gravity flows at the head of the Eel submarine canyon, northern California margin. *J. Geophys. Res. Ocean.* **2004**, *109*, C03019. [CrossRef]
26. Xu, J.; Sequeiros, O.E.; Noble, M.A. Sediment concentrations, flow conditions, and downstream evolution of two turbidity currents, Monterey Canyon, USA. *Deep Sea Res. Part I* **2014**, *89*, 11–34. [CrossRef]
27. Bailey, L.P.; Clare, M.A.; Rosenberger, K.J.; Cartigny, M.J.; Talling, P.J.; Paull, C.K.; Gwiazda, R.; Parsons, D.R.; Simmons, S.M.; Xu, J. Preconditioning by sediment accumulation can produce powerful turbidity currents without major external triggers. *Earth Planet. Sci. Lett.* **2021**, *562*, 116845. [CrossRef]
28. Azpiroz-Zabala, M.; Cartigny, M.J.; Talling, P.J.; Parsons, D.R.; Sumner, E.J.; Clare, M.A.; Simmons, S.M.; Cooper, C.; Pope, E.L. Newly recognized turbidity current structure can explain prolonged flushing of submarine canyons. *Sci. Adv.* **2017**, *3*, e1700200. [CrossRef]
29. Descy, J.P.; Darchambeau, F.; Lambert, T.; Stoyneva-Gaertner, M.P.; Bouillon, S.; Borges, A.V. Phytoplankton dynamics in the Congo River. *Freshw. Biol.* **2017**, *62*, 87–101. [CrossRef]
30. Barats, A.; Féraud, G.; Potot, C.; Philippini, V.; Travi, Y.; Durrieu, G.; Dubar, M.; Simler, R. Naturally dissolved arsenic concentrations in the Alpine/Mediterranean Var River watershed (France). *Sci. Total Environ.* **2014**, *473*, 422–436. [CrossRef]
31. Khripounoff, A.; Crassous, P.; Bue, N.L.; Dennielou, B.; Jacinto, R.S. Different types of sediment gravity flows detected in the Var submarine canyon (northwestern Mediterranean Sea). *Prog. Oceanogr.* **2012**, *106*, 138–153. [CrossRef]
32. Palanques, A.; de Madron, X.D.; Puig, P.; Fabres, J.; Guillén, J.; Calafat, A.; Canals, M.; Heussner, S.; Bonnin, J. Suspended sediment fluxes and transport processes in the Gulf of Lions submarine canyons. The role of storms and dense water cascading. *Mar. Geol.* **2006**, *234*, 43–61. [CrossRef]
33. Halbwachs, M.; Sabroux, J.C.; Grangeon, J.; Kayser, G.; Tochon-Danguy, J.C.; Felix, A.; B'eard, J.C.; Villevieille, A.; Vitter, G.; Richon, P. Degassing the "killer lakes" Nyos and Monoun, Cameroon. *Eos Trans. Am. Geophys. Union* **2004**, *85*, 281–285. [CrossRef]
34. Sparks, R.S.J.; Bonnecaze, R.T.; Huppert, H.E.; Lister, J.R.; Hallworth, M.A.; Mader, H.; Phillips, J. Sediment-laden gravity currents with reversing buoyancy. *Earth Planet. Sci. Lett.* **1993**, *114*, 243–257. [CrossRef]
35. Cantelli, A.; Johnson, S.; White, J.; Parker, G. Sediment sorting in the deposits of turbidity currents created by experimental modeling of explosive subaqueous eruptions. *J. Geol.* **2008**, *116*, 76–93. [CrossRef]
36. Gladstone, C.; Pritchard, D. Patterns of deposition from experimental turbidity currents with reversing buoyancy. *Sedim* **2010**, *57*, 53–84. [CrossRef]
37. Hizzett, J.L.; Hughes Clarke, J.E.; Sumner, E.J.; Cartigny, M.; Talling, P.; Clare, M. Which triggers produce the most erosive, frequent, and longest runout turbidity currents on deltas? *GeoRL* **2018**, *45*, 855–863. [CrossRef]
38. Kineke, G.; Woolfe, K.; Kuehl, S.; Milliman, J.D.; Dellapenna, T.; Purdon, R. Sediment export from the Sepik River, Papua New Guinea: Evidence for a divergent sediment plume. *Cont. Shelf. Res.* **2000**, *20*, 2239–2266. [CrossRef]
39. Steel, E.; Buttles, J.; Simms, A.R.; Mohrig, D.; Meiburg, E. The role of buoyancy reversal in turbidite deposition and submarine fan geometry. *Geo* **2017**, *45*, 35–38. [CrossRef]
40. Steel, E.; Simms, A.R.; Warrick, J.; Yokoyama, Y. Highstand shelf fans: The role of buoyancy reversal in the deposition of a new type of shelf sand body. *Bulletin* **2016**, *128*, 1717–1724. [CrossRef]
41. Lu, G.; Wells, M.; van Strygen, I.; Hecky, R.E. Intrusions of sediment laden rivers into density stratified water columns could be an unrecognized source of mixing in many lakes and coastal oceans. *Sedim* **2022**, *69*, 2228–2245. [CrossRef]
42. Morton, B.; Taylor, G.I.; Turner, J.S. Turbulent gravitational convection from maintained and instantaneous sources. *Proc. R. Soc. Lond. Ser. A Math. Phys. Sci.* **1956**, *234*, 1–23. [CrossRef]
43. Jacobson, M.; Testik, F.Y. Turbulent entrainment into fluid mud gravity currents. *Environ. Fluid Mech.* **2014**, *14*, 541–563. [CrossRef]
44. Han, D.; He, Z.; Lin, Y.T.; Wang, Y.; Guo, Y.; Yuan, Y. Hydrodynamics and sediment transport of downslope turbidity current through rigid vegetation. *Water Resour. Res.* **2023**, *59*, e2023WR034421. [CrossRef]

45. He, Z.; Zhao, L.; Lin, T.; Hu, P.; lv, Y.; Ho, H.-C.; Lin, Y.-T. Hydrodynamics of gravity currents down a ramp in linearly stratified environments. *J. Hydraul. Eng.* **2017**, *143*, 04016085. [CrossRef]
46. Carter, L.; Gavey, R.; Talling, P.J.; Liu, J.T. Insights into submarine geohazards from breaks in subsea telecommunication cables. *Oceanography* **2014**, *27*, 58–67. [CrossRef]
47. Talling, P.J.; Clare, M.L.; Urlaub, M.; Pope, E.; Hunt, J.E.; Watt, S.F. Large submarine landslides on continental slopes: Geohazards, methane release, and climate change. *Oceanography* **2014**, *27*, 32–45. [CrossRef]
48. Tanaka, M.; Girard, G.; Davis, R.; Peuto, A.; Bignell, N. Recommended table for the density of water between 0 C and 40 C based on recent experimental reports. *Metro* **2001**, *38*, 301. [CrossRef]
49. Kneller, B.; Nasr-Azadani, M.M.; Radhakrishnan, S.; Meiburg, E. Long-range sediment transport in the world's oceans by stably stratified turbidity currents. *J. Geophys. Res. Ocean.* **2016**, *121*, 8608–8620. [CrossRef]
50. Giorgio Serchi, F.; Peakall, J.; Ingham, D.; Burns, A. A unifying computational fluid dynamics investigation on the river-like to river-reversed secondary circulation in submarine channel bends. *J. Geophys. Res. Ocean.* **2011**, *116*. [CrossRef]
51. Sequeiros, O.E.; Spinewine, B.; Beaubouef, R.T.; Sun, T.; García, M.H.; Parker, G. Characteristics of velocity and excess density profiles of saline underflows and turbidity currents flowing over a mobile bed. *J. Hydraul. Eng.* **2010**, *136*, 412–433. [CrossRef]
52. Hiscott, R.N. Loss of capacity, not competence, as the fundamental process governing deposition from turbidity currents. *J. Sediment. Res.* **1994**, *64*, 209–214. [CrossRef]
53. Sumner, E.J.; Amy, L.A.; Talling, P.J. Deposit structure and processes of sand deposition from decelerating sediment suspensions. *J. Sediment. Res.* **2008**, *78*, 529–547. [CrossRef]
54. Johnson, H.P.; Miller, U.K.; Salmi, M.S.; Solomon, E.A. Analysis of bubble plume distributions to evaluate methane hydrate decomposition on the continental slope. *Geochem. Geophys. Geosyst.* **2015**, *16*, 3825–3839. [CrossRef]
55. Phrampus, B.J.; Hornbach, M.J. Recent changes to the Gulf Stream causing widespread gas hydrate destabilization. *Nature* **2012**, *490*, 527–530. [CrossRef]
56. Dorrell, R.; Peakall, J.; Sumner, E.; Parsons, D.; Darby, S.; Wynn, R.; Özsoy, E.; Tezcan, D. Flow dynamics and mixing processes in hydraulic jump arrays: Implications for channel-lobe transition zones. *Mar. Geol.* **2016**, *381*, 181–193. [CrossRef]
57. Meiburg, E.; Kneller, B. Turbidity currents and their deposits. *AnRFM* **2010**, *42*, 135–156. [CrossRef]
58. Turner, J. Turbulent entrainment: The development of the entrainment assumption, and its application to geophysical flows. *JFM* **1986**, *173*, 431–471. [CrossRef]
59. Imran, J.; Khan, S.M.; Pirmez, C.; Parker, G. Froude scaling limitations in modeling of turbidity currents. *Environ. Fluid Mech.* **2017**, *17*, 159–186. [CrossRef]
60. Akan, A.O.; Iyer, S.S. *Open Channel Hydraulics*; Butterworth-Heinemann: Oxford, UK, 2021.
61. Task Committee on Hydraulic Modeling of the Environmental; Water Resources Institute of ASCE. *Hydraulic Modeling: Concepts and Practice*; ACSE: Reston, VA, USA, 2000.
62. Marino, M.; Stagnitti, M.; Stancanelli, L.M.; Musumeci, R.E.; Foti, E. Dynamics of wave-supported gravity currents in intermediate water. *Cont. Shelf. Res.* **2023**, *267*, 105082. [CrossRef]
63. Monaghan, J. Gravity currents and solitary waves. *Phys. D Nonlinear Phenom.* **1996**, *98*, 523–533. [CrossRef]
64. Tian, Z.; Liu, C.; Jia, Y.; Song, L.; Zhang, M. Submarine trenches and wave-wave interactions enhance the sediment resuspension induced by internal solitary waves. *J. Ocean Univ. China* **2023**, *22*, 983–992. [CrossRef]
65. Falcini, F.; Fagherazzi, S.; Jerolmack, D. Wave-supported sediment gravity flows currents: Effects of fluid-induced pressure gradients and flow width spreading. *Cont. Shelf. Res.* **2012**, *33*, 37–50. [CrossRef]
66. Tian, Z.; Huang, J.; Xiang, J.; Zhang, S.; Wu, J.; Liu, X.; Luo, T.; Yue, J. Interaction between internal solitary waves and the seafloor in the deep sea. *Deep. Undergr. Sci. Eng.* **2024**, *3*, 149–162. [CrossRef]
67. Piper, D.J.; Cochonat, P.; Morrison, M.L. The sequence of events around the epicentre of the 1929 Grand Banks earthquake: Initiation of debris flows and turbidity current inferred from sidescan sonar. *Sedim* **1999**, *46*, 79–97. [CrossRef]
68. Rothwell, R.; Thomson, J.; Kähler, G. Low-sea-level emplacement of a very large Late Pleistocene 'megaturbidite'in the western Mediterranean Sea. *Nature* **1998**, *392*, 377–380. [CrossRef]
69. Talling, P.; Wynn, R.; Masson, D.; Frenz, M.; Cronin, B.; Schiebel, R.; Akhmetzhanov, A.; Dallmeier-Tiessen, S.; Benetti, S.; Weaver, P. Onset of submarine debris flow deposition far from original giant landslide. *Nature* **2007**, *450*, 541–544. [CrossRef]
70. Milliman, J.D.; Syvitski, J.P. Geomorphic/tectonic control of sediment discharge to the ocean: The importance of small mountainous rivers. *J. Geol.* **1992**, *100*, 525–544. [CrossRef]
71. Kao, S.; Dai, M.; Selvaraj, K.; Zhai, W.; Cai, P.; Chen, S.; Yang, J.; Liu, J.; Liu, C.; Syvitski, J. Cyclone-driven deep sea injection of freshwater and heat by hyperpycnal flow in the subtropics. *Geophys. Res. Lett.* **2010**, *37*, L21702. [CrossRef]

water

MDPI

Article

Investigating the Element Geochemical Behavior and Provenance of Surface Sediments in the Offshore Area of Sierra Leone, Africa: Insights from Major and Trace Elements

Mengying Hu [1], Yingtao Zhu [2,*], Wenliang Chen [3], Bin Yu [2,*], Pengpeng Zhang [1], Chuanqi Hu [2] and Ruijia Jin [2]

1 Institute of Geophysical and Geochemical Exploration, Chinese Academy of Geological Sciences, Langfang 065000, China; humengying@mail.cgs.gov.cn (M.H.); zhangpengpeng@mail.cgs.gov.cn (P.Z.)

2 Tianjin Research Institute for Water Transport Engineering, Ministry of Transport, Tianjin 300456, China; huchuanqi96@163.com (C.H.); ruijia_jin@163.com (R.J.)

3 Qingdao Boyan Marine Environmental Science & Technology Co., Ltd., Qingdao 266114, China; cwlouc_love@126.com

* Correspondence: zhuyt@tiwte.ac.cn (Y.Z.); 13821810298@163.com (B.Y.)

Abstract: The element geochemical behavior and provenance of marine sediments are of significance to understanding the oceanic material cycle. Here, we tested the grain size and major and trace elements of 35 surface sediments in the offshore area of Sierra Leone, analyzed the content characteristics and controlling factors of the elements, discussed the material source of the sediments, and made a comparative study with the sediments in the offshore area of China. The results show that sandy silt is the main sediment type in the research area, and the average sediment mean grain size (Mz) is 4.15Φ. The content of Ca in the samples is the highest among the major elements (except Si), with an average of 5.1%. The content of Sr is the highest among the trace elements (except Ti, P, and Mn), with an average of 378.2 μg/g. The results of correlation analysis and factor analysis show that there are three main sources of sediments in the research area, namely, terrigenous weathering products, ilmenite-dominated ore, and oceanic biochemical substances. Compared with the sediments in China offshore, the sediments in the study area are more affected by marine biochemistry and have special ore input characteristics.

Keywords: offshore area of Sierra Leone; surface sediments; major and trace elements; element geochemical behavior

1. Introduction

The offshore of the continental shelf is an important hub connecting the land and the deep sea in the regional position, and the region is significantly affected by the land–sea interaction [1]. These unique factors make it play a vital role in the study of the global material cycle process [2,3]. Therefore, the discussion of sediment sources and sinks in the offshore of the continental shelf has become the focus of researchers [4–9]. There are numerous studies only on the offshore areas of China and surrounding countries, such as the Yellow River Delta [10,11], the Yangtze River Delta [12,13], the Pearl River Delta [14], and the Bengal sedimentary fan [15,16], which have all attracted much attention.

Marine sediment, classified by formation mode, mainly comes from cosmic, rocky, biological, or aquatic components [17]. In a series of processes of weathering, transportation, sedimentation, resuspension, and redeposition, the content of major and trace elements will change with it, so the content and combination characteristics of major and trace elements in sediments record a lot of information about the source-to-sink process and sedimentary environment [18,19]. The elements (e.g., Al, Fe, Mg, K, Na, etc.) are mainly controlled by rock-forming minerals, and their combination characteristics can reflect the composition of the parent rock in the source area [20–23]. The elements Ca and Sr are not only controlled by the source rock, but are also important indicators of biological processes in the ocean [24–26]. Elements, such as Cr, Cu, Mo, Fe, Mn, etc., are sensitive to the changes in redox conditions in the sedimentary environment, and are often used to explore the changes of redox state in the marine sedimentary environment [27–31]. Over the years, relatively stable elements in the surface environment (such as Ti, Al, Sc, Th, etc.) and their ratios have also become important indicators for exploring the source of marine sediments [32–34]. In addition, human exploitation of the ocean will also cause changes in the element contents in marine sediments, especially some heavy metals. Some metals are toxic to the environment (such as Co, Ni, Cd, Cu, As, Hg, Pb, etc.). It is of great significance to study the background concentration of and variation in these metal elements for marine environmental protection and development [35–38].

Although a large number of related studies have been carried out on the geochemical characteristics and provenance of marine sediments, these studies have some drawbacks. These are due to funding and other reasons: either the research area is large, but the data volume is small, or the data volume is large, but the area is small. And there is a large blank area of basic survey data in this research. Therefore, if we want to understand the characteristics of global marine sedimentation, we can only work together and enrich the basic survey data step by step. From this scientific perspective, we choose an offshore area with relatively poor water system development and relatively disadvantaged economic development. Sierra Leone is located on the west coast of Africa, bordering Guinea in the north and northeast, Liberia in the East and Southeast, and the Atlantic Ocean in the West and south. The terrain of Sierra Leone is characterized as high in the East and low in the West. It gradually transitions from a plain about 70 km wide along the coast to a plateau on the eastern border. Sierra Leone has a tropical monsoon climate and is one of the countries with the most rainfall in West Africa. In order to reveal the elemental geochemical behavior of the sediments in the offshore area of Sierra Leone, we tested the grain size and major and trace elements of 35 surface sediments in the research area, analyzed the content characteristics and controlling factors of the elements, discussed the material source of the sediments, and made a comparative study with the characteristics of the sediments in the offshore area of China. We aim to fill some gaps in the study of elemental chemistry in sediments. The research area and sampling stations are shown in Figure 1.

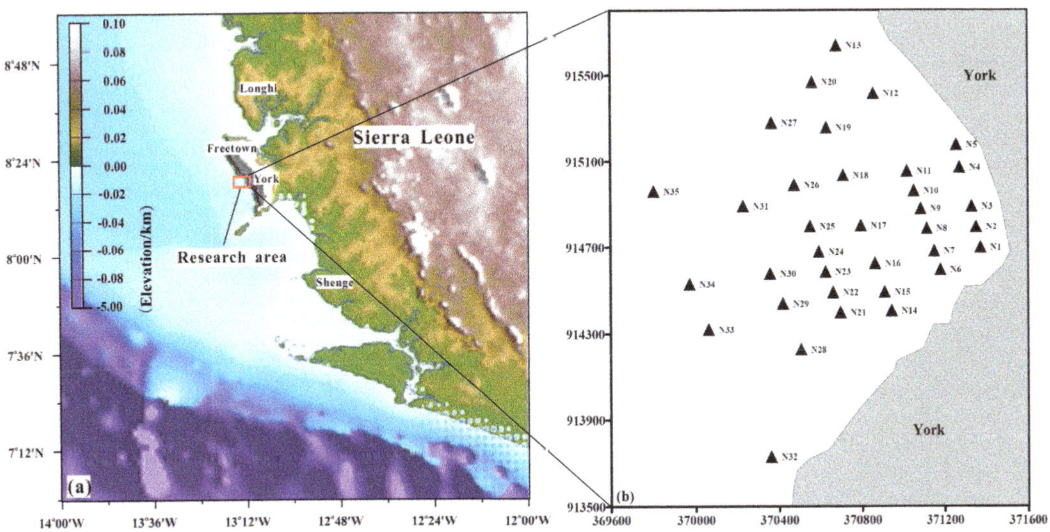

Figure 1. Location map of research area (**a**) and sampling sites (**b**). The coordinate system of (**b**) is CGCS2000, and the central longitude is 12° W.

2. Materials and Methods

2.1. Sample Collection

The samples were collected from a western area offshore of Freetown, Sierra Leone (Figure 1a). A clam-type sediment sampler was used to collect these sediment samples at a depth of <20 cm. The sampling sites are shown in Figure 1b.

2.2. Grain Size Analyses

The particle size measurement of sediment was carried out using a laser particle size analyzer (Mastersizer 2000, Malvern Panalytical, Westborough, MA, USA), whose measurement range and resolution are 0.017–2000 μm and 0.1Φ, respectively. Organic matter and calcium components were removed by using hydrogen peroxide and hydrochloric acid before measurement. Afterwards, samples were rinsed several times with deionized water, then separated with sodium hexametaphosphate, and finally tested on the machine. The grain size parameters were calculated based on the moment method [39]. The measurements of these samples were completed at the Institute of Geophysical and Geochemical exploration, Chinese Academy of Geological Sciences.

2.3. Major and Trace Element Analyses

The content of major and trace elements in the sediments was tested by two instruments, with the major elements measured by ICP-OES (Inductively Coupled Plasma–Optical Emission Spectrometry, SPECTRO Analytical Instruments, Kleve, Germany) and the trace elements measured by ICP-MS (Inductively Coupled Plasma–Mass Spectrometry, Agilent Technologies, Santa Clara, CA, USA). After low-temperature drying, the sediment samples were ground to 200 mesh. After grinding, 0.05 g of the sample was weighed and placed in a polytetrafluoroethylene digestion tank, and then digested by adding twice-distilled hydrofluoric acid and nitric acid (volume ratio was 1:1). After digestion, the sample was placed in a 190 °C oven for 48 h, and then cooled and evaporated on an electric heating plate. Afterwards, about 3 mL of 50% HNO3 was added to the sample container, and then it was placed in a 150 °C oven to continue dissolving the sample for more than 8 h. After reaching a certain volume, we waited for testing. We conducted repeated analysis of several samples and standard sample (GSD-9) analysis to ensure testing accuracy and precision (relative error < 2%). These testing experiments were completed at the Institute of Geophysical and Geochemical Exploration, Chinese Academy of Geological Sciences.

3. Results

3.1. Grain Size Characteristics

Based on Folk and Ward [40], these samples can be divided into five types: sandy silt, silty sand, sand, gravelly muddy sand, and gravelly sand (Figure 2). And sandy silt is the main sediment type. From the nearshore to offshore, the content of clay and silt gradually increases, while the content of sand gradually decreases (Figure 3). The contents of the three particle size components in the sediment vary greatly, among which the average content of sand is 53.31% (varying from 12.6% to 99.47%), the average content of silt is 41.57% (varying from 0.53% to 77.66%), and the average content of clay content is 4.66% (varying from 0 to 9.79%) (see details in Table 1).

The average mean grain size (Mz) of the samples is 4.15Φ (varying from 0.96 to 6.03Φ). The average sorting coefficient (δ) is 1.75 (varies from 0.57 to 2.74). The average skewness (Sk) is 0.8 (varies from −1.37 to 1.91). The average kurtosis (Ku) is 2.32 (varies from 0.72 to 3.34). Overall, these four parameters increase gradually from the nearshore to offshore area (Figure 4).

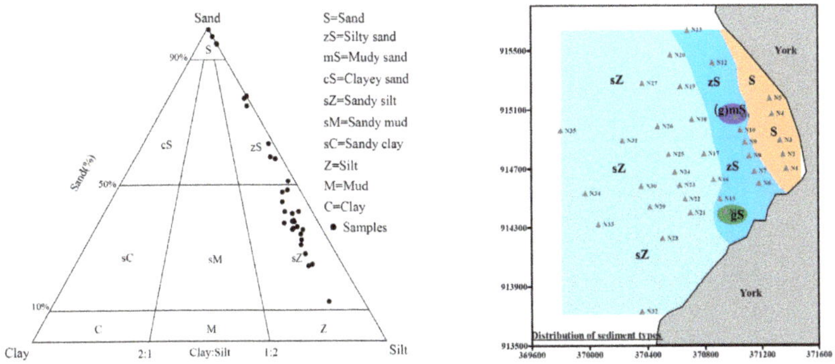

Figure 2. Grain size classification and sediment type distribution. (N11 (gravelly muddy sand) and N14 (gravelly sand) stations are not shown in the left figure).

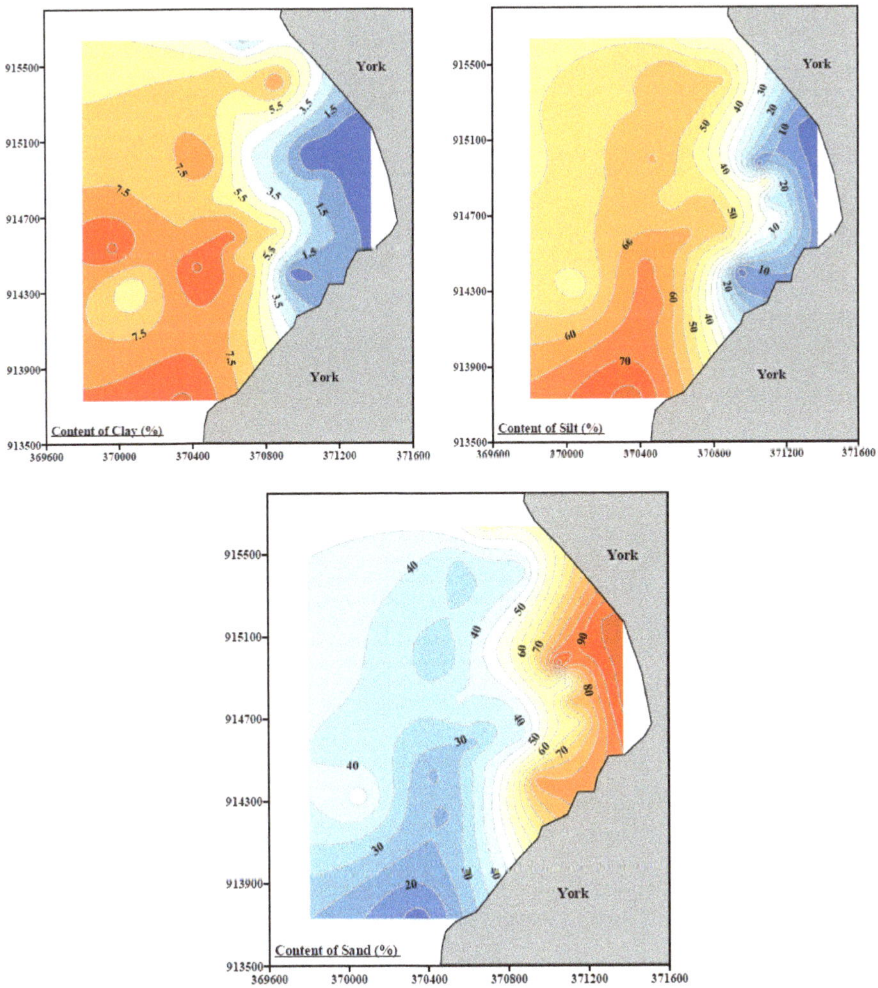

Figure 3. Content distribution of particle size component.

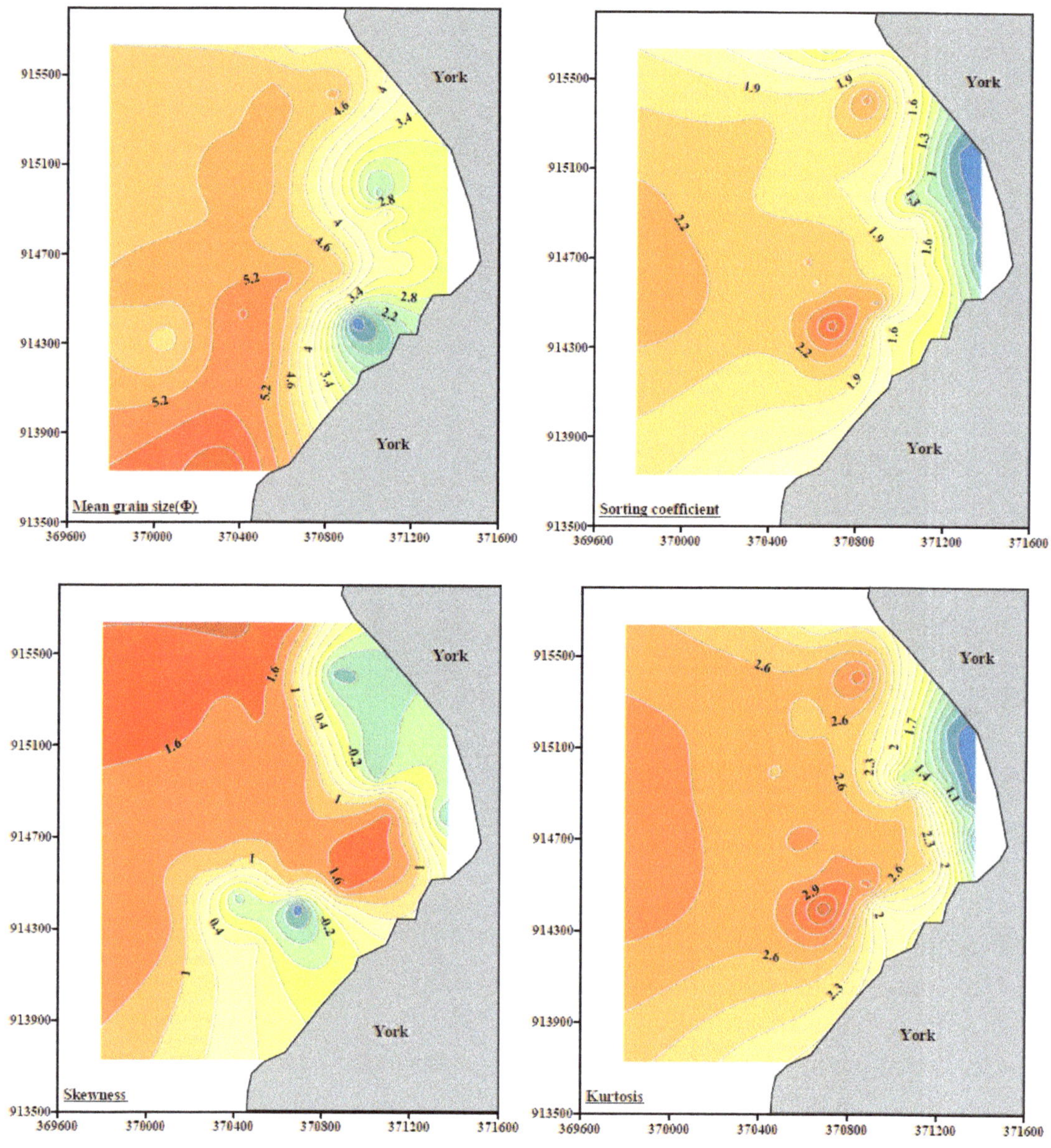

Figure 4. Distribution of particle size parameter.

Table 1. Particle size parameters and fraction content of sediments (n = 35).

	Content of Fractions (%)			Particle Size Parameters			
	Clay	Silt	Sand	Mz (φ)	δ	Sk	Ku
Max	9.79	77.66	99.47	6.03	2.74	1.91	3.34
Min	0.00	0.53	12.60	0.96	0.57	−1.37	0.72
Ave	4.66	41.57	53.31	4.15	1.75	0.80	2.32
STD	3.55	22.50	25.10	1.13	0.60	0.96	0.72
C. V. (%)	76.08	54.14	47.08	27.30	33.99	120.36	31.29

Note: "Ave" means "average", "STD" means "standard deviation", and "C. V." means "coefficient of variation".

3.2. Major and Trace Element Contents

The statistics of the content of constant and trace elements in sediments are presented in Table 2. Unlike in China's coastal areas, the content of Ca is the highest among the major elements, with an average of 5.1% (varying from 0.7% to 18.4%), except for silicon. The content of Al ranks only second, with an average of 4.8% (varying from 1.7% to 10.1%). Like the major elements, the trace element content of sediments in the research area is also significantly different from that in China's coastal areas. The content of Sr in the sediments is the highest, with an average value of 378.2 μg/g (varying from 63.5 to 1566.1 μg/g), among all trace elements (except for Ti, P, Mn). The Ba content is the second highest, with an average of 81.2 μg/g (varying from 6.8 to 137.1 μg/g).

Table 2. The major and trace element content in sediments of research area and the offshore area of China.

Elements	Research Area					China Sea				
	Max	Min	Ave	STD	C V (%)	South China Sea	East China Sea	South Yellow Sea	North Yellow Sea	Bohai Sea
Al *	10.1	1.7	4.8	1.8	38.3	6.5	7.2	7.1	6.8	6.5
Mg *	1.6	0.3	1.1	0.3	25.1	1.1	1.5	1.1	1.4	1.4
Fe *	4.7	1.8	3.4	0.6	17.8	2.9	3.9	3.0	3.1	3.1
K *	0.6	0	0.3	0.1	49.2	1.7	2.2	2.3	2.1	2.1
Ca *	18.4	0.7	5.1	3.4	66	3.1	3.2	1.8	2.6	4.8
Na *	2.2	0.3	0.9	0.5	50	1.2	1.4	2.1	1.6	0.0
Ti	28,064.3	3124.5	8551.8	5714.1	66.8	3840.0	2838.0	3540.0	3598.0	3960.0
P	1221.8	330	670.9	229.3	34.2	480.3	279.4	523.9	586.0	393.0
Mn	1159.2	227.2	541.1	187.6	34.7	542.3	565.5	774.6	637.0	**
Sr	1566.1	63.5	378.2	266.9	70.6	209.7	159.9	**	194.4	**
Ba	137.1	6.8	81.2	35.9	44.2	**	426.9	**	457.0	**
Li	105.7	8.9	47.1	20.6	43.0	**	**	39.3	38.4	**
Zn	101.9	10.1	42.5	16.6	39.1	71.1	86.3	**	80.6	**
Sc	12.9	3	8.1	2.5	30.4	**	**	**	11.7	**
Co	17.9	3.4	8.1	2.7	33.6	**	14.9	12.0	13.2	**
Rb	25.4	0.7	9.6	5.8	60.1	**	**	**	113.2	**
Th	14.6	1.9	8.3	3.3	40	**	**	**	**	**

Note: The unit for elements with "*" is "wt.%" and that of the others is "μg/g". The data of the South China Sea are from [41], the data of the East China Sea are from [42], the data of the South Yellow Sea are from [43], the data of the North Yellow Sea are from [44], and the data of the Bohai Sea are from [45]. "**" means no data.

Figure 5 shows the distribution of major and trace elements of the sediments. Except for Ca, Ti, Sr, and Mn, the content of most elements shows a decreasing trend from the nearshore to offshore area. The contents of Ca are relatively stable except for a few stations. The distribution patterns of Ti and Sr are similar, with some stations showing high content and others showing little change. The Mn content in the sediment of the research area does not change significantly, and there is no obvious distribution pattern (Figure 5).

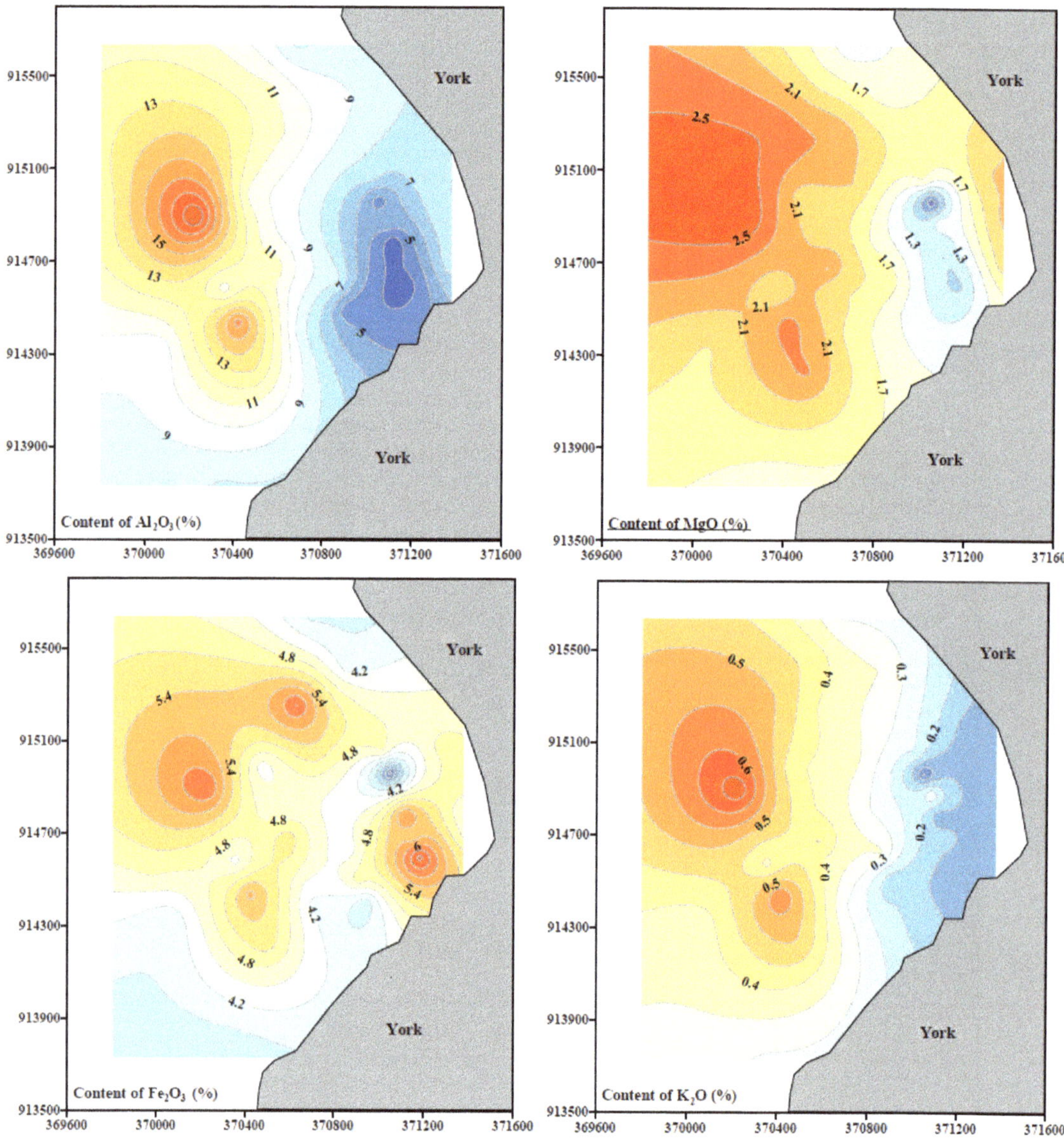

Figure 5. *Cont.*

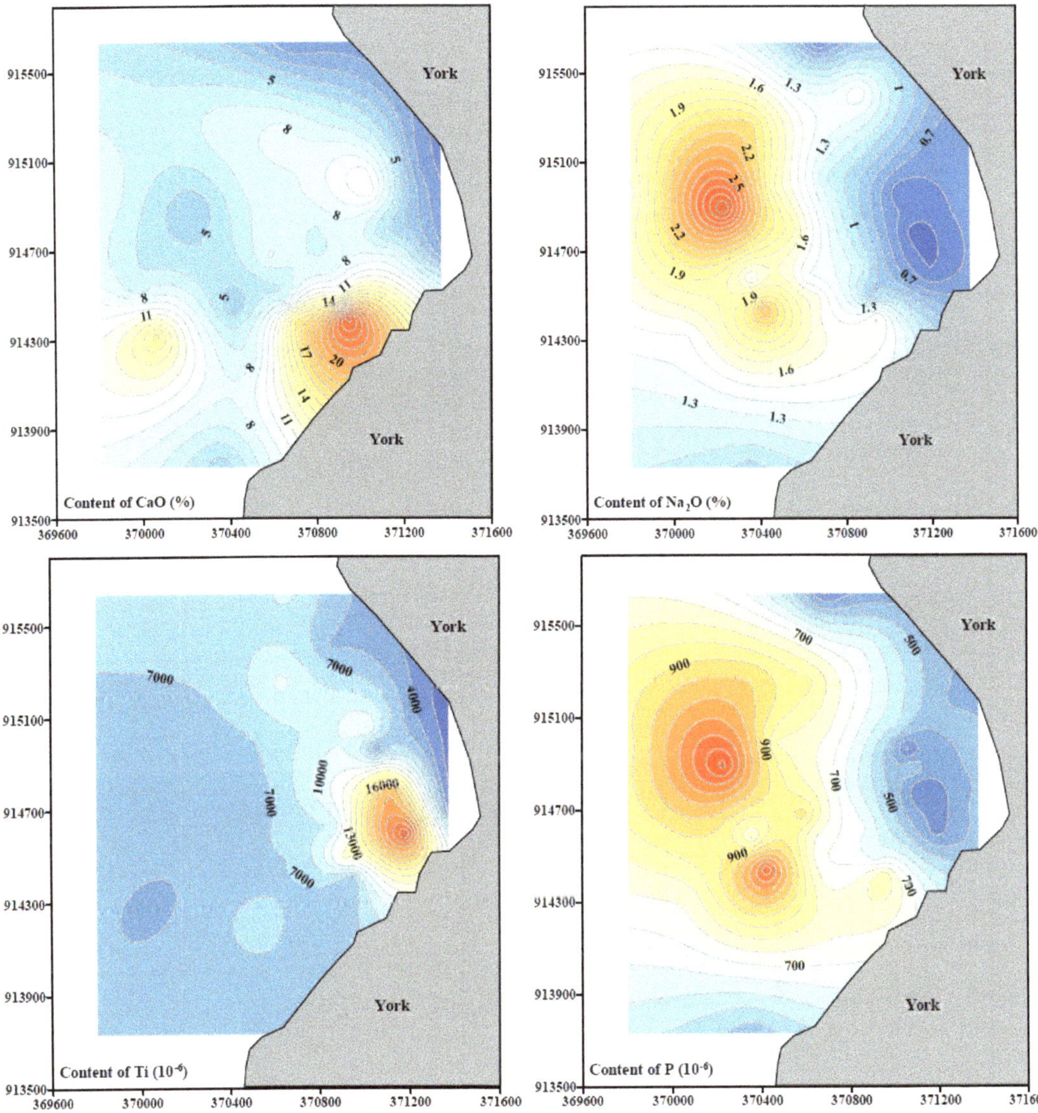

Figure 5. *Cont.*

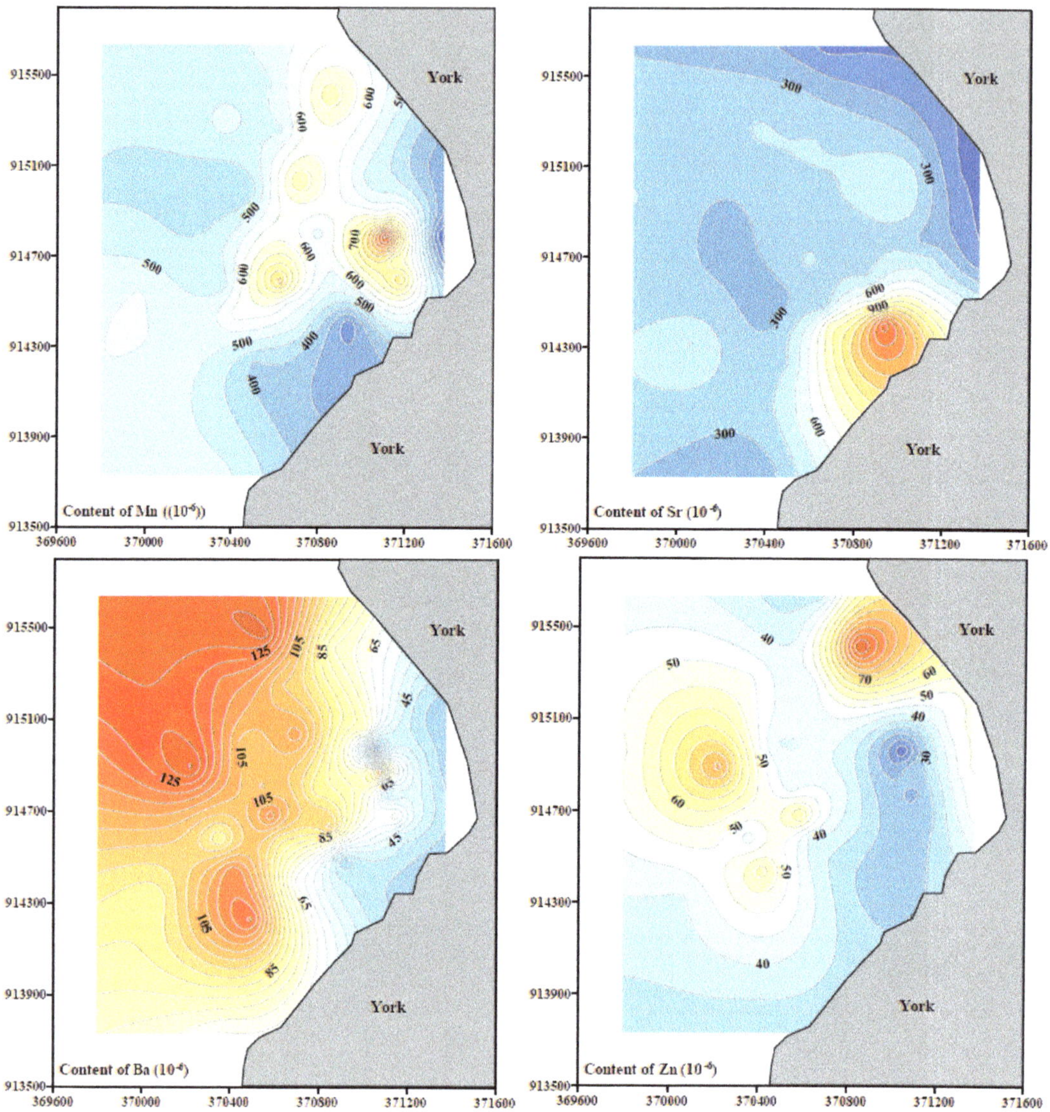

Figure 5. *Cont.*

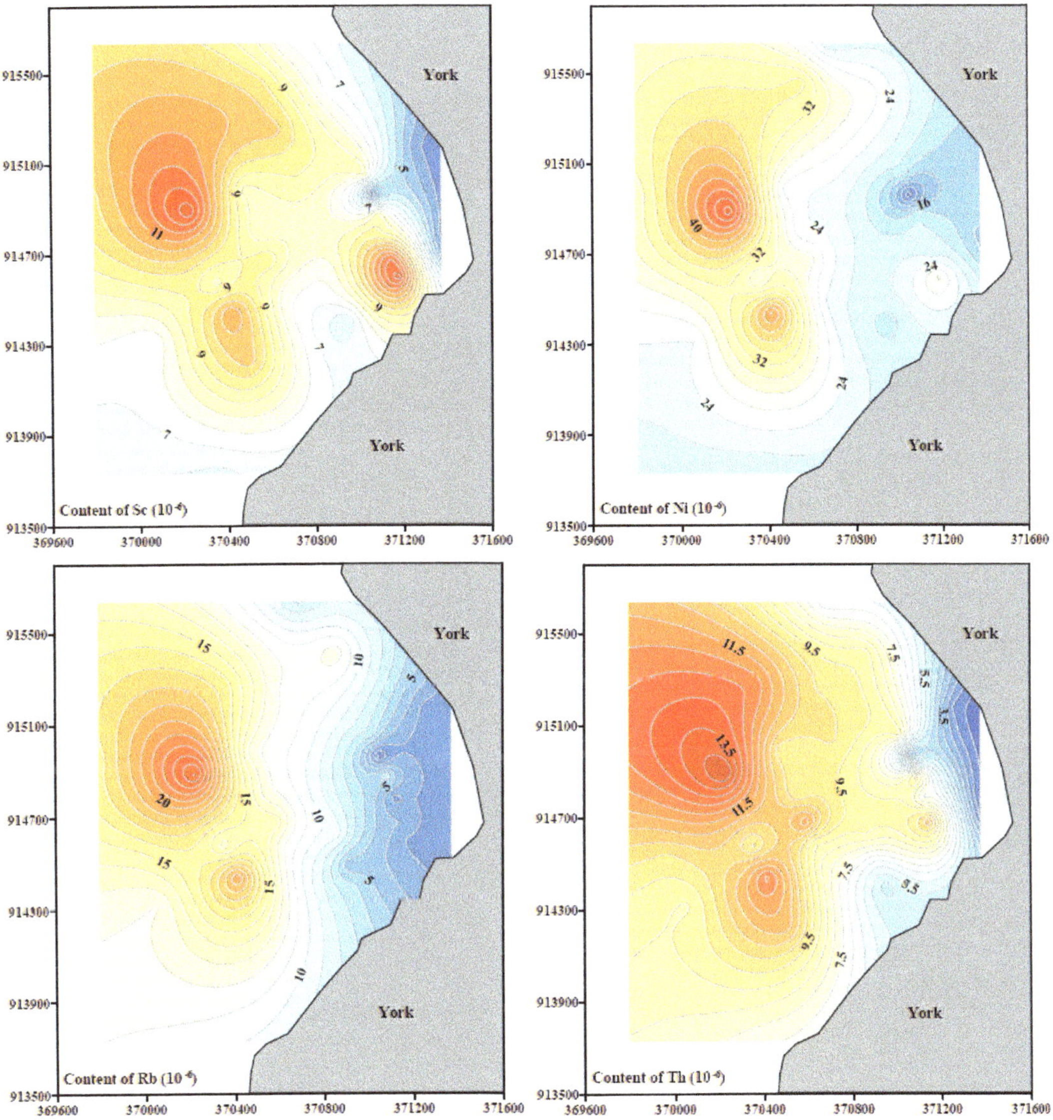

Figure 5. Distribution of major and trace elements of sediments in research area.

4. Discussion

4.1. Correlation Analysis

We use software (SPSS 25.0) to analyze the correlation of element contents and grain size parameters of the sediment samples. Table 3 shows the correlation coefficients between all indicators. Mean grain size (Mz) is highly correlated with clay components (>8Φ) and silt components (4Φ~8Φ), with correlation coefficients (r) of 0.91 and 0.93, respectively, and is highly negatively correlated (r = −0.91) with sand components (−1Φ~4Φ). The correlation coefficient (r) between sediment element content and Mz varies greatly, from −0.3 to 0.77. The content of most elements (except Ca, Ti, and Sr) is positively correlated with Mz, which indicates that the elements are more enriched in fine particle components.

Due to the large amount of SiO_2 in the sand component, the content of other elements is relatively diluted.

Table 3. Correlation coefficients between some indicators of samples (n = 35).

	Mz	Clay	Silt	Sand	Al	Mg	Fe	K	Ca	Na	Ti	P	Mn	Sr	Ba	Li	Zn	Sc	Co	Rb	Th
Mz	1.00																				
Clay	0.91	1.00																			
Silt	0.93	0.91	1.00																		
Sand	−0.91	−0.93	−0.99	1.00																	
Al	0.67	0.68	0.59	−0.60	1.00																
Mg	0.50	0.48	0.38	−0.39	0.84	1.00															
Fe	0.17	0.07	0.13	−0.10	0.29	0.41	1.00														
K	0.77	0.81	0.80	−0.82	0.91	0.71	0.28	1.00													
Ca	−0.30	−0.02	−0.05	−0.02	−0.23	−0.24	−0.22	−0.03	1.00												
Na	0.57	0.73	0.61	−0.66	0.86	0.68	0.21	0.88	0.18	1.00											
Ti	−0.14	−0.21	0.00	0.03	−0.48	−0.49	0.50	−0.23	0.13	−0.35	1.00										
P	0.53	0.68	0.58	−0.63	0.86	0.74	0.31	0.88	0.22	0.95	−0.32	1.00									
Mn	0.19	0.18	0.28	−0.25	−0.18	−0.32	0.33	0.05	−0.03	−0.10	0.61	−0.10	1.00								
Sr	−0.34	−0.04	−0.08	0.00	−0.23	−0.21	−0.16	−0.04	0.96	0.19	0.16	0.24	−0.08	1.00							
Ba	0.72	0.66	0.77	−0.76	0.74	0.57	0.27	0.91	−0.11	0.63	−0.07	0.66	0.13	−0.13	1.00						
Li	0.68	0.74	0.61	−0.64	0.96	0.84	0.29	0.90	−0.10	0.91	−0.45	0.90	−0.14	−0.10	0.66	1.00					
Zn	0.43	0.43	0.29	−0.30	0.67	0.63	0.28	0.53	−0.35	0.55	−0.38	0.52	−0.05	−0.32	0.37	0.69	1.00				
Sc	0.46	0.44	0.59	−0.57	0.41	0.24	0.70	0.63	0.08	0.47	0.58	0.50	0.45	0.10	0.67	0.41	0.19	1.00			
Co	0.07	0.00	0.18	−0.15	−0.11	−0.16	0.75	0.09	0.05	−0.04	0.90	0.01	0.54	0.10	0.20	−0.10	−0.09	0.81	1.00		
Rb	0.75	0.82	0.77	−0.80	0.91	0.69	0.28	0.98	0.00	0.93	−0.26	0.91	0.05	−0.01	0.81	0.94	0.59	0.61	0.07	1.00	
Th	0.68	0.69	0.80	−0.80	0.62	0.42	0.46	0.86	0.09	0.68	0.25	0.70	0.37	0.07	0.84	0.64	0.34	0.89	0.49	0.83	1.00

The correlation between the elements of typical terrigenous weathering products (Al, Mg, K, Na, Rb, etc.) is high (r > 0.84), which indicates the terrigenous characteristics of sediments in the research area. The correlation between the content of Ca and terrestrial weathering products elements (Al, Mg, etc.) is very poor, while the correlation between Ca and Sr content is very high (r = 0.96), indicating that Ca and Sr have homologous characteristics, and are likely to be affected by marine biochemistry. Although Fe is also a typical terrigenous weathering product element, its content is not highly correlated with other terrigenous representative elements (Al, Mg, K, Na, Rb, etc.), indicating that terrigenous input is not the main source of Fe. Although there is a positive correlation between Fe and Mn, the correlation is not high (r = 0.33), indicating that Fe-Mn nodules have a limited impact on the content of Fe in sediments of the research area. Among all elements, only Fe has a relatively high correlation with Ti (r = 0.5), indicating that Fe may be affected by the input of ilmenite. In conclusion, Fe may be carried in by multiple sources. The correlation between Mn and all elements is not high, but the correlation between Mn and Ti is the largest, reaching 0.61. This indicates that the Mn content may be controlled by symbiosis with ilmenite. In conclusion, the materials in the research area may be the result of the joint action of terrigenous weathering products, ilmenite, and marine biochemistry. See Figure 6.

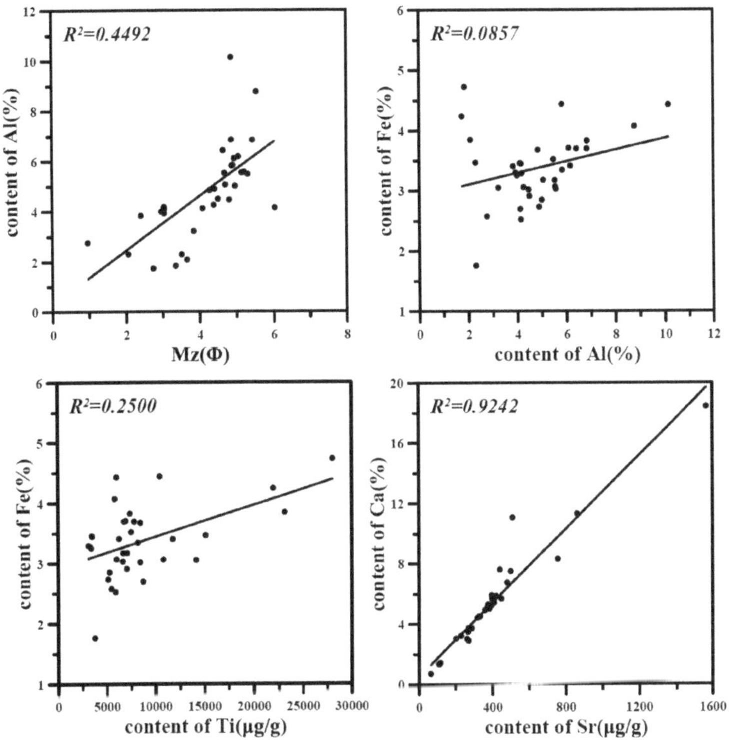

Figure 6. Correlations of content of Al, Fe, Ti, Ca, Sr, and Mz.

4.2. Factor Analysis and Provenance Discussion

To discuss the controlling factors, R-type factor analysis is conducted on the element contents in the samples by the statistical software (SPSS 25.0). Based on the method and covariance matrix, common factors (eigenvalues >1) are extracted. In this analysis, the Kaiser–Meyer–Olkin value and p-value of Bartlett's spherical test are 0.60 ($\geq$0.5) and 4.2×10^{-184} (<0.001), respectively. The values of these two statistical parameters indicate that the analysis method is appropriate and the statistical results have a high level of confidence [44]. We extract three common factors from the statistical data, accounting for 86.19% of the total variance (Table 4). Therefore, we believe that the contents of major and trace elements of sediment in the research area are mainly controlled by three factors.

Factor 1 accounts for 48.69% of the total variance, and we believe it is the determining factor for element content in the sediments. In Factor 1, most element contents have a high load. In particular, the load factor of typical terrestrial weathering product elements (Al, Na, K, Rb, etc.) reaches more than 0.9. Although Fe and Ti are also the main elements in terrigenous weathering products, the load factor of Fe in Factor 1 is only 0.30, and the load factor of Ti is negative (−0.38). The results show that Fe is not only affected by terrigenous weathering products, but also by other sources, and Ti is even diluted by terrigenous weathering products. To sum up, Factor 1 represents the input of terrigenous weathering products.

Factor 2 accounts for 23.14% of the total variance with high load factors of Fe, Ti, Mn, etc. The load factor of Ti in Factor 2 is even as high as 0.90, which is in sharp contrast to the load factor (−0.38) in Factor 1, indicating that Factor 2 may be related to the genesis of Ti-related ores. As we all know, Sierra Leone is a country rich in minerals, of which iron ore is its largest output [46]. However, the load factor of Fe in Factor 2 is not high enough compared with Ti, and the source of iron ore cannot represent Factor 2. Ilmenite often exists in iron ore in the form of accessory minerals, so the input of ilmenite may represent Factor

2. At the same time, Fe^{2+} and Mn^{2+} in ilmenite can be completely replaced by isomorphism, which also explains why the Mn element load in Factor 2 is as high as 0.72. In conclusion, Factor 2 can be interpreted as ore input dominated by ilmenite.

Table 4. Results of factor analysis (n = 35).

Elements	Factor 1	Factor 2	Factor 3	Elements	Factor 1	Factor 2	Factor 3
Al	0.95	−0.09	−0.22	Sr	0.00	0.00	0.97
Mg	0.82	−0.17	−0.28	Ba	0.79	0.29	−0.07
Fe	0.30	0.71	−0.27	Li	0.97	−0.08	−0.11
K	0.96	0.15	0.00	Zn	0.64	−0.09	−0.42
Ca	−0.01	−0.01	0.98	Sc	0.52	0.83	0.11
Na	0.94	−0.04	0.20	Co	−0.02	0.96	0.07
Ti	−0.38	0.90	0.15	Rb	0.97	0.12	0.03
P	0.95	0.00	0.23	Th	0.75	0.57	0.12
Mn	−0.11	0.72	−0.04				
Variance/%	48.69	23.14	14.36	Variance/%	48.69	23.14	14.36

Factor 3 explains 14.36% of the total variance with high load factors of Ca and Sr. The load factors of Ca and Sr in Factor 3 are as high as 0.98 and 0.97, respectively, while the load factors of other elements are very low (<0.3), which indicates that Factor 3 may represent the input of marine biochemical substances. Sr and Ca belong to the second main group of elements, with similar chemical properties and frequent isomorphic substitution. After the death of marine organisms, the organic matter of the corpses is completely decomposed, and only the biological shells or bones rich in Ca and Sr are crushed under the physical action of marine dynamics (waves and currents). The deposition of these substances increases the contents of Ca and Sr in marine sediments. In conclusion, Factor 3 represents the input of marine biochemical substances.

4.3. Comparison of Element Contents in Sediments from the Research Area and the Offshore Area of China

To better understand the elemental geochemical behavior of the sediments in the research area, we select the surface sediment elements of five typical areas in China offshore to compare with them. They are located in the South China Sea (taking the northern coastal waters of the South China Sea as an example) [41], the East China Sea (taking the coastal waters of Zhejiang Province as an example) [42], the Southern Yellow Sea (taking the Jiaozhou Bay as an example) [43], the Northern Yellow Sea (taking the Weihai Bay as an example) [44], and the Bohai Sea (taking the Laizhou Bay as an example) [45]. Through comparison, we find that the contents of Al, K, and Na in the study area are significantly lower than those in the coastal waters of China, but the contents of Fe, Ca, Ti, and Sr are significantly higher (Table 3, Figure 7). This feature corresponds to the analysis results of provenance in the research area. Compared with West Africa, China's river system is developed, and the river sediment flux into the sea is larger [47]. The terrigenous materials carried by rivers into the sea make the content of terrigenous elements (Al, K, Na, etc.) in the sediments of offshore China relatively high. On the contrary, the continental shelf width of the research area is smaller, the water depth is larger, and the ecosystem is more developed compared with China [48]. In addition, the input of terrigenous sediments in the study area is relatively small, which also leads to the relative increase in marine biological materials. This phenomenon is embodied in the high content of Ca and Sr. At the same time, Sierra Leone's rich ore resources (especially ilmenite, a bymineral of iron ore) also have a great impact on the supply of sediments. The content of Fe and Ti in the sediments of the research area is much higher than that in the sediments of offshore China. The element content of sediments in the research area is very different from that in the offshore of China, but its unique sediment source characteristics have great reference value for the elemental geochemistry of sediments in the offshore of China and even the whole world. See Figure 8.

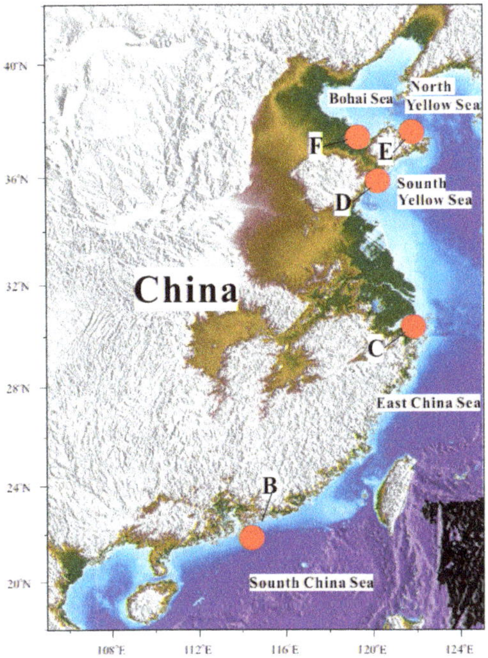

Figure 7. Selected locations in China offshore area.

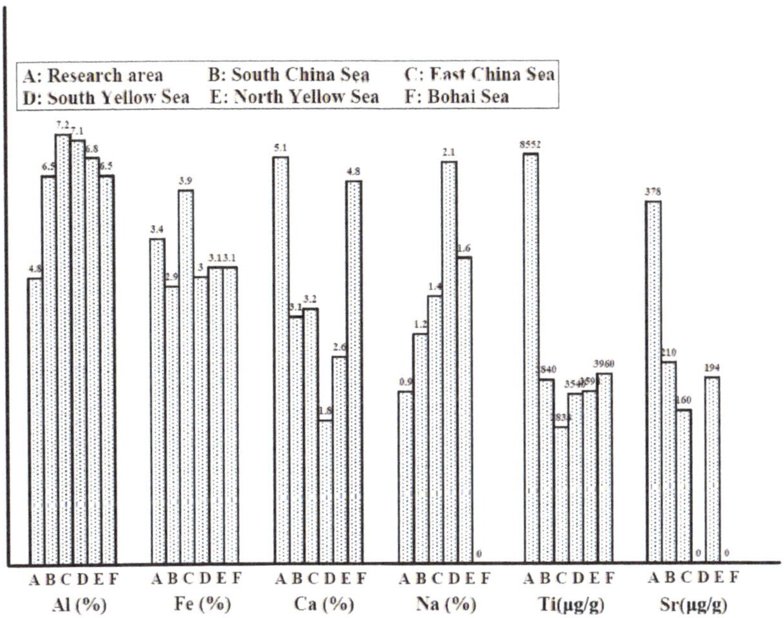

Figure 8. Comparison of element contents in sediments of different regions ("0" means no data).

5. Conclusions

To fill some gaps in the study of elemental chemistry, we chose less popular areas to carry out this supplementary research. Based on 35 surface sediments in the offshore area of Sierra Leone, we tested the grain size and major and trace elements, analyzed the content characteristics and controlling factors of the elements, discussed the material source of the sediments, and made a comparative study with the characteristics of the sediments in the offshore area of China. The results show that sandy silt is the main sediment type in the research area, and the average sediment mean grain size (Mz) is 4.15Φ. The content of Ca in the sediments is the highest, with an average of 5.1%. The content of Sr is the highest among the trace elements (except Ti, P, and Mn), with an average of 378.2 μg/g. There are three main sources of sediments in the research area, namely, terrigenous weathering products, ilmenite-dominated ore, and oceanic biochemical substances, based on factor analysis. The element content of sediments in the research area is very different from that in the offshore of China, but its unique sediment source characteristics have great reference value for the elemental geochemistry of sediments in the offshore of China and even the whole world.

Author Contributions: Conceptualization, M.H. and Y.Z.; methodology, M.H., Y.Z. and P.Z.; formal analysis, M.H.; investigation, B.Y. and R.J.; resources, Y.Z., B.Y. and R.J.; data curation, M.H. and C.H.; writing—original draft preparation, M.H.; writing—Review and Editing, Y.Z., W.C. and C.H.; visualization, M.H. and Y.Z.; supervision, B.Y.; project administration, M.H. and Y.Z.; funding acquisition, M.H. and Y.Z. All authors have read and agreed to the published version of the manuscript.

Funding: This research was funded by (1) China's National Key R&D Program (No. 2022YFB3207400), (2) the basic scientific research business cost project of the Institute of Geophysical and Geochemical Exploration, Chinese Academy of Geological Sciences (No. AS2022J07), and (3) Research Funds for the Central Universities (No. TKS20230205).

Data Availability Statement: Data are contained within the article.

Acknowledgments: We thank Mingyang Liu and Bo Liu from Tianjin Research Institute for Water Transport Engineering, Ministry of Transport, for their assistance with paper writing.

Conflicts of Interest: Author Wenliang Chen was employed by the company Qingdao Boyan Marine Environmental Science & Technology Co., Ltd. The remaining authors declare that the research was conducted in the absence of any commercial or financial relationships that could be construed as a potential conflict of interest.

References

1. Martin, J.; Zhang, J.; Shi, M.; Zhou, Q. Actual flux of the Huanghe (yellow river) sediments to the Western Pacific Ocean. *Neth. J. Sea Res.* **1993**, *31*, 243–254. [CrossRef]
2. Yang, S.; Jung, H.; Lim, D.; Li, C. A review on the provenance discrimination of sediments in the Yellow Sea. *Earth Sci. Rev.* **2003**, *63*, 93–120. [CrossRef]
3. Lim, D.; Kim, J.; Xu, Z.; Jung, H.; Yoo, D.; Choi, M.; Kim, S. Quantitative reconstruction of Holocene sediment source variations in the Yellow and northern East China Seas and their forcings. *Mar. Geol.* **2020**, *430*, 106345. [CrossRef]
4. Li, T.; Rao, W.; Wang, S.; Mao, C. Identifying the clay-size sediment provenance of the radial sand ridges in the southwestern yellow sea using geochemical and Sr-Nd isotopic tracers. *Mar. Geol.* **2023**, *455*, 106957. [CrossRef]
5. Xu, F.; Hu, B.; Dou, Y.; Liu, X.; Wan, S.; Xu, Z.; Tian, X.; Liu, Z.; Yin, X.; Li, A. Sediment provenance and paleoenvironmental changes in the northwestern shelf mud area of the South China Sea since the mid-Holocene. *Cont. Shelf Res.* **2017**, *144*, 21–30. [CrossRef]
6. Zhu, X.; Shen, C.; Zhou, R.; Xu, J.; Zhao, J.; Wang, L.; Ge, X. Paleogene sediment provenance and paleogeographic reconstruction of the South Yellow Sea basin, East China: Constraints from detrital zircon U Pb geochronology and heavy mineral assemblages. *Palaeogeogr. Palaeoecl.* **2020**, *553*, 109776. [CrossRef]
7. Li, M.; Zhu, S.; Ouyang, T.; Tang, J.; He, C. Magnetic fingerprints of surface sediment in the Bohai Sea, China. *Mar. Geol.* **2020**, *427*, 106226. [CrossRef]
8. He, C.; Ouyang, T.; Li, M.; Zhu, S.; Yu, M.; Peng, S.; Zhu, Z.; Wang, Y.; Chen, H.; Tian, C. Nonnegligible contribution of terrigenous sediment inputs from local small watersheds to west regions of the Pearl River Estuary, northern South China Sea shelf. *Mar. Geol.* **2024**, *475*, 107369. [CrossRef]

9. Li, M.; Zeng, Y.; Liu, S.; Tang, M.; Chen, L.; Chen, J.; Amorosi, A.; Ouyang, T. Provenance and sediment dispersal in Pearl River Estuary, southern China unraveled by magnetic properties. *Mar. Geol.* **2024**, *478*, 107431. [CrossRef]
10. Yan, J.; Geng, J.; Su, F. Estimation of the ecosystem service value of the yellow river delta-Laizhou Bay coastal zone considering regional differences and social development. *Environ. Manag.* **2024**, *74*, 192–205. [CrossRef]
11. Lin, C.; Bao, R.; Zhu, L.; Hu, R.; Ji, J.; Yu, S. Surface sediment erosion characteristics and influencing factors in the subaqueous delta of the abandoned yellow river estuary. *Mar. Geol.* **2024**, *468*, 107219. [CrossRef]
12. Wang, F.; Zhang, W.; Nian, X.; Roberts, A.; Zhao, X.; Shang, Y.; Dong, Y. Magnetic evidence for yellow river sediment in the late Holocene deposit of the Yangtze River Delta, China. *Mar. Geol.* **2020**, *427*, 106274. [CrossRef]
13. Saito, Y.; Yang, Z.; Hori, K. The Huanghe (Yellow River) and Changjiang (Yangtze River) deltas: A review on their characteristics, evolution and sediment discharge during the Holocene. *Geomorphology* **2001**, *41*, 219–231. [CrossRef]
14. Wei, X.; Wu, C.; Ni, P.; Mo, W. Holocene delta evolution and sediment flux of the Pearl River, southern China. *J. Quaternary Sci.* **2016**, *31*, 484–494. [CrossRef]
15. Weber, M.E.; Wiedicke, M.H.; Kudrass, H.R.; Hübscher, C.; Erlenkeuser, H. Active growth of the Bengal Fan during sea-level rise and highstand. *Geology* **1997**, *25*, 315–318. [CrossRef]
16. Sun, X.; Liu, S.; Li, J.; Zhang, H.; Shi, X. Major and trace element compositions of surface sediments from the lower Bengal Fan: Implications for provenance discrimination and sedimentary environment. *J. Asian Earth Sci.* **2019**, *184*, 104000. [CrossRef]
17. Yang, S.; Li, C.; Jung, H.; Lee, H. Discrimination of geochemical compositions between the Changjiang and the Huanghe sediments and its application for the identification of sediment source in the Jiangsu coastal plain. *China Mar. Geol.* **2002**, *186*, 229–241. [CrossRef]
18. Liu, J.; Saito, Y.; Kong, X.; Wang, H.; Zhao, L. Geochemical characteristics of sediments as indicators of post-glacial environmental changes off the Shandong peninsula in the Yellow Sea. *Cont. Shelf Res.* **2009**, *29*, 846–855. [CrossRef]
19. Zhu, Y.; Bao, R.; Zhu, L.; Jiang, S.; Chen, H.; Zhang, L.; Zhou, Y. Investigating the provenances and transport mechanisms of surface sediments in the offshore muddy area of Shandong Peninsula: Insights from REE analyses. *J. Mar. Syst.* **2022**, *226*, 103671. [CrossRef]
20. Cheng, P. Sediment Characteristics and Transport Processes of Fine-Grained Material over the Northern Yellow Sea. Ph.D. Thesis, Institute of Oceanology, Qingdao, China, 2000.
21. Lim, D.; Jung, H.; Choi, J. REE partitioning in riverine sediments around the Yellow Sea and its importance in shelf sediment provenance. *Mar. Geol.* **2014**, *357*, 12–24. [CrossRef]
22. Lan, X.; Chen, X.; Mi, B.; Li, R.; Qin, Y.; Wang, Z. Distribution pattern and source of major and trace elements in the central North Yellow Sea since late Pleistocene. *Mar. Geol. Quat. Geol.* **2015**, *35*, 1–10. (In Chinese)
23. Fang, H.; Huang, P.; Zhou, Y.; Zhang, J.; Li, A.; Yan, J. Distribution and controlling factors of the major elements in surface sediments of the North Yellow Sea. *Mar. Sci.* **2015**, *39*, 108–116. (In Chinese)
24. Li, J.; Liu, S.; Feng, X.; Sun, X.; Shi, X. Major and trace element geochemistry of the mid-Bay of Bengal surface sediments: Implications for provenance. *Acta Oceanol. Sin.* **2017**, *36*, 82–90. [CrossRef]
25. Yang, S.; Youn, J. Geochemical compositions and provenance discrimination of the central south Yellow Sea sediments. *Mar. Geol.* **2007**, *243*, 229–241. [CrossRef]
26. Risgaard-Petersen, N.; Revil, A.; Meister, P.; Nielsen, L. Sulfur, iron-, and calcium cycling associated with natural electric currents running through marine sediment. *Geochim. Cosmochim. Acta* **2012**, *92*, 1–13. [CrossRef]
27. Calvert, S.; Pedersen, T. Geochemistry of recent oxic and anoxic marine sediments: Implications for the geological record. *Mar. Geol.* **1993**, *113*, 67–88. [CrossRef]
28. Dean, W.; Gardner, J.; Piter, D. Inorganic geochemical indicators of glacial-interglacial changes in productivity and anoxia on the California continental margin. *Geochim. Cosmochim. Acta* **1997**, *61*, 4507–4518. [CrossRef]
29. Moskovchenko, D.; Soromotin, A.; Khoroshavin, V.; Prikhodko, N.; Kirillov, V.; Koveshnikov, M.; Krylova, E.; Krasnenko, A.; Pechkin, A. Climate change and the Ob River: A reassessment of major and trace element fluxes to the Arctic Ocean. *Water* **2024**, *16*, 2112. [CrossRef]
30. Grosbois, C.; Desmet, M.; Zhang, M.; Gassama, N.; Battaglia, F. Trace element contamination in one of the Yangtze tributaries (Hunan, China)—Source review and potential release from sediments. *Water* **2021**, *13*, 271. [CrossRef]
31. Bilinski, H. Geochemical fractionation and risk assessment of potentially toxic elements in sediments from Kupa River, Croatia. *Water* **2020**, *12*, 2024. [CrossRef]
32. Cho, Y.; Lee, C.; Cho, M. Geochemistry of surface sediments off the southern and western coasts of Korea. *Mar. Geol.* **1999**, *159*, 111–129. [CrossRef]
33. Lim, D.; Jung, H.; Choi, J.; Yang, S.; Ahn, K. Geochemical compositions of river and shelf sediments in the Yellow Sea: Grain-size normalization and sediment provenance. *Cont. Shelf Res.* **2006**, *26*, 15–24. [CrossRef]
34. Rao, W.; Mao, C.; Wang, Y.; Su, J.; Balsam, W.; Ji, J. Geochemical constraints on the provenance of surface sediments of radial sand ridges off the Jiangsu coastal zone, east China. *Mar. Geol.* **2015**, *359*, 35–49. [CrossRef]
35. Sundaray, S.; Nayak, B.; Lin, S.; Bhatta, D. Geochemical speciation and risk assessment of heavy metals in the river estuarine sediments—A case study: Mahanadi basin, India. *J. Hazard. Mater.* **2011**, *186*, 1837–1846. [CrossRef]
36. Liu, Y.; Kuang, W.; Xu, J.; Chen, J.; Sun, X.; Lin, C.; Lin, H. Distribution, source and risk assessment of heavy metals in the seawater, sediments, and organisms of the Daya Bay, China. *Mar. Pollut. Bull.* **2022**, *174*, 113297. [CrossRef]

37. Liu, T.; Zhu, L.; Bao, R.; Hu, R.; Jiang, S.; Zhu, Y.; Song, J. Hydrodynamically–driven distribution and remobilization of heavy metals in surface sediments around the coastal area of Shandong Peninsula, China. *Sci. Total Environ.* **2023**, *857*, 159286. [CrossRef]
38. Lin, H.; Lan, W.; Feng, Q.; Zhu, X.; Zhao, B. Pollution and ecological risk assessment, and source identification of heavy metals in sediment from the Beibu Gulf, South China Sea. *Mar. Pollut. Bull.* **2021**, *168*, 112403. [CrossRef]
39. McManus, J.; Berelson, W.; Klinkhammer, G.; Johnson, K.; Rushdi, A. Geochemistry of barium in marine sediments: Implications for its use as a paleoproxy. *Geochim. Cosmochim. Acta* **1999**, *62*, 3453–3473. [CrossRef]
40. Folk, R.; Ward, A. Brazos River bar: A study in the significance of grain size parameters. *J. Sediment. Res.* **1957**, *27*, 3–26. [CrossRef]
41. Zhao, J. Elements Geochemistry Spatial Multiscale Variability of Surface Sediments and Its Mechanism in Northwest of the South China Sea. Ph.D. Thesis, China University of Geosciences, Wuhan, China, 2016.
42. Dou, Y. Characteristics of Sediment Granularity, Element Geochemistry and Their Significance for Identifying Sedimentary Environment in the Contiguous Sea Areas of Changjiang River Estuary. Master's Thesis, First Institute of Oceanography SOA, Qingdao, China, 2007.
43. Wang, Q.; Wang, J.; Dong, J.; Zou, L.; Wang, R.; Xie, Y.; Xu, M.; Xie, X.; Zhou, X. Geochemical characteristics and controlling factors of surface sediments in Jiaozhou Bay. *Mar. Geol. Front.* **2023**, *39*, 8–19. (In Chinese)
44. Zhu, Y.; Feng, X.; Zhu, L.; Zhong, W. Origin and geochemistry of surface sediments in the mud deposit area off shore the Shandong Peninsula, China. *J. Oceanol. Limnol.* **2021**, *39*, 483–499. [CrossRef]
45. Lei, Z.; Liu, R.; Hu, R.; Qiu, J.; Zhu, L.; Zhu, F.; Zhang, X. Controlling factors and distribution of geochemical characteristics of the surface sediments in the Yellow River Delta. *Mar. Geol. Quat. Geol.* **2022**, *42*, 104–118. (In Chinese)
46. Jiang, J.; Hu, P.; Zhang, H.; Chen, X.; Wang, J.; Xiang, W. The iron ore resource feature and metallogenic regularity in western Africa. *Geol. Bull. China* **2023**, *42*, 1276–1290.
47. Li, Y.; Yan, H.; Rong, Y.; Lin, J.; Xue, W. The distribution characteristics of mercury species in the estuaries of the ten main rivers in China and flux of mercury from rivers to the ocean. *Acta Sci. Circumst.* **2022**, *42*, 323–331. (In Chinese)
48. Zhang, Z. Evaluation of Sustainable Utilization of Fishery Resources in the Coast of West Africa and Its Development Countermeasures. Ph.D. Thesis, Shanghai Ocean University, Shanghai, China, 2024.

water

MDPI

Article

Adaptive Penetration Unit for Deep-Sea Sediment Cone Penetration Testing Rigs: Dynamic Modeling and Case Study

Yusen Zhu [1], Zhiqiang Zhang [1], Xiuqing Yang [2], Zihang Fei [2], Lei Guo [1,*], Gang Xue [1] and Yanjun Liu [3]

1 Institute of Marine Science and Technology, Shandong University, Qingdao 266237, China; 202236984@mail.sdu.edu.cn (Y.Z.); zhangzhiqiang1695@163.com (Z.Z.); xuegangzb@163.com (G.X.)
2 College of Environmental Science and Engineering, Ocean University of China, Qingdao 266100, China; yxqself@163.com (X.Y.); f1079860937@sina.com (Z.F.)
3 Key Laboratory of High Efficiency and Clean Mechanical Manufacture, Ministry of Education, School of Mechanical Engineering, Shandong University, Jinan 250061, China; lyj111ky@163.com
* Correspondence: 201894900036@sdu.edu.cn

Abstract: The reliability and continuity of data are key issues in deep-sea sediment cone penetration testing. Cone penetration testing employs static force to uniformly insert rods into sediment, a process crucial for assessing its mechanics and layering. The clamping manipulator can perform this operation while accommodating sediment sensors of varying types and sizes. However, its requirement to reset post-penetration creates zero-velocity points that diminish test continuity and should be minimized. To address these limitations, this paper proposes a load-adaptive sediment rig that minimizes zero-velocity points, ensures data continuity, and contributes to sedimentology research. This paper analyzes the mechanical properties and layering patterns of sediment, along with the interaction mechanisms between sediment and mechanical structures. Subsequently, a mechanical structure–sediment-integrated model with adaptive control logic is established. Finally, real sediment data are introduced into the physical model for simulation experiments. The simulation results demonstrate that the load-adaptive rig reduces data breakpoints by 50% and increases the maximum single penetration stroke to 1.8 m. Additionally, the load-adaptive rig provides redundancy between penetration force and stroke, automatically reducing penetration force for greater stroke when encountering low-strength sediments and, conversely, sacrificing part of the stroke for greater force. These improvements significantly enhance the continuity of in situ detection data of sediment.

Keywords: sediment test; seabed cone penetration test; hydraulic energy; modeling and simulation; case study

1. Introduction

Cone penetration testing (hereinafter referred to as CPT) is the most commonly used in situ testing method for the mechanical properties of seabed sediments [1]. By pressing the cone rod with specific sensors into the sediment at a constant penetration rate, parameters such as cone resistance, sleeve friction, and pore water pressure are measured, and the relationship between the CPT data and the sediment mechanical parameters is established by a semi-theoretical and semi-empirical method, so as to indirectly obtain the physical and mechanical properties of the sediment layer [2–4]. With the continuous development of sensor technology, the CPT technique can provide more reliable and multi-parameter test data by replacing and developing different probes and sensors, so this in situ testing method has great expandability [5].

Cone penetration testing has been applied to sediment testing since the 1960s [6]. Due to its characteristics of its real-time, efficient, and accurate determination of the engineering properties of submarine soils, it has unparalleled advantages in the comprehensive evaluation of submarine engineering geology [7,8]. Some countries regard cone penetration technology as an essential part of marine engineering geological surveys, and it has been applied in the following areas: (1) To identify sediment types, delineate strata, and determine the vertical stratification of sediment conditions; (2) To estimate the physical and mechanical properties of soil to reflect the engineering properties of subsea soil, such as estimating soil shear strength and the overconsolidation ratio; (3) Piezocone penetration tests (CPTUs), building on traditional CPT, can measure pore water pressure, providing a more accurate reflection of soil liquefaction trends and electrical resistivity [9–11]. A number of reliable commercial seabed-based CPT rigs are now available on the market, such as the ROSON series, the Neptune series, the Manta series, and the Seacalf series [12–14]. These CPT rigs can operate in water depths of up to 3000 meters and are powered by electrical energy transmitted from surface vessels via umbilical cables [15]. Depending on the penetration method, these devices can be divided into two categories: friction wheel and manipulator [16,17]. One of the most important features of this equipment, represented by the Neptune 3000, is the use of pairs of rolling metal wheels to extrude and straighten flexible rods, using friction to achieve penetration of the rods [18,19]. Another common type of rig uses a clamping manipulator for the penetration of the rod [20,21]. During the penetration operation, the upper movable clamping manipulator is in a clamped state, while the lower fixed manipulator is in a released state [22,23]. At this time, the driving hydraulic cylinder piston rod extends, transmitting through the steel cable and pulley block, which drives the movable manipulator on the sliding seat to clamp the rod and slowly and uniformly penetrate into the soil. After a single penetration is completed, the lower fixed manipulator clamps the rod, the upper movable one releases, the penetration cylinder piston rod retracts, and the upper movable manipulator resets. This is one penetration cycle [24]. Repeating the above cycle completes the full penetration of the rod.

The penetration of the rod is subjected to continuous resistance exerted by the soil layer, which usually fluctuates dramatically due to the complexity of the formation [25,26]. In order to collect high-quality raw data in this case, it is required that the penetration speed of the rod with the sensor is kept constant during the test, e.g., 2 cm/s [27–29]. The above-mentioned Neptune 3000 can easily achieve continuous penetration of tens of meters in a single operation due to the use of flexible rods that can be coiled and stowed as well as the metal wheel penetration device. However, due to the poor adaptability of the metal wheels to the diameter of the rods, the cross-sectional area of the tens of meters of the rods of the Neptune 3000 is kept to 2.5 cm^2, which means that the types of sensors that can be accommodated in such a small space are very limited [30,31]. Therefore, the disadvantages of this type of CPT rig are the small variety of parameters acquired and the weak representativeness of the data. Another type of rig uses a clamping manipulator to achieve the penetration of the rod. Although the flexible opening and closing range of its manipulator can accommodate rods of different diameters, after completing a single penetration, the manipulator needs to release its clamp and reset to proceed to the next penetration cycle. During the resetting process, the speed of the rod is 0, so the working continuity of such equipment is relatively poor. In addition, the maximum penetration force and the depth of a single penetration of this equipment are fixed. When encountering soil layers with low resistance (low-load conditions), it is difficult to fully utilize the rig.

Based on the analysis above, to address the problems of clamping manipulators, this paper proposes a load-adaptive CPT rig with the aim of achieving gear shifting according to dynamic loads and obtaining the maximum single penetration stroke under load conditions based on the clamping manipulator-type CPT. The rig uses a pulley assembly to multiply the displacement of the driving hydraulic cylinder, and the pulleys that make up this assembly can switch between fixed pulley and movable pulley states. Therefore, by changing the number of enabled movable pulleys in the pulley assembly, the amplification factor of the hydraulic cylinder displacement can be altered, which, in turn, changes the single penetration stroke of the clamping manipulator. This process is defined as gear shifting in this paper.

The main novelties of this paper are as follows: We propose a load-adaptive CPT rig with a variable pulley block, enhancing stroke under load, minimizing rate zeros, and ensuring data continuity. This paper establishes a mathematical model of the hydraulic cylinder, analyzes its amplitude–frequency characteristics, determines the boundary conditions, and verifies the model through physical simulation. We validate the model using real CPT data, demonstrating the load-adaptive CPT system's superior continuity and efficiency under identical resistance conditions compared to conventional systems. In strata with low resistance, the system automatically shifts to a higher gear, applying minimal yet sufficient force for maximum penetration, which ensures data reliability and continuity while optimizing energy efficiency through force conservation.

The paper is organized as follows: Section 2 introduces the structural design of the load-adaptive CPT rig and the shifting method of the penetration unit; Section 3 details the development and subsequent validation of the system's mathematical model; and Section 4 tests the CPT rig utilizing actual CPT data and, subsequently, analyzes the resulting test outcomes.

2. Structure of the Load-Adaptive CPT Rig

Figure 1 describes the general structure of the load-adaptive CPT rig, and Figure 2 shows the specific structure of the penetration unit. During the penetration test, the active clamping manipulator is in the clamping state, and the lower fixed clamping manipulator is in the loosening state; at this time, the piston rod of the hydraulic cylinder extends and drives the active clamping manipulator to clamp the rod slowly and evenly into the soil layer by means of the transmission of the steel cable and pulley group. After the single penetration is completed, the lower fixed clamping manipulator clamps the rod, the upper movable clamping manipulator is loosened, the piston rod of the penetration cylinder is retracted, the upper movable clamping manipulator is reset, and the above cycle is repeated until the rod penetrates to the specified depth.

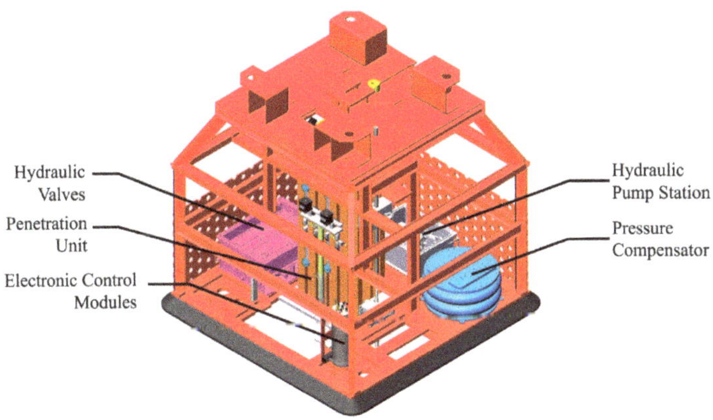

Figure 1. Diagram of the load-adaptive CPT rig.

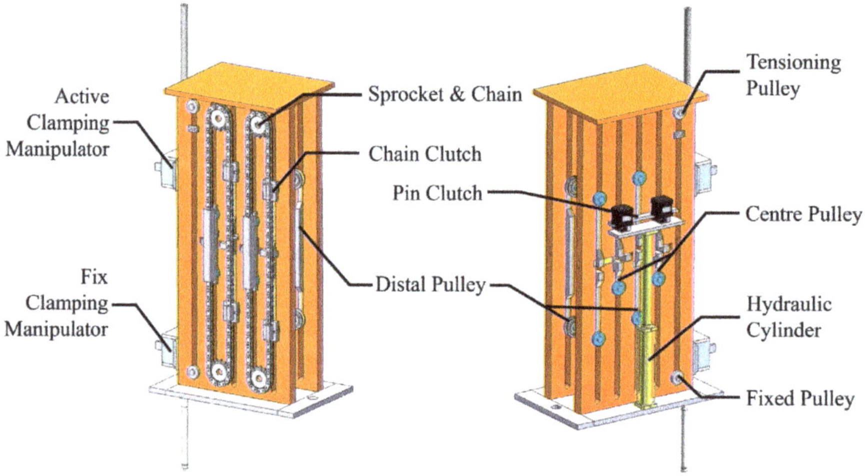

Figure 2. Diagram of the penetration unit.

Figures 2 and 3 depict the structure and operational mechanism of the penetration unit. The unit's central components are the pulley block and steel cable. Within the unit, pulleys are categorized into two types: central pulleys and distal pulleys. For instance, Blocks B and D in Figure 3 are central pulleys, while Blocks A and C are distal pulleys. As observed in Figure 4, the pulleys are mounted on a bracket to constitute a pulley block, ensuring synchronized velocity among pulleys within the same block. Furthermore, the steel cable's end is secured to a metal slider, enabling the cable to move in unison with the slider's motion. The pulley and metal slider are both connected to the hydraulic cylinder via a clutch, remaining stationary when the clutch is disengaged. Upon the pin clutch's activation, pushing the pin into the designated groove, the central pulley and metal slider are actuated by the hydraulic cylinder, initiating movement. Concurrently, the central pulley propels the sprocket chain assembly into motion; as the chain clutch secures the chain, the distal pulley synchronizes with the chain, moving in unison. These clutch mechanisms enable the control and switching of the pulley between two states in both fixed and dynamic pulleys.

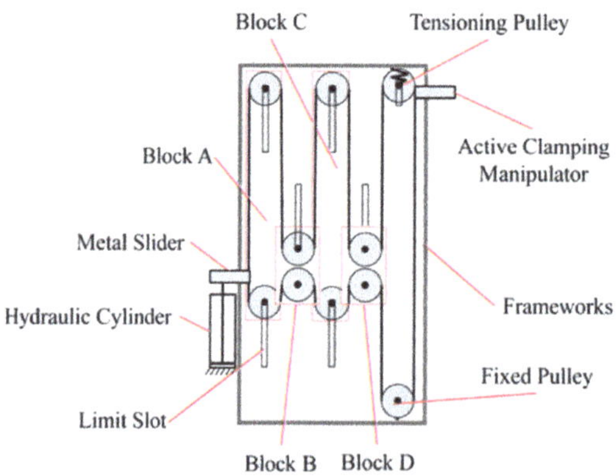

Figure 3. Schematic diagram of the transmission.

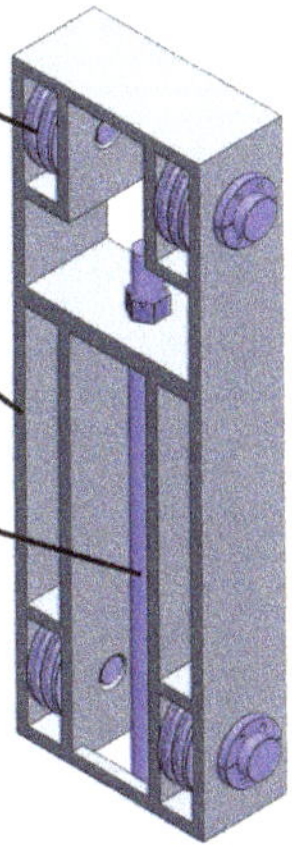

Figure 4. Diagram of the composition of the pulley block.

When the hydraulic cylinder directly actuates the metal slider, the cylinder's stroke is transferred to the clamping manipulator at a 1:1 ratio. According to the principle of the dynamic pulley, the stroke is doubled for each dynamic pulley engaged by the hydraulic cylinder. By altering the number of enabled dynamic pulleys, we can adjust the stroke amplification of the hydraulic cylinder, effectively shifting gears. As depicted in Figure 5, during the transition to the first gear, the driving hydraulic cylinder propels the metal slider upward, with the slider's displacement denoted as d, propelled by the driving force F_i from the cylinder. The metal slider is concurrently subjected to the vertical upward force F_i and the vertical downward cable tension T. Thus, when it undergoes uniform linear motion, the following equation holds:

$$F_i = T \tag{1}$$

Subsequently, the penetration force output by the system in the first gear is as follows:

$$F_{o1} = T = F_i \tag{2}$$

Due to the presence of the tensioning wheel, the change in the length of the steel cable is negligible. At this point, the system's output displacement, which is also the penetration travel, is $D = d$. Differentiating this with respect to time yields the relationship between the hydraulic cylinder piston velocity v_{i1} and the output velocity v_o, as follows:

$$v_{i1} = \dot{d} = \dot{D} = v_o \tag{3}$$

Additionally, to maintain the penetration speed, namely, the output speed, at 2 cm/s, the velocity of the hydraulic cylinder piston in the first gear is:

$$v_{i1} = 2\ \text{cm/s} \tag{4}$$

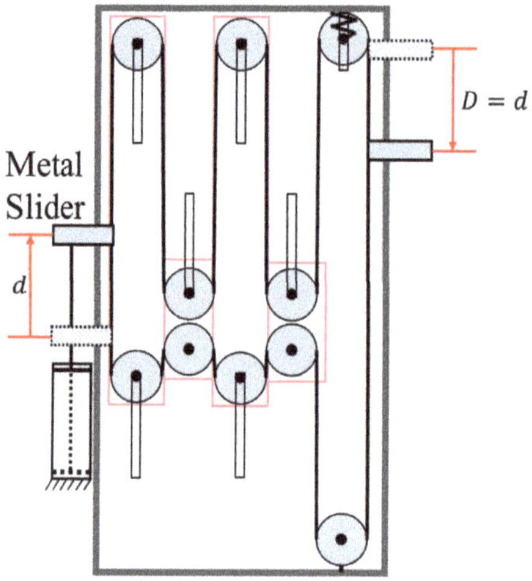

Figure 5. Schematic diagram of 1st gear movement.

As shown in Figure 6, when shifted to the second gear, the driving hydraulic cylinder propels Block B upward, at which point Block B has a displacement of d and is subjected to the driving force F_i from the hydraulic cylinder. For Block B, it simultaneously experiences the vertical upward driving force F_i and the vertical downward cable tension T. According to the principle of the movable pulley, during uniform linear motion, the following equation holds:

$$F_i = 2T \tag{5}$$

$$D = 2d \tag{6}$$

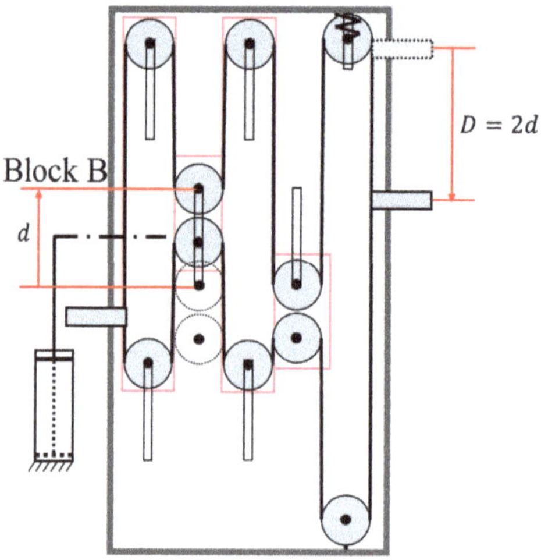

Figure 6. Schematic diagram of 2nd gear movement.

Therefore, the penetration force output by the system in second gear is:

$$F_{o2} = T = \frac{1}{2}F_i \tag{7}$$

The velocity of the hydraulic cylinder piston is:

$$v_{i2} = \dot{d} = \frac{1}{2}\dot{D} = \frac{1}{2}v_o = 1\ \text{cm/s} \tag{8}$$

As depicted in Figure 7, when shifted to the third gear, the driving hydraulic cylinder simultaneously propels the metal slider and Block B upward. At this juncture, the driving force F_i from the hydraulic cylinder can be divided into two components: the force F_{iS} acting on the metal slider and the force F_{iB} acting on Block B, that is:

$$F_i = F_{iS} + F_{iB} \tag{9}$$

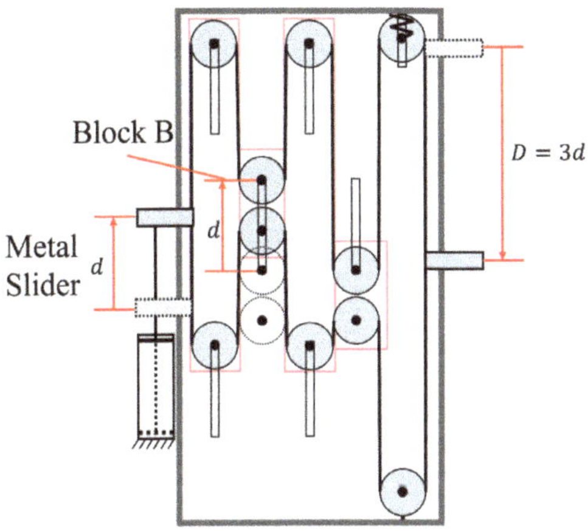

Figure 7. Schematic diagram of 3rd gear movement.

The force conditions for the metal slider and Block B are consistent with Equations (1) and (5), hence:

$$F_i = F_{iS} + F_{iB} = T + 2T = 3T \tag{10}$$

Similarly, the system's output displacement is:

$$D = d + 2d = 3d \tag{11}$$

Thus, in the third gear state, the penetration force and the velocity of the hydraulic cylinder piston are, respectively:

$$F_{o3} = T = \frac{1}{3}F_i \tag{12}$$

$$v_{i2} = \dot{d} = \frac{1}{3}\dot{D} = \frac{1}{3}v_o \approx 0.33\ \text{cm/s} \tag{13}$$

In summary, the motion states and parameters at different gear positions during the penetration process are summarized in Table 1, where "×" indicates that the slider or pulley is in a fixed state, "↑" or "↓" signifies an active state capable of moving in the direction of the arrow, and "×2", "×4", etc., indicates magnification, and "F" is force. It should be noted that as the stroke of the hydraulic cylinder is magnified, the penetration force decreases inversely. Therefore, we need to flexibly shift gears based on the actual penetration resistance to ensure maximum single-penetration travel under load conditions.

Table 1. Motion states and parameters at different gears.

	Gear I	Gear II	Gear III	Gear IV	Gear V	Gear VI	Gear VII	Gear VIII	Gear IX
Pulley Block A	×	×	×	↓	↓	↓	↓	↓	↓
Pulley Block B	×	↑	↑	↑	↑	↑	↑	↑	↑
Pulley Block C	×	×	×	×	×	×	×	↓	↓
Pulley Block D	×	×	×	×	×	↑	↑	↑	↑
Stroke Amplification	0	×2	×2	×4	×4	×6	×6	×8	×8
Metal Slider	↑	×	↑	×	↑	×	↑	×	↑
Total Stroke Amplification	×1	×2	×3	×4	×5	×6	×7	×8	×9
Penetration Force	F	$\frac{1}{2}F$	$\frac{1}{3}F$	$\frac{1}{4}F$	$\frac{1}{5}F$	$\frac{1}{6}F$	$\frac{1}{7}F$	$\frac{1}{8}F$	$\frac{1}{9}F$
Hydraulic Cylinder Speed	2 cm/s	1 cm/s	0.67 cm/s	0.5 cm/s	0.4 cm/s	0.33 cm/s	0.29 cm/s	0.25 cm/s	0.22 cm/s
Penetration Speed					2 cm/s				

3. Modeling of the Hydraulic Drive System for the Penetration Unit

3.1. Modeling of Servo Valve Mechatronics Subsystem

This paper selected the Bosch Rexroth 4WSE2EM6-2X15BET type servo valve, and its parameters are shown in Table 2. This section and the following chapters carry out mathematical modeling based on the parameters of this type of servo valve. The mechatronic subsystem of the servo valve is capable of converting an input voltage signal u into a displacement x_v of the servo valve spool. Specifically, the input voltage u is first converted into a current i by a servo amplifier; the current i is then applied to the coil of the torque motor of the mechatronic converter, causing a force to be generated by a permanent magnet, which is ultimately transferred to the servo valve spool through a mechanical feedback and causes it to generate a displacement x_v.

Table 2. Parameters and values of the servo valve.

Parameters	Symbol (Unit)	Values
Flow-pressure coefficient	$K_c(\mathrm{m^5/N{\cdot}s})$	6.11×10^{-12}
Internal leakage coefficient	$C_{ip}(\mathrm{m^5/N{\cdot}s})$	3×10^{-13}
External leakage coefficient	$C_{ep}(\mathrm{m^5/N{\cdot}s})$	0
Equivalent leakage coefficient	$C_{ie}(\mathrm{m^5/N{\cdot}s})$	3.36×10^{-13}
Additional leakage coefficient	$C_f(\mathrm{m^5/N{\cdot}s})$	-3.55×10^{-14}
Natural frequency	W_h (rad/s)	268
Flow gain	K_{sv} ($\mathrm{m^3/s \cdot A}$)	4.38×10^{-4}
Bandwidth of servo valve	ω_{sv} (rad/s)	0.03
Damping ratio of servo valve	ξ_{sv}	0.6

The mathematical model of the servo amplifier is shown below:

$$K_0(S) = \frac{I(S)}{U(S)} \tag{14}$$

where $I(S)$ is the output current, $U(S)$ is the input voltage, and $K_0(S)$ is the gain amplifier.

In engineering applications, it is a common assumption that an electromechanical transducer can be modeled as a second-order system, neglecting the dead zone, and its mathematical representation is shown below 32:

$$\frac{X_v(S)}{I(S)} = \frac{K_{sv}}{\frac{S^2}{\omega_{sv}{}^2} + \frac{2\xi_{sv}S}{\omega_{sv}} + 1} \tag{15}$$

where K_{sv} is the flow gain of servo valve, ω_{sv} is the bandwidth of servo valve, and ξ_{sv} is the damping ratio of servo valve.

3.2. Modeling of Asymmetric Cylinder Controlled by Servo Valve Subsystem

Figure 8 shows the schematic diagram of the asymmetric cylinder controlled by the valve subsystem, and the direction of the arrow shown is the positive direction of each variable. By establishing the subsystem model of the asymmetric cylinder controlled by the servo valve, the quantitative relationship between the spool displacement x_v and the piston displacement y of the hydraulic cylinder can be obtained.

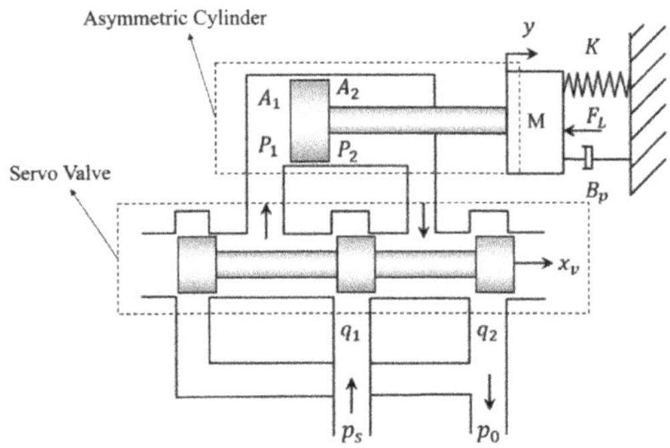

Figure 8. Schematic diagram of asymmetric cylinder controlled by servo valve subsystem.

For the purpose of further analysis, the following assumptions are established: the servo valve is paired with an asymmetric cylinder; the internal and external leakage of the hydraulic cylinder is assumed to be laminar; the connecting pipeline between the servo valve and the hydraulic cylinder is characterized as short and thick, with the dynamic characteristics of the pipeline and pressure losses being disregarded; the supply pressure p_s of the subsystem is considered constant, while the return pressure $p_0 = 0$; and the influences of temperature and liquid compressibility on the entire subsystem are neglected. The following discussion focuses on the process in which the servo valve spool moves to the right and the hydraulic cylinder piston extends to the right [32].

When the servo valve spool is shifted to the right, i.e., $x_v > 0$, the flow equation of the servo valve is:

$$q_1 = wx_vC_d\sqrt{\frac{2}{\rho}(p_s - p_1)} \tag{16}$$

$$q_2 = wx_vC_d\sqrt{\frac{2}{\rho}p_2} \tag{17}$$

where q_1 is the rodless chamber flow rate of hydraulic cylinder, q_2 is the rod-end chamber flow rate of hydraulic cylinder, p_1 is the pressure in the rodless chamber of the hydraulic cylinder, p_2 is the pressure in the rod-end chamber of the hydraulic cylinder, p_s is the pressure of the hydraulic source, C_d is the flow coefficient, w is the area gradient of the servo valve orifice, ρ is the density of hydraulic fluid, and x_v is the spool displacement.

Considering the process of the piston extending after the hydraulic cylinder is subjected to hydraulic pressure, the hydraulic cylinder inlet chamber flow rate q_1 and return chamber flow rate q_2 are:

$$q_1 = \frac{dV_1}{dt} + C_{ip}(p_1 - p_2) + C_{ep}p_1 + \frac{V_1}{\beta_e}\frac{dp_1}{dt} \tag{18}$$

$$q_2 = \frac{dV_2}{dt} + C_{ip}(p_1 - p_2) - C_{ep}p_1 - \frac{V_2}{\beta_e}\frac{dp_2}{dt} \quad (19)$$

where C_{ip} is the internal leakage coefficient of the hydraulic cylinder, C_{ep} is the external leakage coefficient of the hydraulic cylinder, and β_e is the bulk elastic modulus.

Notice that when ignoring the compressibility of the hydraulic fluid and hydraulic cylinder leakage, according to Equations (16)–(19), there are:

$$q_1 = \frac{dV_1}{dt} = A_1\frac{dy}{dt} = wx_vC_d\sqrt{\frac{2}{\rho}(p_s - p_1)} \quad (20)$$

$$q_2 = \frac{dV_2}{dt} = A_2\frac{dy}{dt} = wx_vC_d\sqrt{\frac{2}{\rho}p_2} \quad (21)$$

where A_1 is the effective area of piston side, A_2 is the piston area on the rod side, and y is the piston displacement.

From Equations (20) and (21):

$$n = \frac{q_2}{q_1} = \frac{A_2}{A_1} = \sqrt{\frac{p_2}{p_s - p_1}} < 1 \quad (22)$$

Ideally, the output power N_L of the servo valve and the output power N_{out} of the hydraulic cylinder are equal, i.e.,:

$$N_L = p_Lq_L = F_Lv = N_{out} \quad (23)$$

where p_L is the load pressure, q_L is the load flow, F_L is the external load force, and v is the piston speed.

According to the force analysis of the piston in Figure 8, there is:

$$F_L = p_1A_1 - p_2A_2 \quad (24)$$

The speed of the piston of the hydraulic cylinder is:

$$v = \frac{q_1}{A_1} = \frac{q_2}{A_2} \quad (25)$$

Associating Equations (23)–(25), we obtain:

$$p_Lq_L = (p_1A_1 - p_2A_2)\frac{q_1}{A_1} = \left(p_1 - p_2\frac{A_2}{A_1}\right)q_1 = (p_1 - np_2)q_1 \quad (26)$$

The load pressure is defined as follows:

$$p_L = \frac{F_L}{A_1} = \frac{p_1A_1 - p_2A_2}{A_1} = p_1 - np_2 \quad (27)$$

From Equations (26) and (27):

$$q_L = q_1 \quad (28)$$

Associating Equations (22) and (28), we obtain:

$$p_1 = \frac{n^3p_s + p_L}{1 + n^3}, p_2 = \frac{n^2(p_s - p_L)}{1 + n^3} \quad (29)$$

From Equations (18), (28), and (29):

$$q_L = q_1 = A_1 \frac{d_y}{dt} + C_{ie} p_L + C_f p_s + \frac{V_t}{4\beta_e} \frac{dp_L}{dt} \tag{30}$$

where C_{ie} is the equivalent leakage coefficient, C_f is the additional leakage coefficient, and V_t is the equivalent total volume.

When the piston extends, all output forces and load forces of the hydraulic cylinder should be in a state of force balance, hence, it follows that:

$$p_1 A_1 - p_2 A_2 = p_L A_1 = m_t \frac{d^2 y}{dt^2} + B_p \frac{dy}{dt} + Ky + F_L \tag{31}$$

where m_t is the total mass translated to the piston, B_p is the total viscous damping coefficient, and K is the load spring stiffness.

In order to conduct a linear analysis of the system, nonlinear load forces and the like must be neglected. Therefore, based on the flow characteristics of fluids through small orifices, Equation (16) is linearized under small perturbations, simplifying to:

$$q_L = q_1 = K_q x_v - K_c p_L \tag{32}$$

where K_q is the flow gain, and K_c is the flow-pressure coefficient.

Applying the Laplace transform to Equations (30), (31), and (32), and simplifying, we obtain the following mathematical model with input as the spool displacement and external load and output as the hydraulic cylinder piston displacement:

$$Y(S) = \frac{\frac{K_q}{A_1} X_v - \frac{K_{ce}}{A_1{}^2}(1 + \frac{V_t}{4\beta_e K_{ce}} S) F_L}{S(\frac{S^2}{\omega_h{}^2} + \frac{2\xi_h}{\omega_h} S + 1)} \tag{33}$$

where ω_h is the natural frequency, K_{ce} is the total flow-pressure coefficient ($K_{ce} = K_c + C_{ie}$), and ξ_h is the hydraulic damping ratio ($\xi_h = \frac{K_{ce}}{A_1}\sqrt{\frac{\beta_e m_t}{V_t}} + \frac{B_p}{4A_1}\sqrt{\frac{V_t}{\beta_e m_t}}$). Therefore, the transfer function of the hydraulic cylinder displacement $Y(S)$ to the spool displacement X_v is:

$$\frac{Y(S)}{X_v} = \frac{K_q / A_1}{S(\frac{S^2}{\omega_h{}^2} + \frac{2\xi_h}{\omega_h} S + 1)} \tag{34}$$

The transfer function of the hydraulic cylinder displacement $Y(S)$ to the interference force F_L is:

$$\frac{Y(S)}{F_L} = \frac{-\frac{K_{ce}}{A_1{}^2}(1 + \frac{V_t}{4\beta_e K_{ce}} S)}{S(\frac{S^2}{\omega_h{}^2} + \frac{2\xi_h}{\omega_h} S + 1)} \tag{35}$$

3.3. Modeling of Servo Valve Displacement Sensor Subsystem

As an essential component for the implementation of negative feedback automatic control in servo valves, the displacement sensor transforms the input displacement signal from the hydraulic cylinder piston into a voltage signal, which is then output to the electrical components of the servo valve. The mathematical model of the displacement sensor is:

$$U_f(S) = Y K_f \tag{36}$$

where U_f is the output voltage, K_f is the gain of the displacement sensor and Y is the displacement of hydraulic cylinder piston.

3.4. Modeling of Hydraulic Drive System

By assembling the servo valve mechatronics subsystem, the asymmetric cylinder controlled by the servo valve subsystem, and the servo valve displacement sensor subsystem, the overall mathematical model of the hydraulic transmission system can be obtained. By combining Equations (14), (15), and (34)–(36), the open-loop transfer function of the system is obtained:

$$G(S)H(S) = \frac{K_V}{S(\frac{S^2}{\omega_{sv}{}^2} + \frac{2\xi_{sv}}{\omega_{sv}} S + 1)(\frac{S^2}{\omega_h{}^2} + \frac{2\xi_h}{\omega_h} S + 1)} \tag{37}$$

where K_V is the open-loop amplification factor ($K_V = K_f K_{sv} K_f / A_1$).

Subsequently, modeling was carried out using the Simulink component of MATLAB R2022b, according to Equation (37), and the model created by Simulink is shown in Figure 9.

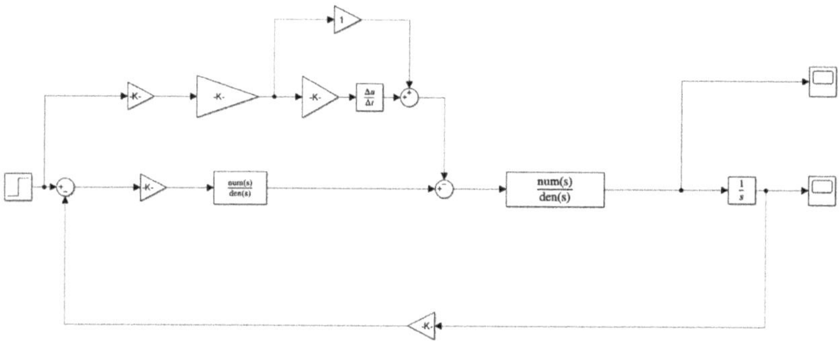

Figure 9. Mathematical modeling of hydraulic transmission system in Simulink.

The parameter values in Table 3 were used in the further analysis of the model.

Table 3. Calculated parameters and values.

Parameters	Symbol (Unit)	Values
External load equivalent mass	m_t(kg)	400
Flow rate	n	0.61
Total flow-pressure coefficient	K_{ce}(m^5/N·s)	6.45×10^{-12}
Hydraulic damping ratio	ξ_h	0.25

The external load equivalent mass in Table 3 is the total mass of the hydraulic cylinder and the pulley block. This is because their motions are translational, and there is no rotational inertia, so we modeled them using 3D modeling software SOLIDWORKS 2022, which, subsequently, showed the total mass. In addition, the flow rate, the total flow-pressure coefficient, and the hydraulic damping ratio were obtained using the calculations of Equations (22) and (33), respectively.

Stability is one of the most critical characteristics of a control system and an essential condition for its normal operation. Any system will deviate from its original equilibrium state under the influence of disturbances, resulting in an initial deviation. Stability refers to the system's ability to return to the original equilibrium state from the initial deviation state after the disturbance has subsided. If the system's dynamic process gradually decays and tends to zero over time, it is considered stable; conversely, if the system's dynamic process diverges over time under the influence of disturbances, the system is deemed unstable. The system's characteristic equation was assessed in the time domain using the Routh stability criterion. If all coefficients of the characteristic equation are positive, the system can be determined to be in a stable state.

According to the formula:

$$2\xi_h\omega_h > K_v \tag{38}$$

Substituting the data gives:

$$2 \times 0.25 \times 268 = 134 > 0.523 \tag{39}$$

Thus, it can be determined that the hydraulic transmission system of the penetration unit is stable and possesses a significant gain margin.

In the frequency domain, system stability was analyzed by plotting the amplitude and phase frequency response curves, with relative stability in the frequency domain referred to as the stability margin, commonly quantified by the phase margin and the gain margin [33]. The condition for system stability typically requires a phase margin between 30 deg and 60 deg and a gain margin greater than 6 dB. Consequently, the stability of the system can be analyzed through its open-loop frequency characteristics [34,35]. Given the open-loop transfer function, the open-loop Bode plot can be obtained using MATLAB/Simulink R2022b simulation analysis to determine the system's gain and phase margins. The Bode plot for Equation (37) is shown in Figure 10:

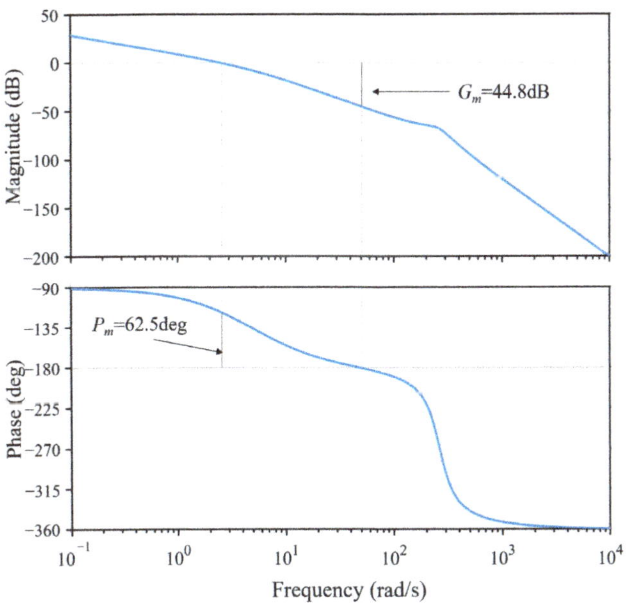

Figure 10. Bode plot for Equation (37).

From Figure 10, it is evident that the system's gain margin is 44.8 dB, and the phase margin is 62.5 degrees, thus meeting the stability requirements and exhibiting excellent stability. Under the condition without a controller, a unit step signal was applied to the system to observe its transient response. The simulation was performed on the model, as shown in Figure 9, and the results are shown in Figures 11 and 12.

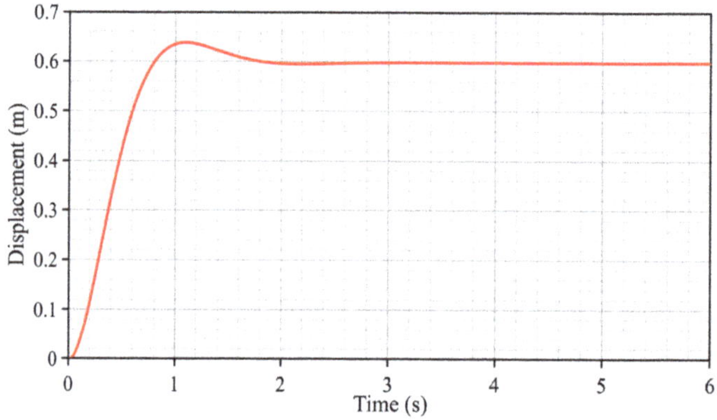

Figure 11. Output displacement of the system under a unit step signal input.

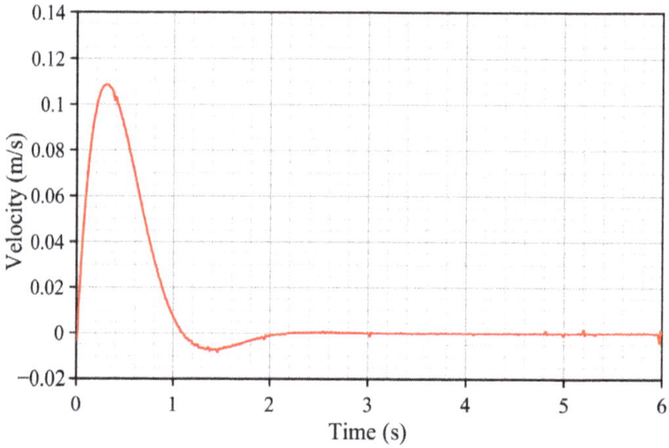

Figure 12. Output velocity of the system under a unit step signal input.

As depicted in Figure 11, upon the application of a unit step input to the system, the displacement response exhibits the following characteristics: the response time is moderate, ranging from 3 to 4 s, yet overshoot is observed, which is detrimental to practical engineering applications. Furthermore, according to the prevailing standards both domestically and internationally, the penetration rate should be controlled within a range of 2 cm/s ± 10%. The velocity of the hydraulic transmission system illustrated in Figure 12 clearly fails to meet these requirements. Consequently, it is imperative to incorporate a controller into the original system for regulation.

The hydraulic transmission system employs a PID (Proportional Integral Derivative) control algorithm, combined with negative feedback control methods, to effectively stabilize the penetration rate under conditions of uncertain soil parameters [36]. The transfer function of the PID control algorithm is represented as follows:

$$\frac{I(s)}{e(s)} = K_p + K_i\frac{1}{s} + K_d s \quad (40)$$

where K_p is the proportional gain, K_i is the integral gain, K_d is the derivative gain, $I(s)$ is the input, and $e(s)$ is the error. The complete model of the hydraulic transmission system after adding the PID controller is shown in Figure 13.

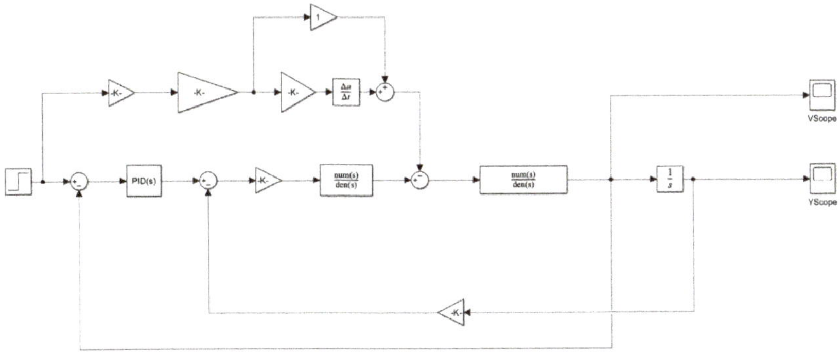

Figure 13. Diagram of complete model of the hydraulic drive system.

3.5. Determination of System Boundary Conditions

In the field of hydraulic transmission, the hydraulic cylinder crawling phenomenon is prevalent, characterized by non-uniform motion, intermittent stops, and erratic speed variations during low-speed operation. This instability and discontinuity in movement arise from unstable flow to or from the actuator, manifesting as intermittent flow disruptions. Particularly at low speeds, the phenomenon can range from subtle vibrations to significant jumps, severely impacting the stability and control precision of the hydraulic system. As detailed in Table 1, while maintaining a constant penetration speed of 2 cm/s, the stroke of the hydraulic cylinder increases with higher gear positions, inversely reducing its movement velocity. By decreasing the hydraulic cylinder's speed, one can identify the onset of sudden velocity changes, indicative of the crawling phenomenon, thus revealing the rig's speed lower limit.

The hydraulic transmission system depicted in Figure 13, augmented with a PID controller, was configured with fixed parameters. By gradually reducing the system's speed, we obtained the outcomes presented in Figure 14. Observations reveal that at a set speed of 0.38 cm/s for the hydraulic cylinder, the crawling phenomenon emerges, failing to satisfy the precision requirements for penetration speed. Consequently, the maximum stroke amplification ratio of the hydraulic cylinder is limited to five, per the velocities detailed in Table 1; thus, the system is capped at a maximum of five gears.

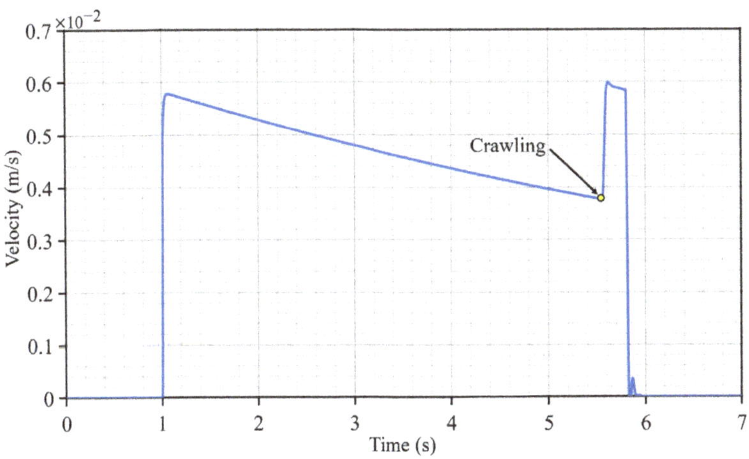

Figure 14. Speed in case of crawling phenomenon of hydraulic cylinder.

4. Discussion

To verify whether the adaptive CPT system can shift gears based on dynamic loads, thereby achieving maximum single-penetration travel under load conditions, a model with time-varying dynamic loads was subjected to simulation experiments to discuss the model's stability under load disturbances. As a reference for calculating the dynamic loads, a set of in situ test data was selected from the reference literature 37. These resistance data were chosen because they represent the typical variation in the sediment resistance during the penetration of the rod.

4.1. Calculation of Dynamic Load

To optimize the design of the transmission system for achieving uniform penetration of the rod, it is essential to analyze the patterns of load variation experienced by the rod during sediment penetration and determine the correlation between the penetration depth of the rod and the penetration load. The mechanics of this penetration are highly complex, with factors influencing the resistance to penetration in marine sediments, including the substrate type, penetration depth, water content, porosity, and wet density.

With the development of domestic and international CPT rigs and the urgent need for marine engineering construction, the use of CPT in marine engineering surveys has increased significantly, generating a substantial amount of measured data. These data provide strong support for studying the correlation between CPT data and soil physical–mechanical parameters. Wei Duan and colleagues [37] conducted a total of 383 seabed CPTU tests at the site of the Hong Kong–Zhuhai–Macao Bridge in the Pearl River Delta using Fugro Seacalf equipment, obtaining the relationship between cone resistance and sleeve friction varying with depth within a 53 m depth range beneath the seabed. The results of their tests are shown in Figure 15.

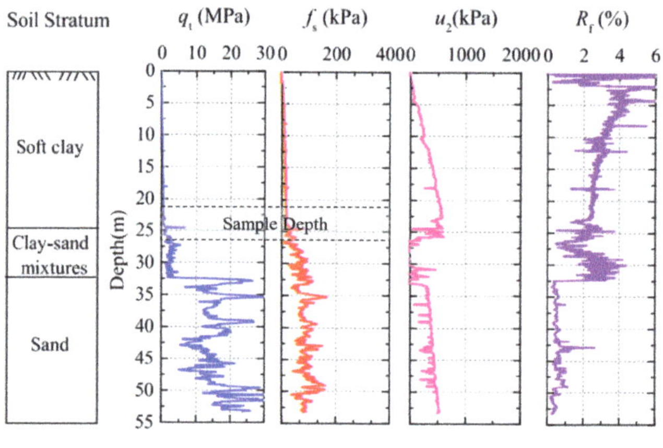

Figure 15. In situ CPT data referenced from Duan W et al. 37.

From the figure, it can be observed that both cone resistance and sleeve friction gradually increase with penetration depth within the same soil layer, with significant changes occurring at soil boundary areas. When the soil transitions from soft clay to clay–sand mixtures, there is a sharp increase in cone resistance, rising from approximately 0.8 MPa to about 8 MPa. Upon reaching the sand layer, the value continues to increase uniformly to over 15 MPa.

Due to the complexity and uncertainty of submarine geological conditions, the rod may penetrate strata of varying hardness during penetration, and the resistance it encounters will change with the depth of penetration and the type of soil layer. Consequently, the load on the hydraulic transmission system varies over time, leading to unstable penetration speeds. In actual penetration processes, the total resistance experienced by the rod, as shown in Figure 16a, primarily consists of the total cone resistance Q_c and the total sleeve friction force F_f, which can be calculated by Equations (41) and (42). In addition, as shown in Figure 16b, since the pore water pressure is mainly horizontal and only a vertical component exists at the tip of the cone, which has a very small effect on the vertical motion of the rod, it is ignored in the following analysis in order to simplify the model:

$$Q_c = q_c \cdot A \tag{41}$$

$$F_f = f_s \cdot A_h \tag{42}$$

where q_c is the cone resistance, f_s is the sleeve friction, A is the area of the tip of the cone, and A_h is the total area of the rod sleeve in contact with the soil.

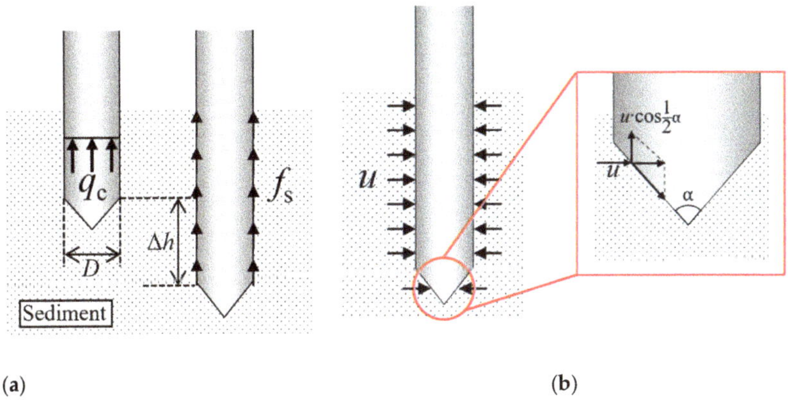

Figure 16. Force analysis diagram of the rod. (**a**) Cone resistance and sleeve friction. (**b**) Pore water pressure.

Thus, the total penetration resistance F_t is obtained:

$$F_t = Q_c + F_s = q_c \cdot A + f_s \cdot A_h = q_c \cdot A + \Sigma f_s \cdot \pi \cdot D \cdot \Delta h \tag{43}$$

where D is the rod diameter ($D \approx 40$ mm), and Δh is the depth of penetration of the rod.

Referencing the in situ test data illustrated in Figure 15, this paper selected a soil sample segment transitioning from soft clay to clay–sand mixtures. Due to the significant changes in the relevant parameters associated with soil type variations, such typical data can effectively test the robustness of the control system. Subsequently, the data were simplified to meet the design and simulation requirements of the penetration unit and hydraulic transmission system, predicting the changes in cone resistance and sleeve friction during the penetration process [38,39]. The organized data are presented in Figure 17, showing the curves of cone resistance and sleeve friction as a function of depth. Following this, using Equation (43), the variation curve of the hydraulic cylinder load force was calculated and is depicted in Figure 18.

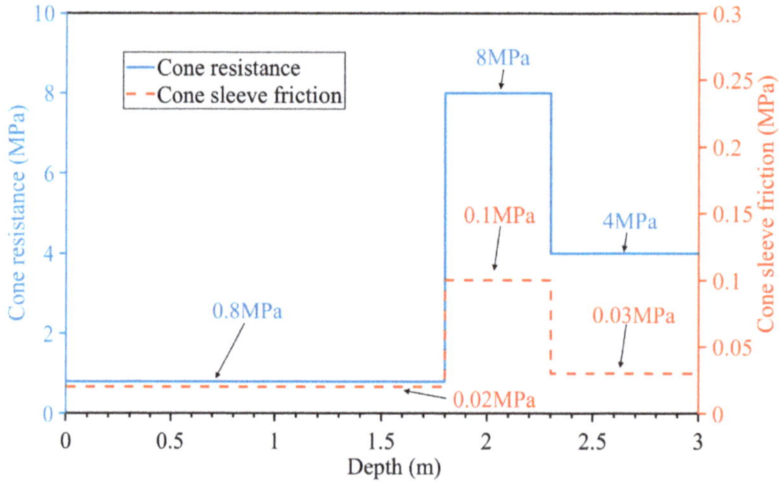

Figure 17. Depth–cone resistance curve and depth–cone sleeve friction curve.

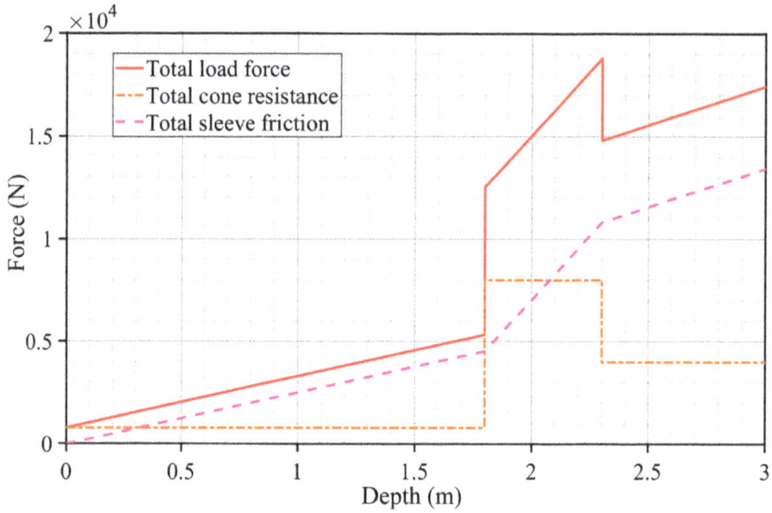

Figure 18. Depth–load force curves.

4.2. Model Consistency Validation and Control Logic Design

To achieve the adaptive function of maximizing single-penetration travel under load conditions, we first utilized the Simscape library in MATLAB/Simulink R2022b to construct a simulation model. This model was then compared with the simulation results of the mathematical model shown in Figure 9 to verify the consistency between the two. Subsequently, the adaptive gear-shifting control logic was programmed using Stateflow.

Simscape, a powerful multi-domain physical system modeling and simulation library introduced by MathWorks, enables the rapid creation of physical simulation models within the Simulink environment. By employing component models based on physical connections and physics-to-Simulink signal conversion modules, Simscape allows physical models to integrate and interact directly with other component libraries, facilitating the modeling, control, and co-simulation of hydraulic systems. The hydraulic transmission system simulation model established using Simscape is depicted in Figure 19.

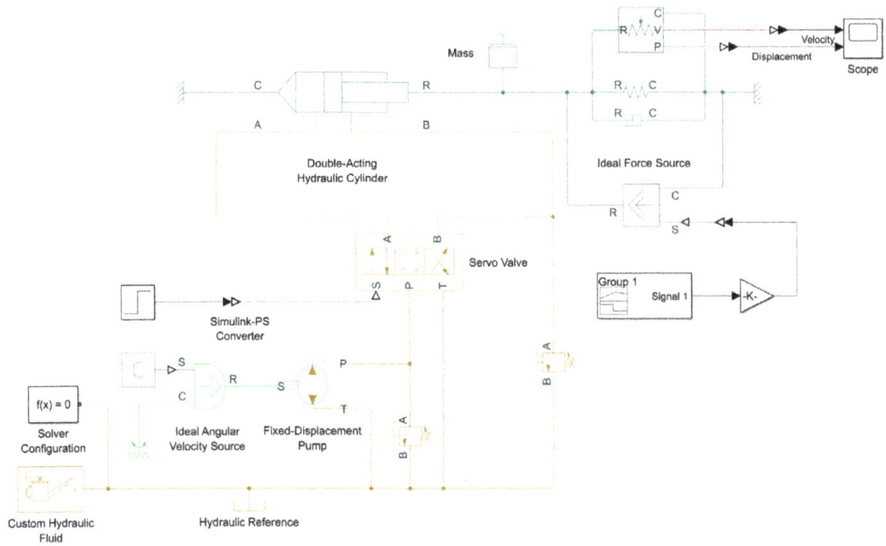

Figure 19. Simulation model of a hydraulic transmission system in Simulink using Simscape.

The parameter values used in the aforementioned physical model are consistent with those in Table 2. The simulation results of this model were compared with the outcomes of the mathematical model presented in Figure 9. As shown in Figure 20, the velocity profiles of the hydraulic cylinder piston essentially overlap, validating the mathematical model.

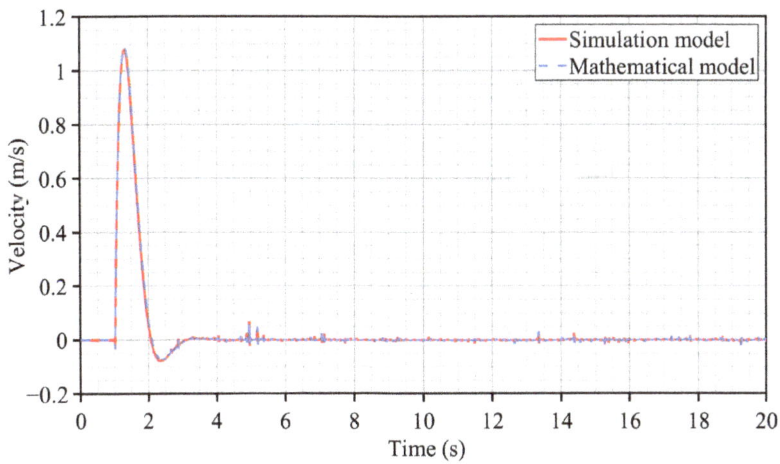

Figure 20. Comparison of the results of simulation model with results of mathematical model.

Subsequently, the load-adaptive shifting module was designed using Stateflow, with the fundamental logic being that during startup, the fifth gear is used for penetration operations. When sensors detect an external load exceeding the nominal load range or when the safety valve in the hydraulic system overflows, the gripping mechanical hand pauses and resets. It then sequentially downshifts until a new gear can overcome the external load, after which operations resume. The following candlestick chart is used to represent the design loads corresponding to the five gears, as shown in Figure 21, where the red upper shadow indicates the maximum load corresponding to the gears, and the real body represents the nominal load.

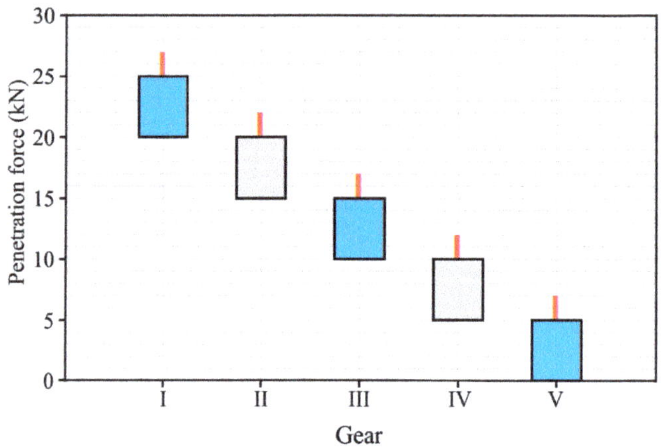

Figure 21. Design loads for each gear.

As can be seen, a load margin of 2 kN is reserved for each gear position, which is a safety redundancy designed to take into account the response time of the system. In addition, the load ranges corresponding to two neighboring gears are partially overlapped, which ensures the movement of the clamping manipulator during the gear shift is as smooth as possible. Next, the designed adaptive gearshift module, PID controller, and simulation model were assembled to form the complete system, as shown in Figure 22.

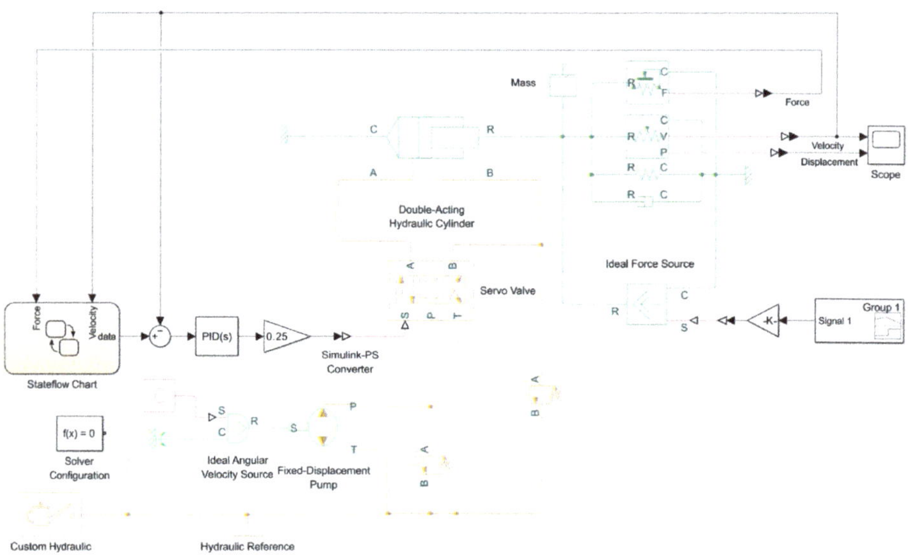

Figure 22. Complete hydraulic transmission system.

4.3. Static Load Testing

In order to test and validate the performance of the PID controller, the system model was simulated after applying 5 kN, 10 kN, 15 kN, 20 kN, and 25 kN static loads sequentially. The simulation results are shown in Figure 23.

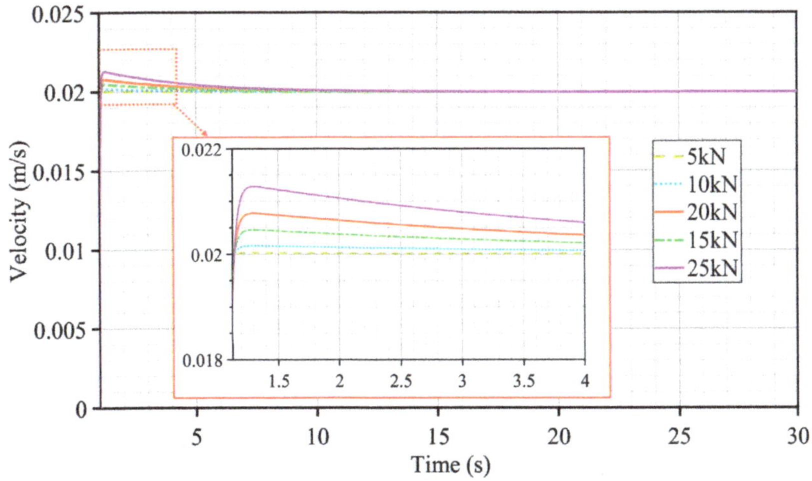

Figure 23. Velocity of the system obtained by simulation under different static loads.

The results indicate that system velocity oscillations amplify with escalating external loads, peaking at an overshoot when subjected to an external load of 25 kN. However, under PID controller regulation, system velocity remains consistently within the specified tolerance of 2 cm/s ± 10%, satisfying the design specifications.

4.4. Dynamic Load Testing

The dynamic load presented in Figure 18 was applied to both the adaptive CPT system designed in this study and the standard CPT system (disabled adaptive functions, with a fixed single penetration stroke of 0.6 m), with a target penetration depth of 3 m set for the simulation. As shown in Figure 24, the figure indicates that the adaptive CPT system, when set to a penetration depth of 3 m, performed a total of three penetration operations, taking 170 s in total, whereas the standard non-adaptive CPT system, due to its fixed 0.6 m stroke, executed six penetration operations, taking 180 s in total. Additionally, the simulation results of the adaptive CPT system were compared with the soil layer data, as depicted in Figure 25. The comparison reveals that for a penetration depth of 0–1.8 m, the soil type is soft clay with minimal resistance, prompting the system to automatically shift to the fifth gear, achieving a single penetration stroke of 1.8 m. When the depth is between 1.8 and 2.3 m, the system encounters the boundary between soft clay and clay–sand mixtures, where the penetration resistance peaks at 18,000 N, causing the system to switch to the second gear for penetration. For depths between 2.3 and 3 m, the soil type transitions entirely to clay–sand mixtures, with resistance slightly decreasing and stabilizing, leading the system to shift to the third gear to complete the remaining penetration.

Tests found that load-adaptive CPT reduced data breakpoints by 50% from four to two for the same external load compared to traditional CPT. In addition, in the depth range of 0 to 1.8 m, the adaptive CPT system obtained a maximum single penetration stroke of 1.8 m due to the smaller external load. In summary, the load-adaptive CPT system flexibly shifts gears in response to external loads, achieving maximum single-penetration strokes and overcoming load forces more efficiently than conventional CPT systems. It demonstrates superior penetration continuity and operational efficiency, requiring fewer penetration attempts under identical resistance conditions.

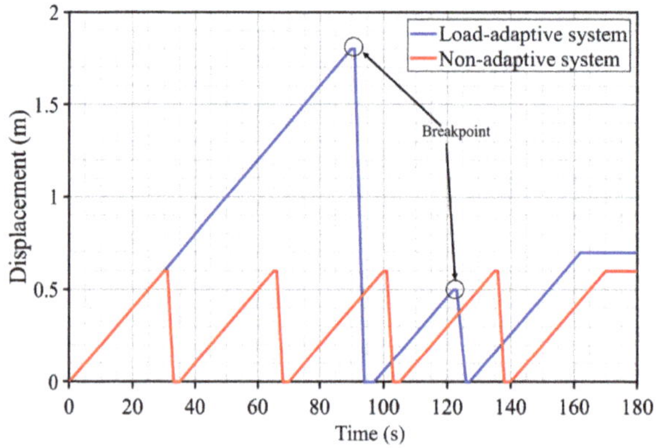

Figure 24. Comparison between load-adaptive CPT system and non-adaptive CPT system.

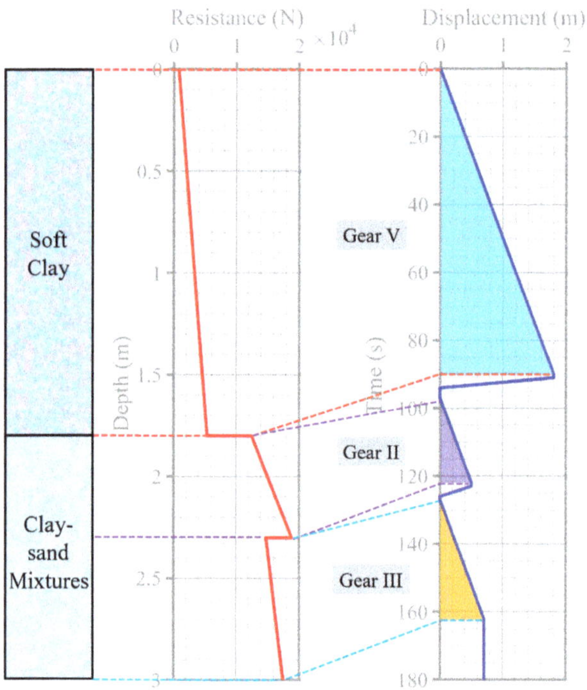

Figure 25. Simulation results of load-adaptive CPT system compared with soil resistance.

5. Conclusions

This paper addresses issues with the sediment cone penetration test by proposing a novel load-adaptive sediment CPT rig. This rig is designed to maximize single-penetration strokes under conditions that overcome sediment resistance, minimize zero-rate conditions, and ensure data continuity. The rig's penetration unit, driven by a hydraulic cylinder and pulley-block system, amplifies the penetration stroke. The clamping manipulator's flexibility in accommodating rods and sediment sensors of various diameters enhances the CPT rig's environmental adaptability, enriches test data, and better supports sedimentology research.

This paper analyzed the mechanical properties and layering patterns of sediment, along with the interaction mechanisms between sediment and mechanical structures. Subsequently, a mathematical model of a mechanical structure–sediment-integrated system was established using MATLAB/Simulink R2022b software. We also confirmed the model's boundary conditions through a comprehensive analysis of sediment characteristics and the hydraulic system. To facilitate system function design, we developed a physical simulation model, conducted comparative tests between the physical and mathematical models, and further designed a load-adaptive control logic to form a complete model.

This study simulated and tested the mechanical structure–sediment-integrated model using real sediment test data from the Pearl River Delta estuary (the Hong Kong–Zhuhai–Macao Bridge site). The results indicate that the rig in this paper exhibits superior continuity and operational efficiency under identical penetration resistance conditions when compared to conventional CPT rigs. In low-resistance sediment layers, it automatically shifts to a higher gear, employing minimal penetration force for maximum single-stroke penetration while ensuring data continuity and conserving energy. Conversely, when encountering abrupt sediment resistance fluctuations, the rig downshifts to deliver increased penetration force, and the hydraulic system stabilizes the rod's speed to maintain accuracy. Thus, the novel load-adaptive CPT rig and its mechanical structure–sediment model emerge as an innovative option for subsea geological investigations and sedimentology research.

Author Contributions: Conceptualization, L.G. and Y.L.; methodology, G.X.; data curation, Z.Z.; writing—original draft preparation, Z.Z. and Y.Z.; writing—review and editing, G.X. and Y.Z.; supervision, Z.F.; project administration, X.Y. All authors have read and agreed to the published version of the manuscript.

Funding: This research was funded by the Laoshan Laboratory, grant number LSKJ202203500; the National Natural Science Foundation of China (NSFC), grant number No. 42477153; National Key Research and Development Program of China, grant number 2024YFF0507000; the Project of Sanya Yazhou Bay Science and Technology City, grant number SCKJ-JYRC-2023-03; Science and Technology special fund of Hainan Province, grant number DSTIC-CYCJ-2022010.

Data Availability Statement: Data are contained within the article.

Acknowledgments: I would like to thank Lei Guo and Shuting Huang for their guidance and everyone who contributed to this article.

Conflicts of Interest: The authors declare no conflict of interest.

References

1. Wang, J.; Tang, G.; Huang, J.X. Analysis and modelling of a novel hydrostatic energy conversion system for seabed cone penetration test rig. *Ocean. Eng.* **2018**, *169*, 177–186. [CrossRef]
2. Douglas, B.J. Soil classification using electric cone penetrometer. In Proceedings of the Sympsium on Cone Penetration Testing and Experience, Geo-Technical Engineering Division, ASCE, St. Louis, MO, USA, 26–30 October 1981.
3. Wahl, D.A.J. *Implementation of Variable Rate Cone Penetration Testing: An Experimental Field Study*; University of California: Davis, CA, USA, 2012.
4. Ghose, R.; Goudswaard, J. Integrating S-wave seismic-reflection data and cone penetration test data using a multiangle multiscale approach. *Geophysics* **2004**, *69*, 440–459. [CrossRef]
5. White, D.J. CPT equipment: Recent advances and future perspectives. *Cone Penetration Test.* **2022**, *2022*, 66–80.
6. Ji, F.D.; Jia, Y.G.; Liu, X.L.; Guo, L.; Zhang, M.S.; Shan, H.X. In situ measurement of the engineering mechanical properties of seafloor sediment. *Mar. Geol. Quat. Geol.* **2016**, *36*, 191–200.
7. Luo, T.; Song, Y.; Zhu, Y.; Liu, W.; Liu, Y.; Li, Y.; Wu, Z. Triaxial experiments on the mechanical properties of hydrate-bearing marine sediments of South China Sea. *Mar. Pet. Geol.* **2016**, *77*, 507–514. [CrossRef]
8. Ma, H.; Zhou, M.; Hu, Y.; Hossain, M.S. Effects of cone tip roughness, in-situ stress anisotropy and strength inhomogeneity on CPT data interpretation in layered marine clays: Numerical study. *Eng. Geol.* **2017**, *227*, 12–22. [CrossRef]
9. Lunne, T.; Robertson, P.K.; Powell, J.J.M. Cone penetration testing in geotechnical practice. *Eng. Geol.* **1999**, *50*, 219–220.

10. Robertson, P.K. Soil classification using the cone penetration test. *Can. Geotech. J.* **1990**, *27*, 151–158. [CrossRef]
11. Yin, K.S.; Zhang, L.M.; Wang, H.J.; Zou, H.F.; Li, J.H. Marine soil behaviour classification using piezocone penetration test (CPTu) and borehole records. *Can. Geotech. J.* **2021**, *58*, 190–199. [CrossRef]
12. Wang, C.; Guo, L.; Jia, L.; Sun, W.; Xue, G.; Yang, X.; Liu, X. Development and application of a 3,000-m Seabed Cone Penetration Test and Sampling System based on a hydraulic drive. *Front. Mar. Sci.* **2024**, *11*, 1377405. [CrossRef]
13. Storteboom, O.; Woollard, M.; Verhagen, J. Efficiency examined of hands-free Cone Penetration Testing using the SingleTwist™ with COSON. In *Cone Penetration Testing 2022*; CRC Press: Boca Raton, FL, USA, 2022; pp. 230–235.
14. Shoukat, G.; Michel, G.; Coughlan, M.; Malekjafarian, A.; Thusyanthan, I.; Desmond, C.; Pakrashi, V. Generation of Synthetic CPTs with Access to Limited Geotechnical Data for Offshore Sites. *Energies* **2023**, *16*, 3817. [CrossRef]
15. Lunne, T. The CPT in offshore soil investigations–a historic perspective. JK Mitchell Lecture. In Proceedings of the 2nd International Symposium on the Cone Penetration Test, CPT, Huntington Beach, CA, USA, 9–11 May 2010; p. 10.
16. Randolph, M.F. New tools and directions in offshore site investigation. *Aust. Geomech. J.* **2016**, *51*, 81–92.
17. Robertson, P.K. Cone penetration test (CPT)-based soil behaviour type (SBT) classification system—An update. *Can. Geotech. J.* **2016**, *53*, 1910–1927. [CrossRef]
18. Lu, Y.; Duan, Z.; Zheng, J.; Zhang, H.; Liu, X.; Luo, S. Offshore cone penetration test and its application in full water-depth geological surveys. In *IOP Conference Series: Earth and Environmental Science*; IOP Publishing: Bristol, UK, 2020; Volume 570, p. 042008.
19. Qiao, H.; Liu, L.; He, H.; Liu, X.; Liu, X.; Peng, P. The practice and development of T-bar penetrometer tests in offshore engineering investigation: A comprehensive review. *J. Mar. Sci. Eng.* **2023**, *11*, 1160. [CrossRef]
20. McCALLUMA, A. Cone penetration testing (CPT) in Antarctic firn: An introduction to interpretation. *J. Glaciol.* **2014**, *60*, 83–93. [CrossRef]
21. McCALLUMA, A. A brief introduction to cone penetration testing (CPT) in frozen geomaterials. *Ann. Glaciol.* **2014**, *55*, 7–14. [CrossRef]
22. Ghanizadeh, A.R.; Aziminejad, A.; Asteris, P.G.; Armaghani, D.J. Soft Computing to predict earthquake-induced soil liquefaction via CPT results. *Infrastructures* **2023**, *8*, 125. [CrossRef]
23. Cui, Y.; Guo, L.; Liu, T.; Yang, Z.; Ling, X.; Yang, X.; Lu, K.; Xue, G. Development and application of the 3000 m-level multiparameter CPTu in-situ integrated test system. *Mar. Georesources Geotechnol.* **2023**, *41*, 400–411.
24. Robertson, P.K.; Cabal, K.L. *Guide to Cone Penetration Testing for Geotechnical Engineering*; Gregg Drilling & Testing: Signal Hill, CA, USA, 2015.
25. Du, Y.; Zhu, L.; Zou, H.; Zhang, L.; Cai, G.; Liu, S. Evaluation of CPTU-based soil classification charts for offshore sediments in Pearl River Delta, China. In Proceedings of the Geo-Congress 2020, Minneapolis, Minnesota, 25–28 February 2020; American Society of Civil Engineers: Reston, VA, USA, 2020; pp. 633–639.
26. Khosravi, A.; Martinez, A.; DeJong, J.T. Discrete element model (DEM) simulations of cone penetration test (CPT) measurements and soil classification. *Can. Geotech. J.* **2020**, *57*, 1369–1387. [CrossRef]
27. Jia, R.; Hino, T.; Chai, J.; Yoshimura, M. Interpretation of density profile of seabed sediment from nuclear density cone penetration test results. *Soils Found.* **2013**, *53*, 671–679. [CrossRef]
28. Fudong, J.; Yonggang, J.; Xiaolei, L. Development and application of new offshore CPT equipment. *Coast. Eng.* **2016**, *35*, 1–9.
29. Krage, C.P.; DeJong, J.T.; Schnaid, F. Estimation of the coefficient of consolidation from incomplete cone penetration test dissipation tests? *J. Geotech. Geoenvironmental Eng.* **2015**, *141*, 06014016. [CrossRef]
30. de Lange, D.A.; van Duinen, T.A.; Peters, D.J. Large diameter cone penetrometers: What is an appropriate location for the transition to the rod diameter. In *Cone Penetration Testing 2022*; CRC Press: Boca Raton, FL, USA, 2022; pp. 127–132.
31. Niazi, F.S.; Mayne, P.W. Cone penetration test based direct methods for evaluating static axial capacity of single piles. *Geotech. Geol. Eng.* **2013**, *31*, 979–1009. [CrossRef]
32. Zhang, S.; Li, S.; Minav, T. Control and performance analysis of variable speed pump-controlled asymmetric cylinder systems under four-quadrant operation. *Actuators* **2020**, *9*, 123. [CrossRef]
33. Ma, Y.; Gu, L.C.; Xu, Y.G.; Shi, L.C.; Wang, H.T. Research on control strategy of asymmetric electro-hydraulic servo system based on improved PSO algorithm. *Adv. Mech. Eng.* **2022**, *14*, 16878132221096226. [CrossRef]
34. Krishnamurthy, V.; Seshadri, V. Model reduction using the Routh stability criterion. *IEEE Trans. Autom. Control.* **1978**, *23*, 729–731. [CrossRef]
35. Kim, S.D.; Cho, H.S.; Lee, C.O. Stability analysis of a load-sensing hydraulic system. *Proc. Inst. Mech. Eng. Part A Power Process Eng.* **1988**, *202*, 79–88. [CrossRef]
36. Margolis, D.L.; Hennings, C. Stability of hydraulic motion control systems. *ASME. J. Dyn. Sys., Meas., Control.* **1997**, *119*, 605–613. [CrossRef]
37. Duan, W.; Congress, S.S.C.; Cai, G.; Puppala, A.J.; Dong, X.; Du, Y. Empirical correlations of soil parameters based on piezocone penetration tests (CPTU) for Hong Kong-Zhuhai-Macau Bridge (HZMB) project. *Transp. Geotech.* **2021**, *30*, 100605. [CrossRef]

38. Cui, G.; Li, Y.; Pei, W. The effect of natural water content on the penetration resistance of seabed sediment. *Hydrogr. Surv. Charting* **2005**, *6*, 51–53.
39. Xue, G.; Liu, Y.; Guo, L.; Liu, B. Optimization on motion-robust and energy-saving controller for hydraulic penetration system of seabed equipment. *Proc. Inst. Mech. Eng. Part M J. Eng. Marit. Environ.* **2021**, *235*, 792–808. [CrossRef]

water

MDPI

Article

Analysis of Shoreline Change in Huizhou–Shanwei Region (China) from 1990 to 2023

Sizheng Li [1,2,3], Feng Gui [4], Jirong Feng [4], Yang Wang [1,2,3], Yanwei Song [1,2,3], Wanhu Wang [1,2,3,*] and Cong Lin [1,2,3,*]

1 Haikou Research Center of Marine Geology, China Geological Survey, Haikou 571100, China; zhengsl@mail.cgs.gov.cn (S.L.); wangyang01@mail.cgs.gov.cn (Y.W.); songyanwei@mail.cgs.gov.cn (Y.S.)
2 Haikou Key Laboratory of Marine Contaminants Monitoring Innovation and Application, Haikou 571127, China
3 Innovation Base for Island Reef Spatial Resource Investigation, Monitoring, and Technology Utilization, Haikou 571127, China
4 School of Marine Science and Technology, Zhejiang Ocean University, Zhoushan 316022, China; fgui@zjou.edu.cn (F.G.)
* Correspondence: wangwanhu@mail.cgs.gov.cn (W.W.); lincong0621@163.com (C.L.)

Abstract: The dynamic change in the shorelines reflects an important sign to the socio-economic development of coastal areas. The Huizhou–Shanwei region of China has experienced rapid socio-economic development over the past 33 years. The study of the dynamic change in the shorelines in this region can provide basic data support for the marine environmental protection and regional development planning in this region. Based on Landsat RS (remote sensing) images from 1990 to 2023, this study obtained the length and structure data of the shorelines in eight periods by manual visual interpretation. DSAS (Digital Shoreline Analysis System) and other methods were also used to calculate indices such as EPR (End Point Rate) and fractal dimension of the shorelines The results show that, during 33 years, the length of the shorelines increased 15.83 km, with an average growth rate of 0.48 km/y; the value of the intensity of change in the shorelines was 0.08%; the average EPR was 3.66 (m/y), and the artificiality index of the shorelines increased from 0.2895 to 0.4295; the greatest intensity of change was in the estuarine shorelines, with an intensity of change of −2.69%. The overall change in the fractal dimension of the shorelines was small, both between 1.0395 and 1.0673; the shorelines became slightly more curved. As far as the influencing factors are concerned, the influence of the natural environment is a long process, and human activities are more capable of changing the length and shape of the shorelines in a short period of time, with factors such as the degree of economic development having a greater impact on the shorelines.

Keywords: Huizhou–Shanwei; spatio-temporal analysis; DSAS; EPR; fractal dimension

1. Introduction

The coastal zone is an important site for human socio-economic development, while the shorelines act as demarcation lines between the land and sea [1,2], reflecting changes induced by the natural environment and human intervention (i.e., socio-economic development) [3–5]. Dramatic changes in the shorelines have been associated with a wide range of problems in the coastal zone, such as deterioration of seawater quality, ecological imbalance, etc. In this context, there is an urgent need to strengthen the protection, management, and sustainable use of the shorelines and related to coastal zone resources [6]. Therefore, monitoring and rationally analyzing the changes in the length and type of the shoreline is a highly important task. The use of remote sensing images to extract shorelines and analyze their spatial and temporal changes has become a major part of shorelines research. As Wu [7] (2014) demonstrated, the structural changes to the shoreline of the Chinese mainland over the past 70 years have been remarkably substantial. The prevailing trend has been an increase in artificial features along the shorelines. However, the rate of change in the shoreline and its causes vary significantly between different regions. Zhang [8] used remote sensing images to analyze the changes in the shorelines in the southern part of the Yellow River estuary, and concluded that the shorelines in the southern part of the Yellow River Delta mainly undergo gradual erosion. Zhao [9] (2023) concluded that the shorelines of central Jiangsu have been continuously advancing seaward for many years due to sedimentation and beach reclamation. It is evident that the coast of the poldered land area is distinguished by its high level of activity. Xia Wang [10] argues that in the course of Ningbo's development, natural erosion or siltation is often disturbed due to activities such as urban development, agricultural development, and port construction. Mshelia [11] analyzed the fluctuation of the Durban shoreline from 1990 to 2023 using Lansat images and DSAS to assess erosion and accretion patterns. Adenugb [12] found that human activities have contributed to the urbanization of the island city, resulting in a shoreline retreat of about 56% of its total length. Although analysis of shoreline changes is critical, few studies have examined changes along the Huizhou–Shanwei shoreline.

The Huizhou–Shanwei region is located in the southeastern part of Guangdong Province and is also an important zone for the development of the marine economy in Guangdong Province [13,14]. The region has many characteristics, such as curved coasts, extensive mudflats, and abundant biological life. The management, protection and scientifically efficient use of its shorelines resources are conducive to the economic growth and the coordinated development of coastal ecology and marine economy in the Huizhou–Shanwei region [15]. However, there are fewer studies related to the shoreline changes in this region.

This study identifies the shoreline length and structural changes in Huizhou–Shanwei over the past 33 years using Landsat remote sensing images by visual interpretation. It then analyses the changes and explains the driving factors. The result provides scientific evidence for the healthy management and development of the coastal zone in Huizhou–Shanwei region.

2. Overview of the Study Area

The Huizhou–Shanwei region is located in the southeastern part of China's Guangdong Province, in a subtropical oceanic monsoon climate zone, with a warm climate, abundant rainfall, and precipitation concentrated in the summer months. The large number of rivers and lakes within the region can be seen in Figure 1. The summer and autumn seasons are heavily affected by typhoons, and there is a clear seasonal shift in wind direction, with the prevailing wind direction being easterly throughout the year. The region is dominated by low hills with a mountain range trend that slopes from northeast to southwest. Small areas of floodplain are distributed along the coast. The sand content of the rivers is moderately high. The intensity of stream erosion in the mountainous areas is high during the rainy season, but is easily influenced by hydraulic engineering [16]. The region plays an important role in the development of the interface between the Pearl River Delta (PRD) and the Chaoshan region, and is an important corridor connecting the PRD and the Chaoshan region. With the reform and opening up, the region has taken over the economic and industrial spillover functions of Shenzhen and Hong Kong. The population has also increased significantly, and socio-economic progress has continued. This has led to significant changes in the construction of the coastal zone area and the development and utilization of the shoreline. However, with the expansion of human development and urbanization, the region is also facing problems such as water pollution and changes in the structure of the shorelines.

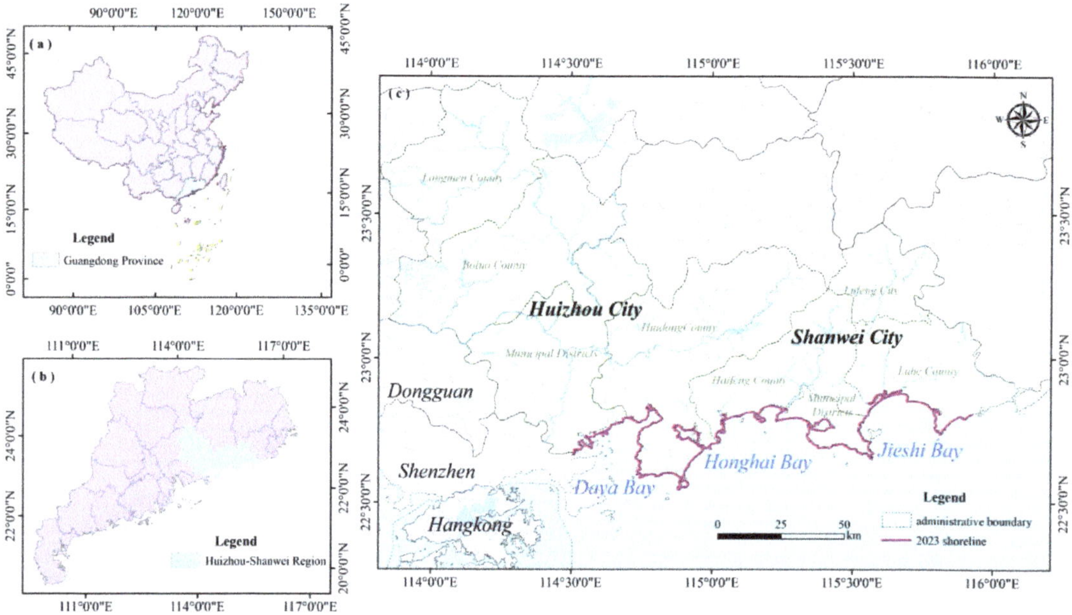

Figure 1. Location of study area. (**a**) Location map of Huizhou–Shanwei. (**b**) Huizhou–Shanwei in Guangdong Province. (**c**) Overview map of the study area.

3. Data Sources and Methodolgy

3.1. Data Sources

Remote sensing images related to the study area are Landsat (https://www.usgs.gov/ (accessed on 1 May 2024)). In the interpretation of the shoreline from the Landsat images, we also used historical images from Google Earth as a reference for some areas of uncertainty. The socio-economic data of Huizhou–Shanwei region are mainly sourced from the statistical yearbooks and bulletins of the relevant regions. The basic information of Landsat images used is shown in Table 1.

Table 1. The basic information of remote sensing images in this study.

Landsat Scene Identifier	Satellite and Sensor Identifier	Path/ Row	Spatial Resolution	Date	Cloud Cover
LT51210441990327BJC00	LANDSAT5 TM	121/44	30 m	23 November 1990	0
LT51210441995085CLT03	LANDSAT5 TM	121/44	30 m	26 March 1995	0
LT51210442000259BJC00	LANDSAT5 TM	121/44	30 m	15 September 2000	0
LT51210442005064BJC00	LANDSAT5 TM	121/44	30 m	5 March 2005	0
LE71210442010342EDC00	LANDSAT7 ETM	121/44	30 m	8 December 2010	0
LC81210442015220LGN01	LANDSAT8_OLI_TIRS	121/44	30 m	8 August 2015	5.96
LC81210442020106LGN00	LANDSAT8_OLI_TIRS	121/44	30 m	15 April 2020	4.88
LC91210442023106LGN00	LANDSAT9_OLI_TIRS	121/44	30 m	16 April 2023	0.58

3.2. Shorelines Extraction and Validation

There has been some ambiguity in the definition of the natural shorelines, and in the context of marine management, it is generally regarded as the natural shorelines when there are no artificial non-permeable structures in the intertidal zone up to above the high tide line of the mean high tide, and when the coast is maintained in its natural state [17]. On the contrary, if there are artificial non-permeable dykes such as seawalls, wave protection dykes, erosion protection dykes, etc., between the intertidal zone and the high tide line of the mean high tide, they are regarded as artificial shorelines. The shoreline in this study refers to the multi-year mean high tide level.

After acquiring the remote sensing images of the study area, the remote sensing data were subjected to FLASSH atmospheric correction and radiometric calibration and debarring operations in ENVI 5.6. To minimize errors caused by inconsistencies in the imagery from year to year, the other data were corrected for ensemble accuracy using the 2023 remote sensing images, and the ground control points were also selected for error control. After this step, a Modified Normalized Difference Water Index (*MNDWI*) is operated on the imagery [18]. This index not only enhances the grey-level gradient at the land-water boundary, but also optimizes the distinction between water bodies and non-water bodies. The improved normalized water body index is calculated as follows:

$$MNDWI = \frac{Green - SWIR}{Green + SWIR} \tag{1}$$

where *Green* represents the green light band in the image and *SWIR* represents the short infrared band in the image.

In addition, manual visual interpretation is still the current remote sensing interpretation method with high accuracy. In order to obtain the shorelines of the study area, this study established the shorelines interpretation signs [19] (Table 2) by combining the actual characteristics of the area, the information of the image's hue, texture, geomorphology, and the characteristics of the surrounding features after the operation of *MNDWI*.

Table 2. Types and their description.

Type	Photo	Description
artificial	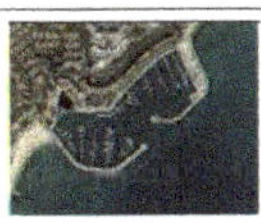	It is grayish-white and linear, with a straight waterside line; farmed areas or salt flats are regularly blocky.
biogenic		Shorelines with predominantly mangrove-growing and other marine forested shores are imaged in patches.
sandy		Smooth water's edge; bright white, even tone in areas not reached by the tide, darker tone in areas wetted by the tide; strips of beach with clear boundaries to land vegetation and sea water
muddy		Darker color tone, dense vegetation on one side, sparse or no vegetation on the other, generally with tidal flume development.
rocky	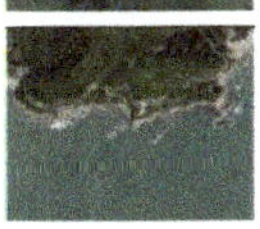	It has typical textural features. The rocks are light-toned and striped on remote sensing images, and the coast is often dotted with reefs, boulders, sea cliffs and other landforms.
estuarine		Located at the mouth of the Sea River, it is irregular in shape and undergoes a marked change in hue.

Subsequently, the shoreline is interpreted and corrected by means of manual visual interpretation. Thereafter, the processed images and the interpreted coastlines are superimposed in ArcGIS 10.7 for the purposes of analysis and mapping. In instances where the clarity of the coastline interpretation was deemed to be questionable, further refinement and improvement was conducted by referring to Google Earth images of similar dates.

3.3. Shoreline Artificial Index

The artificial intensity (frequency of occurrence) of the shorelines indicates the degree of transformation of the natural shoreline into an artificial shoreline, which is expressed by the shorelines artificial index I_A (the proportion of artificial shorelines to the total amount of shorelines in a certain coastal sector) [20]. The formula is as follows:

$$I_A = \frac{N}{L}, \tag{2}$$

where (I_A) represents the artificial shoreline index; (N) is the artificial shoreline length; (L) *is* the total length of shoreline.

The greater the I_A, the greater the degree of shorelines artificiality.

3.4. Intensity of Shoreline Change

In order to quantify the degree of annual average change in shorelines length over a given period of time [19], the intensity of shoreline length change from year i to year j (LCI_{ij}) is used (Equation (3))

$$\mathrm{LCI}_{ij} = \frac{L_j - L_i}{L_i(j-i)} \times 100\% \tag{3}$$

where (L_i, L_j) are the lengths of the shorelines in km for the corresponding years (i,j), respectively.

The greater the absolute value of LCI_{ij}, the greater and more drastic the change in shorelines length in the study area over the time period.

3.5. Baseline Method for Calculating the Rate of Shorelines Change

The baseline method provides good access to the rate at which the shorelines is advancing or receding seaward [21]. In this study, the EPR (End Point Rate) tool in DSAS (Digital shoreline Analysis System) was used to calculate the rate of change in shorelines [22]. The basic principle is to make a baseline to one side of the shorelines, and then make equally spaced vertical lines intersecting the baseline. The rate of change of the endpoints ($E_{i,j}$) from year i to year j, is calculated using the Static tool in DSAS [23]. The formula is as follows:

$$E_{i,j} = \frac{d_j - d_i}{\Delta T_{i,j}}, \tag{4}$$

where ($\Delta T_{i,j}$) is the time interval between years i and j, and d_j and (d_i) is the vertical distance from the shorelines to the baseline from year j to year.

3.6. Shorelines Fractal Dimension

Since non-regular geometries are difficult to describe quantitatively in terms of units such as length and area, fractal theory introduces a more natural language for description, making it easier to understand non-regular geometries [24,25].

Fractal dimension is a parameter used to characterize the complexity and irregularity of fractals, solving the problem of describing and quantifying irregular geometries [26]. The principle of the grid method is to vary the length (ε) of the square grid covering the shorelines, and then count the total number ($N(\varepsilon)$) of corresponding grids. Thence, we can obtain a series of corresponding square grid 'length-totals' of 'ε—$N(\varepsilon)$' [27]. Thus, when the length of the square grid is $\varepsilon_1, \varepsilon_2, \ldots, \varepsilon_k$, the corresponding number of square grids covering the shoreline is $N(\varepsilon_1), N(\varepsilon_2), \ldots, N(\varepsilon_k)$. The formula is as follows:

$$\lg N(\varepsilon_k) = -D\lg\varepsilon + Z \quad (5)$$

where (D) is the fractal dimension of the shorelines to be measured and (Z) is a constant to be determined.

Since the resolution of Landsat is 30 m, the grid edge length should also be a multiple of 30. By combining the existing research results and the situation of the study area, nine cases with grid edge lengths of 30, 60, 90, 120, 150, 210, 240, and 480 m were selected in this study.

4. Results

4.1. Analysis of Changes in Shorelines Length and Structure

According to the results of RS images interpretation, the changes in the length of the shorelines and their type changes in the study area can be obtained (see Figure 1). The statistics of shoreline types in each year using ArcGIS 10.7 are reported in Table 1.

The results of the study show that the shoreline increased from 616.71 km to 632.54 km with an average annual growth rate of 0.48 km. However, the overall trend in the evolution of shoreline length is "decrease-increase-decrease". The places where significant changes have occurred are in Figure 2a–d: the shorelines have become straight in Figure 2a, the shorelines are more curved in Figure 2b, and the two changes in Figure 2c,d are roughly parallel to the shore.

Statistics on the total length of the shorelines in each period show that the overall length of the Huizhou–Shanwei shorelines increased by 15.82 km from 1990 to 2023: the length of the shorelines was 616.71 km in 1990, then gradually decreased to 603.76 km in 2000, and then continued to increase to 634.19 km in 2020, and then decreased to 632.54 km in 2023. As far as shorelines types are concerned, during this 33-year period, there has been an increase in artificial and sandy shorelines and a decrease in other types of shorelines: artificial shorelines have increased by 93.15 km, sandy shorelines have increased by 7.04 km, biogenic shorelines have decreased by 6.79 km, muddy shorelines have decreased by 3.65 km, bedrock shorelines have decreased by 66.78 km, and estuarine shorelines have decreased by 7.14 km. Of these, the three areas with the most significant changes throughout the shorelines are a, b, and c in Figure 2.

Shorelines length change and structural change can only show the change in the length of each type of shoreline, and cannot see the change characteristics between artificial and natural shorelines. And the shorelines artificial index can intuitively reflect the transformation and influence of human activities on the shorelines. It can be seen in Figure 3 and Table 3. From 1990 to 2023, the artificial index of the Huizhou–Shanwei shorelines increased from 0.2895 to 0.4295, which indicates that some of the shoreline was converted to artificial shorelines during this period and the growth was relatively rapid. Prior to 2015, the shorelines labor index consistently showed an increasing trend, and after 2015, the shorelines labor index decreased slightly. This is demonstrated by the fact that in 1990–2000, the artificiality index increased by about 1% per year, a mere 0.0183 over a 10-year period; in 2000–2005, the artificiality index increased rapidly in this period, from 30.78% to 43.56%. In 2005–2015, the shorelines artificiality index is still increasing at this time, but the increase is extremely slow in this period. In 2015–2023, the artificial index decreased by 0.0387.

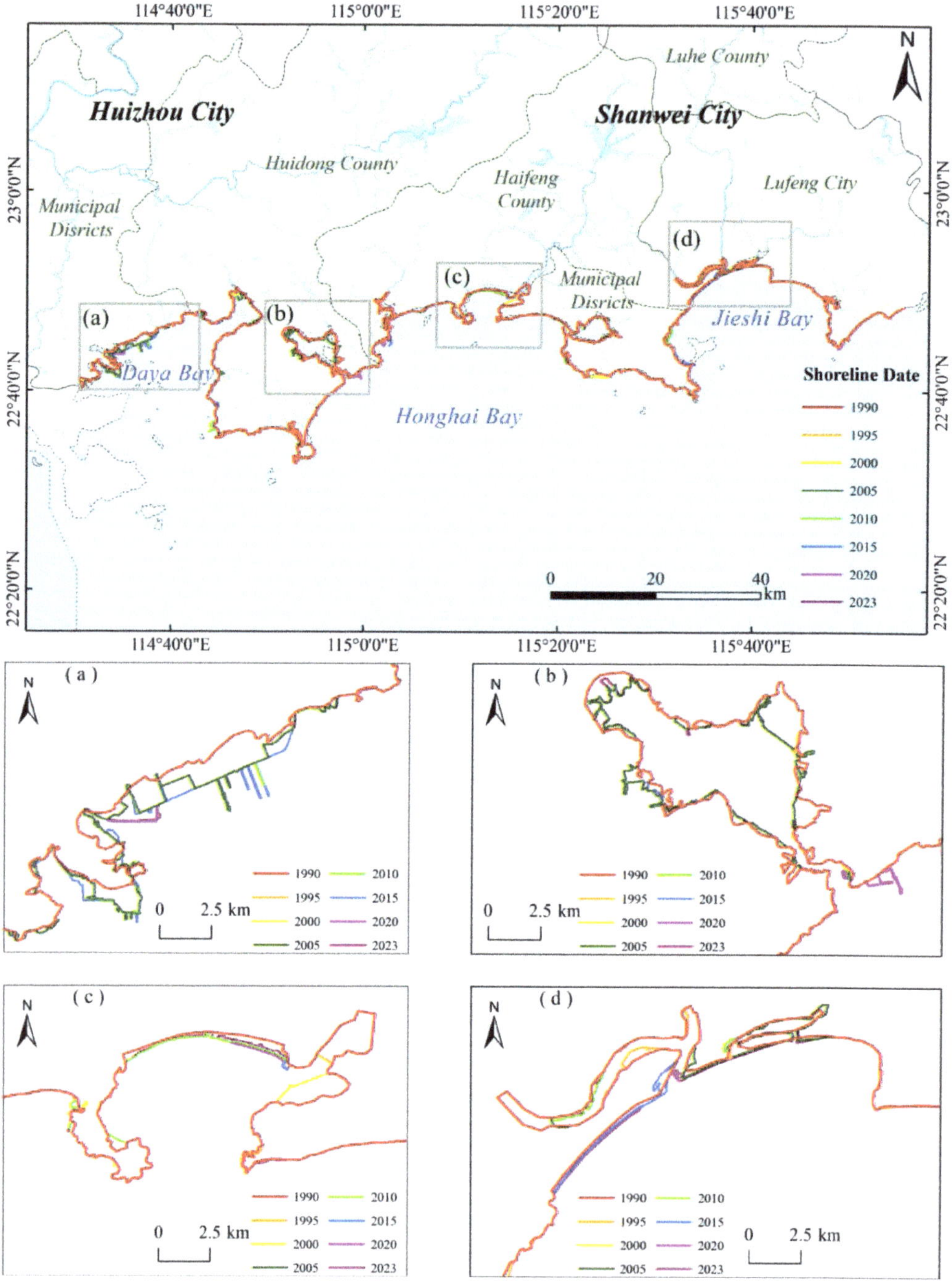

Figure 2. Location of the shorelines in different periods of Huizhou–Shanwei; (**a**) Significant shoreline artificial; (**b**) Shoreline changes of the lagoon; (**c**) Shoreline changes in the northern part of Honghai Bay; (**d**) Shoreline changes in the estuary.

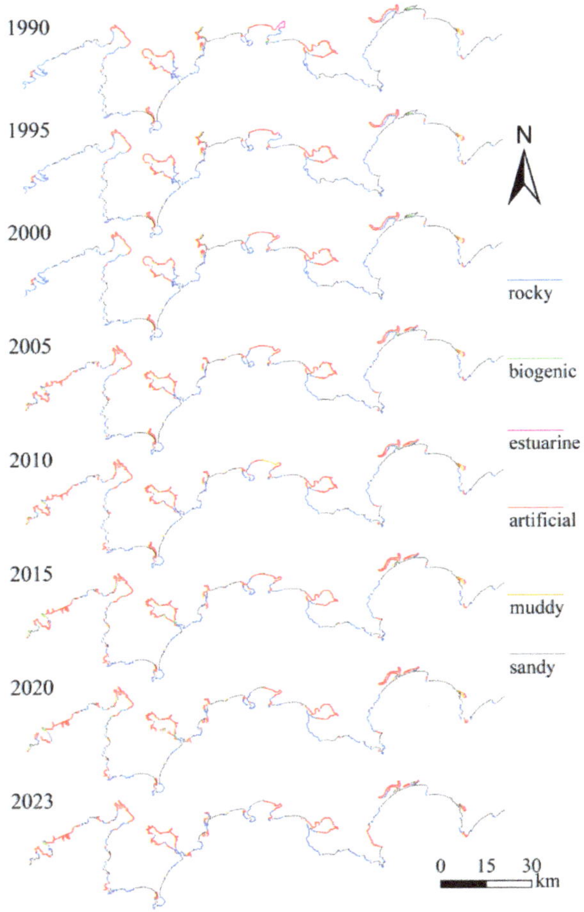

Figure 3. Changes in shoreline types and length in Huizhou–Shanwei.

Table 3. Length (km) and artificial index of each type of shorelines in Huizhou–Shanwei.

Types / Years	Artificial	Biogenic	Sandy	Muddy	Rocky	Estuarine	All	I_A
1990	178.53	9.90	187.69	8.73	223.82	8.04	616.71	0.2895
1995	180.33	7.90	186.38	7.84	223.00	1.38	606.83	0.2972
2000	185.84	7.90	183.81	8.83	216.00	1.38	603.76	0.3078
2005	269.12	7.80	166.24	9.70	163.68	1.35	617.89	0.4356
2010	274.76	7.13	165.93	15.75	159.22	1.35	624.14	0.4402
2015	295.13	13.66	174.40	12.08	133.81	1.28	630.35	0.4682
2020	281.43	12.97	187.26	10.50	140.75	1.28	634.19	0.4438
2023	271.68	3.11	194.72	5.08	157.04	0.90	632.54	0.4295

4.2. Intensity of Shorelines Change

Intensity of shorelines change can be seen in Table 4. In terms of intensity of shorelines change, the overall shoreline intensity of change was 0.08%, which is the smallest change compared to the six shoreline types. However, in terms of phases, in 1990–1995, the estuarine shorelines had the greatest intensity of change at −16.57%; in 1995–2000, the muddy shorelines had the greatest intensity of change at 2.52%; in 2000–2005, the artificial shorelines had the greatest intensity of change at 8.96%; in 2005–2010, the muddy shorelines had the greatest intensity of change at 12.47%; in 2010–2015, the greatest intensity of biological shorelines change was 18.30%; in 2015–2020, the greatest intensity of muddy shorelines change was −2.61%; and in 2020–2023, the greatest intensity of biological shorelines change was −25.35%.

Table 4. Intensity of change of Huizhou–Shanwei shorelines.

Period \ Type	Artificial	Biogenic	Sandy	Muddy	Rocky	Estuarine	Total
1990–1995	0.20%	−4.05%	−0.14%	−2.04%	−0.07%	−16.57%	−0.32%
1995–2000	0.61%	0.00%	−0.28%	2.52%	−0.63%	0.00%	−0.10%
2000–2005	8.96%	−0.25%	−1.91%	1.98%	−4.85%	−0.40%	0.47%
2005–2010	0.42%	−1.71%	−0.04%	12.47%	−0.54%	0.00%	0.20%
2010–2015	1.48%	18.30%	1.02%	−4.66%	−3.19%	−1.14%	0.20%
2015–2020	−0.93%	−1.00%	1.48%	−2.61%	1.04%	0.00%	0.12%
2020–2023	−1.15%	−25.35%	1.33%	−17.22%	3.86%	−9.69%	−0.09%
1990–2023	1.58%	−2.08%	0.11%	−1.27%	−0.90%	−2.69%	0.08%

4.3. Rate of Shorelines Change

The 1990 shoreline was buffered by 1000 m in the landward direction, and the baseline was obtained through adjustment. A 2000 m vertical line was then generated from the baseline to the sea direction, and an equally spaced shoreline vertical section with 500 m was created to generate the baseline. Following the deletion of a number of sections deemed to be unqualified, the total number of sections was reduced to 810. However, subsequent analysis revealed that the number of sections in conformity with the study's requirements was 644.

The EPR of the study area was calculated using the DSAS plug-in and the previous equations, and the spatial and temporal evolution of the Huizhou–Shanwei shorelines was analyzed from 1980 to 2023. The results showed that the average EPR for the entire study area was 3.66 (m/y), with transect number 23 producing the maximum EPR value of 30.21 (m/y) and the transect number 272 producing the minimum EPR value of −42.12 (m/y). There are 350 values of EPR greater than 0.79 values of EPR less than zero, and 215 values of zero. This indicates that the Huizhou–Shanwei shorelines mainly showed spatial seaward advancement during the period 1990–2023, in which the forms of advancement were mainly siltation and reclamation. Combined with Figure 4 and the field study, it appears that the shoreline at sector A is straightening out and advancing seaward, mainly due to harbor construction and engineering works. The area at sector B has an outwardly protruding peninsula, mainly a tourist area, where there is little overall change in the shorelines, and at the left and right ends of sector B are two bays, each with an overall seaward advance in the form of siltation. The shorelines are more variable in sector C, where their western half is advancing seaward in the form of estuarine siltation. The eastern side of the shorelines is overall straighter, advancing seaward in an orderly manner, mainly due to engineering works. The area at sector D is mainly a new tourist area that has been constructed in the last two years, where the original shorelines have been planned and modified artificially, allowing the shorelines to be set back in comparison to 1980.

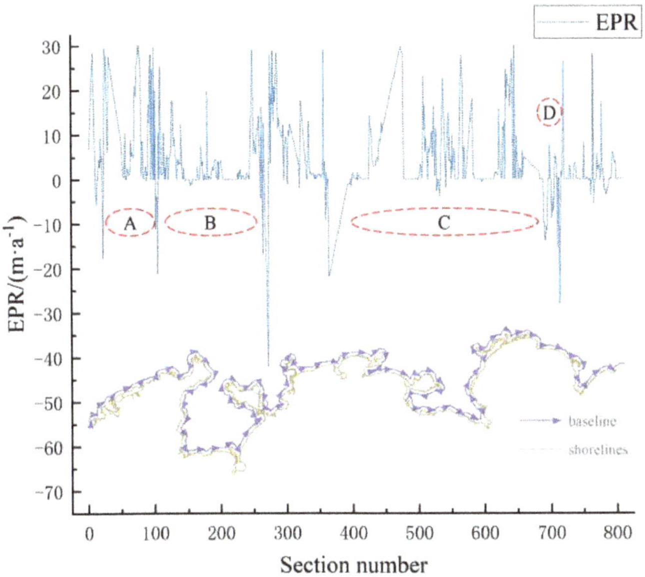

Figure 4. Combination picture of Huizhou–Shanwei EPR changes, baseline and shorelines.

4.4. Shorelines Fractal Dimension Change

Based on the grid side lengths set in the previous section, the number of grids capable of covering the shorelines was calculated for each side length for each year. The linear relationship "grid side length-number of grids" ("ε—$N(\varepsilon)$") was then calculated for each year according to Formula (5). The results are given in Table 5.

Table 5. Indicators of the fractal dimension of the shorelines at each stage of Huizhou–Shanwei.

Year	Equation of Linear Regression	R^2	Fractal Dimension
1990	$y = -1.0395x + 5.9578$	0.9997	1.0395
1995	$y = -1.04x + 5.9517$	0.9997	1.04
2000	$y = -1.0447x + 5.9582$	0.9995	1.0447
2005	$y = -1.0594x + 5.9871$	0.9996	1.0594
2010	$y = -1.0615x + 5.9953$	0.9996	1.0615
2015	$y = -1.0671x + 6.0069$	0.9996	1.0671
2020	$y = -1.0673x + 6.0097$	0.9997	1.0673
2023	$y = -1.0662x + 6.0084$	0.9996	1.0662

In Table 5, x denotes $\lg\varepsilon$ and y denotes $\lg N(\varepsilon_k)$. The calculation results show that the correlation coefficients R^2 between x and y are all above 0.9995, indicating that the fractal nature of Huizhou–Shanwei shorelines in each period exists objectively, and it is possible to use this method to study the shoreline change. The change in fractal dimension could reflect the changing characteristics of shoreline morphology. The larger the value of fractal dimension is, the more zig-zagged the shoreline becomes and the morphology tends to be more complex; on the contrary, the shorelines become smooth and the morphology tends to be simpler.

According to Table 5, the fractal dimension of the Huizhou–Shanwei shoreline has ranged from 1.0395 to 1.0673, with the minimum value of 1.0395 (observed in 1990), and the maximum value of 1.0673 (recorded in 2020). Among them, the fractal dimension of the shorelines has been getting larger between 1990 and 2020, with the largest change from

2000 to 2005, increasing from 1.0447 to 1.0594, an increase of 0.0153. After 2020, the fractal dimension of the shorelines decreased by 0.0011. According to the above results it can be seen thatn in 1990, the Huizhou–Shanwei shoreline has the simplest morphology, and in 2020 the Huizhou–Shanwei shoreline has the most tortuous morphology.

The fractal dimension can accurately reflect the morphological changes in the shoreline [24,26].With the expansion of the urban land scale, the impact of human activity on the shoreline has gradually increased, such as the construction of ports, reclamation and breeding, and the development of tourism. These human interventions will change the original morphology of the shoreline, making the originally smooth shoreline more curved, resulting in an increase in the fractal dimension of the shoreline.

5. Discussion

5.1. Comparative Analysis of Shorelines Variation

As an important indicator of regional development, changes in the shorelines can reflect the speed of the development process of the region to a non-negligible extent [28]. Many research studies have shown that the construction of ports and other infrastructures can seriously affect the type structure of shorelines [23,29]. Huizhou–Shanwei region is distributed in three bays: Daya Bay, Honghai Bay, and Jieshi Bay. There are three bays spread across the region and numerous small harbors within the bays. These harbors are critical nodes within provincial maritime transport networks. The rapid economic development has led to an increase in harbor construction so that the conversion of silty shorelines into artificial ones.

In addition, the area of Figure 2b,c—the estuarine delta areas of the region—are susceptible to wave and tidal influences and are characterized by morphological instability [30]. As planning and modifications have been made to the estuarine areas of the study area, both improvements in river transportation and estuarine environments have resulted in the reduction in estuarine shorelines.

Meanwhile, Yang (2014) [31] pointed out that the shorelines of the southern mainland of China were relatively stable from 1990 to 2010, and the fractal dimension basically did not change much. This coincides with the results of this study [31]. In addition, in previous studies, among the many cities in Guangdong's Greater Bay Area, Huizhou has exhibited a slow growth in shoreline fractal dimension compared to cities like Guangzhou and Shenzhen [24,32]. The present study confirms this view. And this study further concludes that the reason for shoreline stabilization in the region is due to the presence of some sandy and biological shorelines (mainly mangrove reserves) within the region. These areas play an important role in stabilizing the fractal dimension.

5.2. Driving Factor Analysis

5.2.1. Natural Factors

Changes in the shorelines are influenced by a number of factors. For natural shorelines, the location and geology of the shorelines are the main factors affecting its change, and therefore changes in natural shorelines are not likely to be significant in a short period of time [33]. Also, factors affecting shorelines change include river sediment loads, tidal action, etc.

Sandy and bedrock shorelines are more stable in terms of the evolutionary development of the earth itself. The study was carried out in 1990, and there have been no major geological events or earth movement processes in the study area in the past 33 years. Therefore, in terms of the natural evolution of the shorelines during the study period, except for the estuarine shorelines, the shoreline type has not changed much as a result of the changes in the natural environmental factors. For example, in Figure 2c, there is a large river inlet, which was an estuarine shoreline in 1990. After this time, the shoreline was pushed seaward, the length of the shoreline became shorter, and the type of shoreline was converted from an estuary to an artificial shoreline, which led to a change in the depositional environment of the estuary. Affected by the influence of the geostrophic force, deposition began on the right side of the river flow—that is, to the left half of Figure 2c. The shoreline type in the left half of Figure 2c was altered by the process of estuarine water flow. In the aftermath of the changes, artificial shorelines in the left half of the gradually formed through siltation, and then formed a muddy shoreline in 2010.

5.2.2. Socio-Economic Factors

Human socio-economic activities also affect shorelines type changes. Since the reform and opening up, China's economy has developed rapidly. By integrating relevant yearbooks, statistical bulletins and other information, the changes in major socio-economic indicators in the Huizhou–Shanwei region since 1990 can be obtained, as shown in Figure 5:

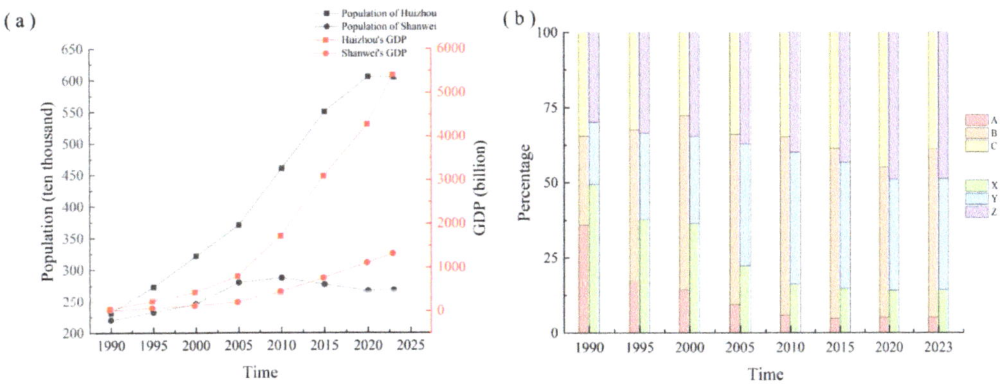

Figure 5. Changes in socio-economic indicators in Huizhou–Shanwei Region; (**a**) Changes in population and GDP; (**b**) Changes in industry proportions in Huizhou and Shanwei cities. **Note:** A: Percentage of tertiary industry in Huizhou. B: Percentage of secondary industry in Huizhou. C: Percentage of primary industry in Huizhou. X: Percentage of primary industry in Shanwei. Y: Percentage of secondary industry in Shanwei. Z: Percentage of tertiary industry in Shanwei.

In 1990, the year-end resident populations of Huizhou City and Shanwei City were 2312.5 thousand and 2199.9 thousand, respectively. By the end of 2022, the year-end resident populations of Huizhou City and Shanwei City will be 6050.2 and 2682.6, respectively, with a significant increase in population. Within the study period, Huizhou City and Shanwei City also experienced a rapid increase in GDP (Gross Domestic Product), and the city's industrial structure is undergoing dramatic changes. In 1990, Huizhou's GDP and Shanwei's GDP was 4.9 billion *yuan* and 2.3 billion *yuan*, respectively. By 2023, Huizhou's GDP was 540.1 billion *yuan* and Shanwei's was 132.2 billion *yuan*. The rapid increase in population, large-scale economic construction, and the increase in the proportion of industry have led to drastic changes in the surface type of the region [29], as well as influencing changes in the shorelines. Accelerated urban construction, more land reclamation activities, and harbor construction are being carried out, which have a great impact on the shorelines, and the proportion of man-made shorelines begins to rise. The most typical area is the NAME (Figure 2a), where the shoreline has experienced a clear seaward movement and became straightened by 2005 due to the construction of ports; hence, the natural shorelines has changed to an artificial one. In Figure 2d, there are numerous small rivers that discharge into the sea draining a very fertile area. The existence of many villages and towns show that people have been working here for a long time, with their activity contributing to shoreline change, as discussed earlier. Nevertheless, from 1980 to 2023, the shoreline in question remained unaltered as an artificial shoreline.

5.2.3. Policy Factors

Policies play an important role in the economic development of a region, and similarly, changes in policies play a decisive role in shorelines changes [2].

In addition, in 2017, the State Oceanic Administration (China) issued *The Measures for the Management of Shorelines Protection and Utilization*, emphasizing that the development and utilization of shorelines resources must be constrained by the carrying capacity of the marine ecosystem, and that marine ecological safety must be ensured as a prerequisite. *The Measures for the Management of shorelines Protection and Utilization* states that, by 2020, the national natural shorelines retention rate will be no less than 35% (excluding island shorelines). For this study area, prior to this document, the artificial index of the shorelines has been increasing, reaching a maximum in 2015 at 0.4682. After this document was published, with the implementation of documents and policies, the artificial shorelines index decreased to 0.4438 in 2020, and the artificial shorelines decreased to 0.4295 in 2023, decreasing to its pre-2005 state. In contrast, the natural retention of the shorelines has increased incrementally, and the length of the natural shorelines has also increased. At present, the natural shoreline retention rate of the region is above this standard, and the overall state is high. As the country requires the development of the marine economy and the development of the coastal zone, the local government should establish a sound system of graded protection of the shoreline, and carry out targeted and hierarchical differentiated management modes for different shorelines.

6. Conclusions

Based on the remote sensing images from 1990 to 2023, the Huizhou–Shanwei shoreline was extracted by visual interpretation. We found that the shoreline increased from 616.71 km to 632.54 km, with an average growth rate of 0.48 km/y. This is associated with a significant reduction in rocky shoreline and increase in artificial shoreline, with the latter becoming increasingly curved. This change is the combined result of natural estuarine processes (i.e., siltation), and human works related to social development and related policies. Therefore, a hierarchical system of shoreline management needs to be established in order to reduce coastal damage and erosion.

Author Contributions: Conceptualization, S.L. and C.L.; methodology, S.L., C.L. and Y.W.; software, S.L., J.F.; validation, S.L., F.G. and J.F.; formal analysis, S.L.; investigation, F.G.; resources, Y.W., Y.S. and C.L.; data curation, S.L., Y.W. and C.L.; writing—original draft preparation, S.L., C.L.; writing—review and editing, S.L., F.G. and Y.S.; visualization, S.L., J.F. and W.W. supervision, F.G. and J.F.; funding acquisition, Y.W. All authors have read and agreed to the published version of the manuscript.

Funding: This research was funded by Science and Technology Innovation Fund of Command Center of Integrated Natural Resources Survey Center (KC20230017), China Geological Survey Projects (DD20230415) and Innovation Foundation of Science and Technology for "Nanhai New Star" Projects (Grant No. NHXXRCXM202353) of Hainan province.

Data Availability Statement: Data are contained within the article.

Acknowledgments: We are deeply grateful for the comments noted by the anonymous reviewers and editors.

Conflicts of Interest: The authors declare no conflicts of interest.

References

1. Zhang, Y.; Hou, X. Characteristics of Coastline Changes on Southeast Asia Islands from 2000 to 2015. *Remote Sens.* **2020**, *12*, 519. [CrossRef]
2. Zheng, Z.; Wu, Z.; Chen, Y.; Yang, Z.; Marinello, F. Exploration of Eco-Environment and Urbanization Changes in Coastal Zones: A Case Study in China over the Past 20 Years. *Ecol. Indic.* **2020**, *119*, 106847. [CrossRef]
3. Almar, R.; Boucharel, J.; Graffin, M.; Abessolo, G.O.; Thoumyre, G.; Papa, F.; Ranasinghe, R.; Montano, J.; Bergsma, E.W.J.; Baba, M.W.; et al. Influence of El Niño on the Variability of Global Shoreline Position. *Nat. Commun.* **2023**, *14*, 3133. [CrossRef] [PubMed]
4. Barbier, E.B. Climate Change Impacts on Rural Poverty in Low-Elevation Coastal Zones. *Estuar. Coast. Shelf Sci.* **2015**, *165*, A1–A13. [CrossRef]
5. Rajasree, B.R.; Deo, M.C.; Sheela Nair, L. Effect of Climate Change on Shoreline Shifts at a Straight and Continuous Coast. *Estuar. Coast. Shelf Sci.* **2016**, *183*, 221–234. [CrossRef]
6. Rangel-Buitrago, N. Human Epoch-Human Responsibility: Rethinking Coastal Zone Management in the Anthropocene. *Ocean Coast. Manag.* **2023**, *244*, 106801. [CrossRef]
7. Wu, T.; Hou, X.; Xu, X. Spatio-Temporal Characteristics of the Mainland Coastline Utilization Degree over the Last 70 Years in China. *Ocean Coast. Manag.* **2014**, *98*, 150–157. [CrossRef]
8. Zhang, X.; Yang, Z.; Zhang, Y.; Ji, Y.; Wang, H.; Lv, K.; Lu, Z. Spatial and Temporal Shoreline Changes of the Southern Yellow River (Huanghe) Delta in 1976–2016. *Mar. Geol.* **2018**, *395*, 188–197. [CrossRef]
9. Zhao, B.; Liu, Y.; Wang, L. Evaluation of the Stability of Muddy Coastline Based on Satellite Imagery: A Case Study in the Central Coasts of Jiangsu, China. *Remote Sens.* **2023**, *15*, 3323. [CrossRef]
10. Wang, X.; Liu, Y.; Ling, F.; Liu, Y.; Fang, F. Spatio-Temporal Change Detection of Ningbo Coastline Using Landsat Time-Series Images during 1976–2015. *ISPRS Int. J. Geo-Inf.* **2017**, *6*, 68. [CrossRef]
11. Mshelia, Z.H.; Amatebelle, E.C.; Belle, J.A. Geospatial Analysis of Shoreline Change of Ethekwini Coastline from 1990–2023. *Sci. Afr.* **2025**, *28*, e02685. [CrossRef]
12. Adenugba, O.; Li, H.; Daramola, S.; Adewale, B.; Gong, Z. Effects of Localized Development on Land Use and Coastline Dynamics: A Focus on Recent Changes along the Lekki Peninsula. *Reg. Stud. Mar. Sci.* **2024**, *78*, 103744. [CrossRef]
13. Sun, J.; Shu, S.; Hu, H.; Deng, Y.; Li, Z.; Zhou, S.; Liu, Y.; Dang, M.; Huang, W.; Hou, Z.; et al. Location Optimization of Unmanned Aerial Vehicle (UAV) Drone Port for Coastal Zone Management: The Case of Guangdong Coastal Zone in China. *Ocean Coast. Manag.* **2025**, *262*, 107576. [CrossRef]
14. Wang, Y.-S.; Lou, Z.-P.; Sun, C.-C.; Sun, S. Ecological Environment Changes in Daya Bay, China, from 1982 to 2004. *Mar. Pollut. Bull.* **2008**, *56*, 1871–1879. [CrossRef]
15. Li, K.; Xu, E. High-Accuracy Continuous Mapping of Surface Water Dynamics Using Automatic Update of Training Samples and Temporal Consistency Modification Based on Google Earth Engine: A Case Study from Huizhou, China. *ISPRS J. Photogramm. Remote Sens.* **2021**, *179*, 66–80. [CrossRef]
16. Liu, X.J.; Xü, J.N.; Kettner, A.J.; Wang, Y.; Yi, J.J. Decadal Shifts and Future Projections of the Han River Delta Coastline. *Mar. Geol.* **2025**, *484*, 107520. [CrossRef]
17. Gairin, E.; Collin, A.; James, D.; Maueau, T.; Roncin, Y.; Lefort, L.; Dolique, F.; Jeanson, M.; Lecchini, D. Spatiotemporal Trends of Bora Bora's Shoreline Classification and Movement Using High-Resolution Imagery from 1955 to 2019. *Remote Sens.* **2021**, *13*, 4692. [CrossRef]
18. Mu, K.; Tang, C.; Tosi, L.; Li, Y.; Zheng, X.; Donnici, S.; Sun, J.; Liu, J.; Gao, X. Coastline Monitoring and Prediction Based on Long-Term Remote Sensing Data—A Case Study of the Eastern Coast of Laizhou Bay, China. *Remote Sens.* **2024**, *16*, 185. [CrossRef]
19. Yang, F.; Zhang, L.; Chen, B.; Li, K.; Liao, J.; Mahmood, R.; Hasan, M.E.; Mamun, M.M.A.A.; Raza, S.A.; Sutrisno, D. Long-Term Change of Coastline Length along Selected Coastal Countries of Eurasia and African Continents. *Remote Sens.* **2023**, *15*, 2344. [CrossRef]
20. Huang, L.; Zhao, C.; Jiao, C.; Zheng, G.; Zhu, J. Quantitative Analysis of Rapid Siltation and Erosion Caused Coastline Evolution in the Coastal Mudflat Areas of Jiangsu. *Water* **2023**, *15*, 1679. [CrossRef]
21. Mansour, N.; Sarhan, T.; El-Gamal, M.; Nassar, K. Assessing the Compatibility of EPR Rates with One-Dimensional Numerical Modelling in Monitoring Shoreline Kinematics along with Supplying near/Long Term Forecasts. *Reg. Stud. Mar. Sci.* **2024**, *71*, 103391. [CrossRef]
22. Rafi, S.; Mourya, N.K.; Balasani, R. Evaluation of Shoreline Alteration Along theJagatsinghpur District Coast, India (1990–2020) Using DSAS. *Ocean Coast. Manag.* **2024**, *253*, 107132. [CrossRef]
23. Rahbani, M.; Ghaderi, D. Long Term Investigation on Shoreline Changes of an Island, Inside a Gulf (Hormuz Island). *Reg. Stud. Mar. Sci.* **2024**, *71*, 103399. [CrossRef]

24. Hu, X.; Wang, Y. Coastline Fractal Dimension of Mainland, Island, and Estuaries Using Multi-Temporal Landsat Remote Sensing Data from 1978 to 2018: A Case Study of the Pearl River Estuary Area. *Remote Sens.* **2020**, *12*, 2482. [CrossRef]
25. Manno, G.; Lo Re, C.; Basile, M.; Ciraolo, G. A New Shoreline Change Assessment Approach for Erosion Management Strategies. *Ocean Coast. Manag.* **2022**, *225*, 106226. [CrossRef]
26. Kappraff, J. The Geometry of Coastlines: A Study in Fractals. *Comput. Math. Appl.* **1986**, *12*, 655–671. [CrossRef]
27. Colak, A.T.I. Geospatial Analysis of Shoreline Changes in the Oman Coastal Region (2000–2022) Using GIS and Remote Sensing Techniques. *Front. Mar. Sci.* **2024**, *11*, 1305283. [CrossRef]
28. Al-Attar, I.M.S.; Basheer, M.A. Multi-Temporal Shoreline Analysis and Future Regional Perspective for Kuwait Coast Using Remote Sensing and GIS Techniques. *Heliyon* **2023**, *9*, e20001. [CrossRef]
29. Yuan, R.; Xu, R.; Zhang, H.; Hua, Y.; Zhang, H.; Zhong, X.; Chen, S. Detecting Shoreline Changes on the Beaches of Hainan Island (China) for the Period 2013–2023 Using Multi-Source Data. *Water* **2024**, *16*, 1034. [CrossRef]
30. Anthony, E.J.; Dussouillez, P.; Dolique, F.; Besset, M.; Brunier, G.; Nguyen, V.L.; Goichot, M. Morphodynamics of an Eroding Beach and Foredune in the Mekong River Delta: Implications for Deltaic Shoreline Change. *Cont. Shelf Res.* **2017**, *147*, 155–164. [CrossRef]
31. Yang, L.; Li, J.; Yuan, Q.; Xu, L.; Lu, X.; Wang, M.; Zhao, S. Spatial-Temporal Changes of Continental Coastline in Southern China. *J. Mar. Sci.* **2014**, *32*, 42–49. [CrossRef]
32. Hu, R.; Yao, L.; Yu, J.; Chen, P.; Wang, D. Remote Sensing of the Coastline Variation of the Guangdong–Hongkong–Macao Greater Bay Area in the Past Four Decades. *J. Mar. Sci. Eng.* **2021**, *9*, 1318. [CrossRef]
33. Gao, W.; Du, J.; Gao, S.; Xu, Y.; Li, B.; Wei, X.; Zhang, Z.; Liu, J.; Li, P. Shoreline Change Due to Global Climate Change and Human Activity at the Shandong Peninsula from 2007 to 2020. *Front. Mar. Sci.* **2023**, *9*, 1123067. [CrossRef]

water MDPI

Article

Assessment of Storm Surge Disaster Response Capacity in Chinese Coastal Cities Using Urban-Scale Survey Data

Li Zhu [1,2,*] and Shibai Cui [1]

1 Department of Architecture, Tianjin University, Tianjin 300072, China; cuicuicui2006@163.com
2 APEC Sustainable Energy Center, Tianjin 300072, China
* Correspondence: zhuli1977@tju.edu.cn; Tel.: +86-17743468411

Abstract

Currently, most studies evaluating storm surges are conducted at the provincial level, and there is a lack of detailed research focusing on cities. This paper focuses on the urban scale, using some fine-scale data of coastal areas obtained through remote sensing images. This research is based on the Hazard–Exposure–Vulnerability (H-E-V) framework and PPRR (Prevention, Preparedness, Response, and Recovery) crisis management theory. It focuses on 52 Chinese coastal cities as the research subject. The evaluation system for the disaster response capabilities of Chinese coastal cities was constructed based on three aspects: the stability of the disaster-incubating environment (S), the risk of disaster-causing factors (R), and the vulnerability of disaster-bearing bodies (V). The significance of this study is that the storm surge capability of China's coastal cities can be analyzed based on the results of the evaluation, and the evaluation model can be used to identify its deficiencies. In this paper, these storm surge disaster response capabilities of coastal cities were scored using the entropy weighted TOPSIS method and the weight rank sum ratio (WRSR), and the results were also analyzed. The results indicate that Wenzhou has the best comprehensive disaster response capability, while Yancheng has the worst. Moreover, Tianjin, Ningde, and Shenzhen performed well in the three aspects of vulnerability of disaster-bearing bodies, risk of disaster-causing factors, and stability of disaster-incubating environment separately. On the contrary, Dandong (tied with Qinzhou), Jiaxing, and Chaozhou performed poorly in the above three areas.

Keywords: storm surge; Chinese coastal cities; disaster response capacity; entropy weighted TOPSIS; grey relational analysis

1. Introduction

Storm surge disasters (SSDs) are a common marine hazard along ocean coastlines, characterized by abnormal fluctuations in sea level caused by strong winds and rapid changes in atmospheric pressure. These events are typically triggered by typhoons and extratropical cyclones, which push seawater toward the shore. As a result, sea levels in affected areas can rise far above normal tidal ranges, leading to severe coastal flooding and infrastructure damage [1–4]. Across the globe, storm surges are widely acknowledged as critical coastal hazards that have caused substantial loss of life, economic disruption, and long-term social impacts. Events such as Hurricane Katrina in 2005 (USA) and Storm Xaver in 2013 (Northern Europe) have revealed how inadequate coastal planning and infrastructure can amplify disaster impacts. In response, countries like the Netherlands have developed comprehensive flood defense systems through the integration of delta engineering and coastal zone governance, setting global benchmarks in surge resilience. Likewise, East Asian nations, such as Japan and South Korea, have implemented mixed strategies combining engineered defenses—like sea barriers—with nature-based approaches, including wetland conservation and restoration. These international approaches demonstrate the value of localized, city-scale disaster preparedness in enhancing resilience to storm surges. By contrast, most Chinese studies to date have focused on macro-scale evaluations, with insufficient attention given to differences in vulnerability and preparedness across individual coastal cities. According to 2020 data, China's coastline is 32,977.34 km long, and the cities that have a coastline are frequently impacted by SSDs due to their unique climate and geographic positioning [5,6]. The geographical positioning of China's coastline renders it susceptible to storm surges. These events have historically caused substantial human casualties and economic losses. Specifically, from 2000 to 2022, SSDs inflicted approximately $32 billion in economic damages and left 888 people missing or dead, while the development of the storm surge early-warning system has resulted in a significant reduction in these casualties. However, the human and economic losses caused by SSDs have increased significantly due to the rapid growth of both, coastal populations and economies [7]. This development highlights the complex interplay between technological advancements in disaster prediction and management, and the challenges posed by economic growth and demographic shifts in coastal areas.

Most contemporary research on SSDs in China focuses on storm surge warning forecasts, estimating the damage caused by storm surges, and assessing storm surge tide levels at different recurrence periods. Some of these studies particularly focus on coastal vulnerability and risk assessment. For example, Shi et al. developed a risk assessment method using the Shanghai area as a case study, and Huang et al. conducted a vulnerability assessment for Chinese coastal areas [3,8]. In disaster studies, the relationship between the disaster-bearing bodies and the cause of the disaster is usually extremely complex, and currently, few comprehensive studies regarding storm surges have been made that are based on the capability of the disaster-bearing bodies. Furthermore, research on the causative mechanisms of storm surges lacks systematicity, and future assessments of SSDs should focus on the refinement of current methodologies. Not only is it necessary to refine the types of losses, but it is also important to segment the objects and areas being evaluated. This assessment process should include a detailed delineation of the natural, social, economic, and environmental characteristics typical to a specific region. Currently, research on vulnerable entities based on SSDs is mainly focused on the provincial level, with a need for more research on the city level. Although Meng et al. conducted a similar vulnerability study that paid particular attention to the impact of storm surges on cities, only 30 coastal cities were included in that sample, meaning that more than 20 Chinese coastal prefecture-level cities were ignored in the results [5]. In the context of urban management concerning SSDs, scholars from various countries have conducted extensive research. Blankespoor et al. quantified the coastal protection services provided by mangroves in mitigating storm surges, underscoring the impact of mangrove loss on storm surge areas [9]. Nguyen et al. assessed storm surge risks in aquaculture in the Northern coastal area of Vietnam, providing a scientific basis for proactive response plans and policy-making to reduce storm surge damage [10]. Xu et al. investigated the joint risk of rainfall and storm surges during typhoons in a coastal city in China, providing insights for urban flood risk assessment and management [11]. However, their study lacks a comprehensive evaluation of coastal cities from multiple perspectives and in a holistic manner. The present study evaluates the prefecture-level cities along China's coastline from various perspectives and gives management advice.

A disaster is recognized when an event exceeds the response capability of the affected entity, highlighting the need for a thorough examination of its response capability. To clarify the current response capabilities of Chinese coastal cities to SSDs, this study presents the following approach. (1) Using the H-E-V framework, the disaster system functional system, and the integration of PPRR theory, a framework is developed to assess the disaster response capability (DRC) of coastal cities against storm surges. (2) This study employs the entropy weight TOPSIS method and the weight rank sum ratio (WRSR) method, in combination with the H-E-V framework, to calculate the DRC scores of coastal cities against storm surges. (3) Given the need for more in-depth research on response capability, this paper analyses the SSD response capability of urban disaster-bearing bodies in order to compare the differing response capacities of various cities to SSDs. The results of this research will be of great significance to the governmental bodies responsible for future urban development and emergency planning.

2. Study Area

China's coastline is characterized by a complex administrative composition that includes 9 coastal provincial regions (Taiwan inclusive), 1 autonomous region, 2 special administrative regions, 55 coastal provincial prefecture-level cities (excluding Hong Kong, Macao, and Taiwan), and 242 coastal districts and counties distributed along China's coastline. This paper is based on the context of Chinese districts, and adopts the city as the primary unit of analysis. A total of 54 provincial prefecture-level cities, excluding Hong Kong, Macao, Taiwan, and Sansha City, were initially included in the scope of this study.

Among these coastal cities, it is commonly accepted that the demarcation between the Qiantang River and Hangzhou Bay stretches from Haining Ganpu (part of Jiaxing) to the Yuyao Xisan Sluice (part of Ningbo). Thus, Hangzhou and Shaoxing are not considered coastal cities. Given that the eight estuaries of the Pearl River serve as the boundary between river and sea, Guangzhou and Dongguan are defined as coastal cities. At the same time, Sansha City, due to its minimal population, is temporarily excluded from this categorization. Following a rigorous selection process, this study identifies a cohort of 52 coastal cities (as depicted in Figure 1).

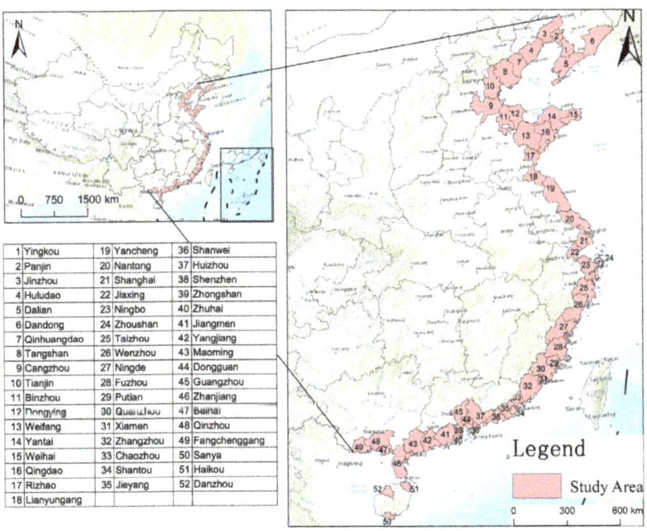

Figure 1. Schematic diagram of the study area.

3. The Construction of Disaster Response Evaluation System for Storm Surges

By making a thorough literature review of both Chinese and foreign research, and taking into account the specific conditions of coastal cities in China, this paper attempts to establish a fine-scale evaluation indicator system of disaster response capability (DRC) with a specific focus on storm surges.

3.1. Principles for Establishing an Indicator System

The establishment of a practical and scientific indicator system must follow some basic principles. Peng et al. proposed that constructing an indicator system should follow the six principles of purpose: purposefulness, completeness, operability, independence, importance, and dynamism [12]. In addition, Lu and Han proposed their own five principles: specific, measurable, achievable, realistic, and time-bound, according to principles previously applied by experts, in conjunction with disaster theory and the specific requirements of this research [13]. Meanwhile, combining the characteristics of SSDs with data gathered from the study area, four principles should be followed to establish an evaluation index system for storm surge response capacity, as proposed in this paper, including. (a) Scientific: Being scientific is the primary basis for evaluating indicator systems. SSDs can cause damage to coastal areas on a wide variety of levels, including people, property, and the environment. Therefore, it is essential that any potential indicators are chosen after a rigorous scientific screening and assessment. (b) Realistic: The aim of the evaluation model is to assess the response of cities to storm surges by identifying and analyzing any weaknesses or potential points of significant damage. Therefore, the indicators used in the system should be both easily accessible and countable, in addition to being descriptive, representative, and easy to conceptualize to ensure the ease of use of the entire system. (c) Generic: The object of evaluation for this indicator system is the city, and many coastal cities are in China. The ability to generalize indicators is among the first issues that should be addressed when establishing an indicator system. While each city has its distinctive features, there are also shared traits that must be identified to establish a standardized indicator system. Recognizing these commonalities is essential to ensure consistency and comparability across different cities. (d) Representative: This paper presents an evaluation model for disaster response capabilities related to SSDs, divided into three dimensions, each containing multiple sub-dimensions. Each sub-system should be broken down into representative indicators for individual evaluation to form a comprehensive evaluation indicator system.

3.2. Theoretical Framework for Storm Surge Response Evaluation in Coastal Cities

3.2.1. H-E-V Framework and Functional System for Disaster Systems

The primary frameworks for assessing urban flood risk are the "Probability–Consequence" model and the "Hazard–Exposure–Vulnerability (H-E-V)" model [14–16]. When comparing these two models, the "H-E-V" framework is comprehensive, clear, and practical, making it popular among scholars and research institutions. For example, the IPCC uses this framework to evaluate urban flood risks [17]. The "H-E-V" framework, commonly applied in flood risk assessments, encompasses three key components: Flood risk = Hazard (H) × Exposure (E) × Vulnerability (V).

According to Peijun Shi, the structural system of a regional disaster system (DS) comprises the disaster-incubating environment (E), disaster-causing factors (F), and disaster-bearing bodies (S), expressed as [18,19]. The system differs from Mileti's approach to structural disaster systems, where these elements are combined into a single environmental system [20]. In 2005, Peijun Shi proposed that the functional system of regional disasters (D_f) consists of the stability of the disaster-incubating environment (S), the risk of disaster-causing factors (R), and the vulnerability of disaster-bearing bodies (V) [20]. This theory resembles Wisner's disaster system, but Wisner emphasizes the interaction between hazard factors and the affected bodies in the disaster system I have checked and revised all [21]. Peijun Shi, on the other hand, believes that hazard factors, affected bodies, and the disaster-incubating environment are equally crucial in the disaster system [18].

Based on previous studies, this research evaluates the disaster response capabilities related to storm surges from three perspectives: the stability of the disaster-incubating environment (S), the risk associated with hazard factors (R), and the vulnerability of the disaster-bearing bodies (V).

3.2.2. Theory of PPRR Model for Crisis Management

Any disaster has a life cycle that goes from inception to development, outbreak, decline, and finally to end. The crisis also has a lifecycle from its occurrence to its development, and traditional crisis management focuses more on post-disaster relief and rehabilitation, with significantly less attention paid to pre-disaster prevention and preparedness [22].

The PPRR model is a widely used theory that originates from crisis management and consists of four phases: Prevention before the crisis (Prevention), preparation before the crisis (Preparation), response during the outbreak (Response), and recovery after the crisis ends (Recovery). These four phases are the generic model of crisis management [23]. These phases correspond to the four disaster response stages, namely prevention, response, relief, and recovery, and thus define the different phases of disaster management and their associated tasks. Disaster prevention capability refers to the ability of the study areas to prevent emergencies and disasters, usually by implementing planning and countermeasures to ensure the safety of property and people. Emphasis is placed on using existing management tools as well as modern science and technology to prevent the occurrence of emergencies or disasters, primarily through the employment of disaster prevention education, increasing awareness of prevention measures, economic investment in disaster prevention, and large-scale disaster prevention projects. Disaster resilience refers to the ability of disaster-bearing bodies to resist a sudden accident or disaster and to maintain their regular function in the event of damage from a disaster. It possesses specific inherent characteristics that are primarily determined by factors such as population demographics, urban spatial organization, and socio-economic factors. The primary manifestation of disaster relief capability lies in the emergency response to disasters, which is usually determined by the government's emergency response capability, emergency resources, transportation, rescue personnel, emergency shelters, and other factors. On the other hand, the strength of a city's recovery capability is mainly influenced by insurance and economic factors [24–28].

3.3. *Construction of an Evaluation Indicator System for Disaster Response Capability*

Building on the theory outlined in the previous section, the assessment system for evaluating disaster response capability to SSDs comprises three components: the vulnerability of the disaster-bearing bodies, the stability of the disaster-incubating environment, and the risk of hazard factors. This study has identified suitable indicators for each component, following the principles of establishing an indicator system. In this study, the construction of the indicator system is grounded in a systematic literature review, and supported by disaster system theory. A wide range of peer-reviewed studies, policy guidelines, and technical reports in the fields of coastal vulnerability, storm surge risk assessment, and urban disaster resilience were reviewed. The following types of indicators were prioritized: (1) those widely used across multiple studies; (2) those reflecting the core dimensions of the H-E-V and PPRR frameworks; and (3) indicator with data availability at the urban scale [5,29–57].

For example, indicators such as "population density", "length of drainage pipelines", and "proportion of elderly population" are frequently used in marine flood vulnerability assessments. "Land use ratio" is recognized as a key variable in studies on environmental stability. Although no studies were found to define "coastline type" as a specific indicator, many have highlighted its role in buffering storm surge impacts. Since this study evaluates urban disaster response from a city-level perspective, "coastline type" is considered an important component of the disaster-incubating environment. The final indicator system should ensure strong scientific validity and applicability to policy.

Regarding the vulnerability of the disaster-bearing bodies, the concept is divided into four stages based on disaster life cycle theory and the PPRR model: disaster prevention capability, disaster resistance capability, disaster relief capability, and recovery capability. In conjunction with the characteristics of the indicators and the four systemic concepts, relevant indicators have been chosen, as shown in Table 1. The stability of the disaster-incubating environment is categorized into natural and human-made environments. The natural environment encompasses various coastline types, each with different abilities to withstand storm surges. Due to varying risks of storm surge damage across different land use types, the human-made environment includes the diverse factors related to land use. The direction definition of indicators is for the subsequent analysis of indicator data using the entropy weight method. Table 2 displays the relevant indicators. Additionally, hazard factors for SSDs generally involve meteorological, hydrological, and geological elements, with the specific indicators listed in Table 3. The theoretical diagram of the evaluation system established based on the H-E-V framework and the PPRR model is displayed in Figure 2.

Table 1. Evaluation indicators for the vulnerability of disaster-bearing bodies.

Segments	Elements	Indicators	Unit	No.	Direction
Disaster prevention capability	Monitoring and early warning capability	Number of tidal stations	piece	V1	+
		Mobile phone subscribers	10,000 people	V2	+
	Education and economic input for disaster prevention	Expenditure on disaster prevention and emergency management	10,000 Yuan	V3	+
		Per capita expenditure on education	Yuan	V4	+
	Environmental factor	Coastline coefficient	km/km^2	V5	−
Disaster resistance	Demographic factors	Relative population density (within 10 km of the coastline)	People/km^2	V6	−
		Proportion of population aged 0~14	%	V7	−
		Proportion of population aged 65 and above	%	V8	−
	Spatial layout	Area of mariculture	Hectare	V9	−
	Infrastructure factors	Road mileage per unit land area	km/km^2	V10	+
		Length of drainage pipelines	km	V11	+
	Economic factors	Fishery output value	Billion Yuan	V12	−
		Proportion of the primary industry at the district (county) level including the coastline	%	V13	−
		Proportion of tourist receipts	%	V14	−
Disaster relief capability	Healthcare infrastructure	Number of hospitals per 10,000 people	pcs	V15	+
		Number of beds per 10,000 people	Bed	V16	+
		Number of doctors per 10,000 people	People	V17	+
	Transport factors	Total highway passenger traffic volume	10,000 people	V18	+
		Total highway freight traffic volume	10,000 tons	V19	+
	Economic factors	Number of shelters	pcs	V20	+

Table 1. *Cont.*

Segments	Elements	Indicators	Unit	No.	Direction
Recovery capability	Basic social security	Proportion of employees covered by basic medical insurance	%	V21	+
	Economic factors	Per capita disposable income of urban permanent residents	Yuan	V22	+
		Per capita disposable income of rural permanent residents	Yuan	V23	+
		Proportion of secondary industry	%	V24	+
		Proportion of tertiary industry	%	V25	+
		Insurance penetration	Yuan/Yuan	V26	+

Table 2. Evaluation indicators for stability of the disaster-incubating environment.

Segments	Elements	Indicators	Unit	No.
Stability of disaster-incubating environment	Coastline Stability	Percentage of estuary shoreline length	%	S1
		Percentage of bedrock shoreline length	%	S2
		Percentage of sandy shoreline length	%	S3
		Percentage of biological shoreline length	%	S4
		Percentage of muddy shoreline length	%	S5
		Percentage of the length of the reclamation and aquaculture shoreline	%	S6
		Percentage of the length of docks, artificial shorelines, and protective embankments	%	S7
	Land Stability	Percentage of farmland area	%	S8
		Percentage of forest area	%	S9
		Percentage of grassland area	%	S10
		Percentage of water area	%	S11
		Percentage of urban and rural, industrial and mining, and residential land area	%	S12
		Percentage of unutilized land area	%	S13

Table 3. Evaluation indicators for the risk of disaster-causing factors.

Segments	Elements	Indicators	Unit	No.	Direction
The risk of disaster-causing factors	Meteorological factors	Annual precipitation	mm	R1	−
		Maximum wind speed	m/s	R2	−
		Days with wind speeds exceeding 10.8 m/s	days	R3	−
	Hydrological factors	The annual average of the highest tide height	mm	R4	−
		Maximum tidal height	mm	R5	−
	Geological factors	Average altitude	m	R6	+
		Average slope	%	R7	+

The indicator framework was designed to align with the thematic priorities of the Chinese "14th Five-Year National Comprehensive Disaster Prevention and Mitigation Plan" issued by the National Disaster Reduction Commission. This includes domains such as monitoring and early-warning systems, infrastructure resilience, emergency response capacity, material reserves, technological support, and community engagement. For instance, indicators related to early-warning stations and emergency shelters directly reflect the Plan's directives on enhancing multi-source monitoring networks and shelter infrastructure. Moreover, The Sustainable Development Goals (SDGs) adopted by the

United Nations in 2015 provide ideas and references for this study. The SDGs provide a unifying framework for achieving human well-being, environmental sustainability, and disaster resilience globally by 2030. A unifying framework is provided to achieve human well-being, environmental sustainability, and disaster resilience globally by 2030. Among the 17 goals, SDG 11 (Sustainable Cities and Communities), SDG 13 (Climate Action), and SDG 1 (End Poverty) explicitly emphasize disaster preparedness, mitigation, and resilience. They integrate the main targets and priorities of the Sendai Framework for Disaster Risk Reduction [57]. Even though the SDGs are primarily intended to apply to the national level, many cities are referring to them and use the Disaster Resilience Scorecard for Cities [58] as instrument for addressing disasters. Both the SDGs and the Disaster Resilience Scorecard for Cities are intended to promote integrated and forward-looking planning, through which the resilience of cities, in terms of infrastructure, social systems, and ecological barriers, can be enhanced to achieve sustainable development.

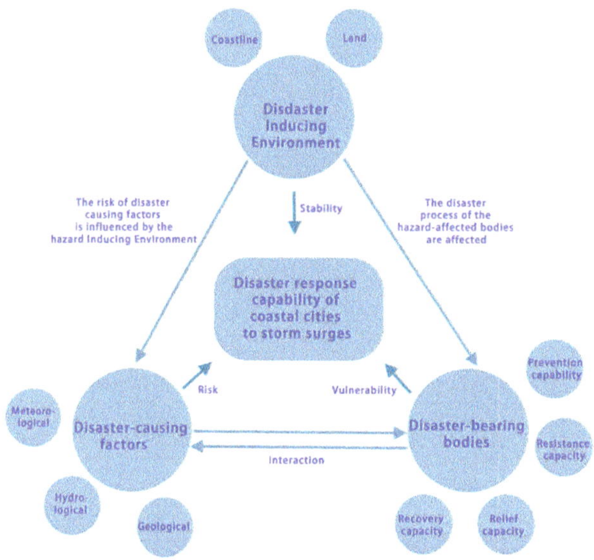

Figure 2. Theoretical map of the evaluation system.

Disaster resilience is considered to be an instrument or precondition facilitating sustainability [59]. A city's ability to effectively recover from extreme events is fundamental to achieving its long-term sustainability. Investments in early warning systems, climate-resilient infrastructure, public health services, and inclusive social protection are not only key aspects of disaster risk management, but are directly linked to the core SDG objectives of eradicating poverty, improving health, and securing livelihoods. Therefore, the systematic assessment of SSD resilience from multidimensional perspectives, including environment, infrastructure, society, and economy, is considered important for ensuring consistency between local adaptation strategies and the global sustainability agenda.

In this study, the concept of SDG objectives is fully referenced in the construction of the SSD response capacity evaluation indicator system. Specifically, indicators such as medical resources, communication access, drainage infrastructure, land use structure, income level, and insurance coverage reflect both the technical preparedness of the city and the broader socio-economic and environmental resilience advocated by the SDG framework. By using an indicator system that matches the SDGs, scientific analysis has been made more systematic and based on a stronger theory. At the same time, the policy relevance of the results has been improved. This helps support the design of more sustainable and flexible coastal management strategies for cities.

Each indicator of vulnerability of disaster-bearing bodies has been carefully selected, with its significance explained below and supported by the relevant literature [5,30,33,34,37–41,44,45,47,49–56]. Tidal detection stations provide critical real-time data for storm surge monitoring and early warning, enabling timely disaster response. A higher number of mobile users indicates a greater capacity for rapid dissemination of information during emergencies. "Mobile phone subscribers" have been regarded as an indicator for communication accessibility and the capacity for rapid information dissemination during emergencies. For instance, regions with higher mobile phone subscription rates often demonstrate better performance in early warning reception, emergency coordination, and public response [53,54]. Higher expenditures on disaster prevention and control indicate stronger institutional capacity for both pre-disaster prevention and post-disaster response. The level of education expenditure reflects public awareness and knowledge of disasters, which indirectly enhances disaster response capacity. The "Coastline Coefficient" indicator represents the abundance of marine resources in terrestrial regions. It is calculated by dividing the coastline length (in kilometers) by the land area (in square kilometers).

The high population density in vulnerable zones increases exposure and response pressure. The proportion of the population aged 0–14 and over 65 represents a group with weaker mobility and greater sensitivity to disasters. The vulnerability of aquaculture areas to storm surge inundation represents potential economic losses, and serves as a test of the resilience of marine economies. Road mileage per unit land area indicates transportation accessibility, which is vital for emergency evacuation and aid delivery. Longer drainage systems suggest better urban flood mitigation capacity. High values of fishery output suggest potential economic vulnerability in coastal areas. A high reliance on fisheries implies greater exposure to marine hazards, such as storm surges. Theoretically, this reflects sectoral dependency vulnerability and socio-ecological sensitivity. Empirical studies have confirmed that cities with higher fishery output tend to suffer more severe economic losses and face longer recovery times after coastal disasters [55,56]. The primary industry is easily affected by climate, and areas with a high proportion have weaker disaster resistance capabilities. Therefore, it is necessary to strengthen the disaster prevention system. The tourism industry is highly sensitive to the environment, and disasters will lead to a rapid decline in the tourism economy, affecting regional economic stability.

Health-related indicators, which include the number of hospitals, beds, and doctors per 10,000 persons, measure the adequacy of the emergency healthcare infrastructure. The passenger and freight transport capabilities determine the efficiency of efficient coordination and evacuation of resources and supply to disaster areas during disasters, and are important indicators of logistical support. The more shelters there are, the stronger the city's ability to accommodate disaster victims during disasters, reflecting the level of emergency reserves. The wide coverage of medical insurance can reduce the burden of post disaster medical care and improve residents' ability to recover after disasters. The per capita disposable income of urban and rural residents reflects household-level adaptive capacity and financial resilience. The secondary industry is greatly affected by infrastructure, and is prone to work stoppages after disasters, while the tertiary industry is highly dependent on the environment and has a significant impact on service interruption after disasters. Its recovery capacity needs to be evaluated. Higher penetration indicates stronger financial resilience and disaster risk transfer capacity.

Among all disaster-bearing body indicators, only population density was calculated specifically within the 10 km buffer zone from the coastline, as the population within this zone is directly exposed to storm surge impacts and plays a passive role in the disaster system, lacking the capacity to support or respond to the event. Other indicators, such as emergency transportation capacity, medical resources, disaster shelters, and institutional mechanisms, reflect the overall DRC at the city level. These resources are not confined to the coastal zone but are spatially distributed across the urban area and can be rapidly mobilized during SSDs to support affected regions. Furthermore, although storm surges mainly affect coastal zones, the ability to respond depends on systems and resources located across the entire city behind coastal zones. Non-coastal parts of the city frequently provide logistical, economic, and administrative support during both the response and the recovery phases. Therefore, using mixed-scale data for the disaster-bearing body indicators provides a more comprehensive reflection of the overall resilience and coordination capacity of the urban system.

The spatial distribution of coastal areas and the types of shorelines are key factors influencing the resilience of coastal cities to storm surge hazards. The importance of land use and shoreline type in storm surge hazards has been noted in several studies [29,32–34,42,43,47]. Coastal land use affects the vulnerability of urban areas and their response strategies. For example, urban construction land contains dense infrastructure, which faces a higher risk of property loss and greater difficulty in evacuation during storm surge events. Agricultural land is considered highly sensitive to seawater intrusion and salinization, which may result in long-term ecological and economic damage. By contrast, areas dominated by wetlands, forests, or mangroves are regarded as natural buffers, through which wave energy is absorbed and the extent of inundation is reduced. Therefore, land use is not only seen as a reflection of regional economic functions but is considered to influence the spatial distribution of disaster resistance and exposure. In this study, the evaluation of land use stability was conducted using the primary classification from the LUCC system developed by the Chinese Academy of Sciences. At the same time, the data were clipped to a 10 km inland buffer from the coastline, following the same rationale as the population indicator, as this area is directly exposed to storm surge impacts. On the other hand, there are significant differences in the resistance of different types of shorelines to storm surge erosion and wave impacts. For instance, rocky shorelines and mangrove shorelines usually have strong natural defenses, while muddy shorelines and aquaculture shorelines are more vulnerable to erosion and overtopping. Therefore, the type of coastline is an important parameter for assessing the stability of disaster incubating environment and developing coastal management strategies.

In summary, the stability of disaster incubating environment is determined by land use and shoreline characteristics, which are regarded as essential components in the construction of an urban storm surge resilience assessment framework.

The reasons for the selection of indicators related to the disaster causal factors are listed, and these indicators are mentioned in the relevant literature [29,31,34–36,45,48].

During storm surge events, heavy rainfall is often observed, which increases coastal flood risk through surface runoff, and worsens urban waterlogging by expanding the inundation area. Therefore, annual precipitation is considered an important indicator. Maximum wind speed is regarded as a major driving factor in the formation of storm surges, as it determines the wind force on the sea surface and influences wave height, tide accumulation, and the destructive power of the hazard. Moreover, the number of days with wind speeds exceeding 10.8 m/s is used to reflect the frequency and persistence of strong wind events. Frequent occurrences are seen as an indication that coastal areas are regularly exposed to wind-driven processes, which may weaken natural defense systems and increase long-term vulnerability. The annual average of the highest tide height and maximum tidal height reflect the tidal level in the city during the year. If the annual mean maximum tide height is high, the buildup of storm surge on top of it is more likely to cause the over-topping of embankments, increasing the risk of urban inundation and infrastructure damage. Maximum tide heights are regarded as representations of the extremes reached by astronomical tides. When they are superimposed on peak storm surges, sea levels are significantly raised, which greatly increases the intensity of the disaster. The average elevation of a city is used to determine its base height relative to sea level. Lower elevations are more susceptible to seawater inundation and flooding, and elevation is regarded as a fundamental indicator in spatial zoning of exposure and geographic sensitivity analyses. Slope is used to assess the rate of surface runoff and the retention time of standing water. Flat areas are seen as more prone to waterlogging, while steep slopes are viewed as favorable for drainage. Therefore, this indicator is considered highly important in the simulation of inundation processes and the planning of drainage systems.

Although some indicators, such as annual precipitation, are not direct causal factors of storm surges, they are still included in this study because they significantly influence disaster response conditions and tend to compound the pressure on emergency response systems during storm surge events. This study does not aim to evaluate the physical severity of storm surges themselves, but rather focuses on assessing the overall disaster response capacity of coastal cities under storm surge scenarios. Therefore, the selection of indicators prioritizes factors that are most likely to influence response effectiveness during SSDs. The inclusion of such compound risk factors reflects a more realistic urban disaster context and supports a more comprehensive resilience evaluation.

Since key data are usually updated every 5 years, most of the data in this study are based on the 2020 data for model demonstration. In this study, the most relevant indicator data for coastal cities comes from the year 2020 for analysis and evaluation. A smaller subset of data, such as those related to coastlines and land use, is taken from the period between 2020 and 2023, as these factors are likely to change slowly. Much of the data on disaster-bearing bodies were obtained from the "China City Statistical Yearbook 2021" and announcements published on the official websites of municipal governments [60]. A smaller portion of the data comes from local government responses and local news extracts obtained through a public application. Data related to disaster factors were sourced from the "Tide Table 2020" published in China and the National Oceanic and Atmospheric Administration (NOAA) of the United States, among others [61]. For the indicators "The annual average of the highest tide height" and "Maximum tidal height", the data were processed using the local mean sea level (MSL) as the vertical reference datum. During data acquisition and calculation, tide levels were adjusted relative to the MSL. The tidal reference datum for each city's observation station is explicitly documented in the "2020 Tide Table" published by Chinese authorities. Coastal data relevant to the disaster-incubating environment were extracted through the GEE platform for remote sensing image analysis, supplemented by manual identification and on-site geolocation verification (Figure 3). Land use data were sourced from the Institute of Geographic Sciences and Natural Resources Research at the Chinese Academy of Sciences and then processed using ArcGIS (Figure 4). Due to space constraints, this paper only presents the visualization of remote sensing analysis of the coastline of Weihai, Zhuhai, and Huizhou, and a visual representation of the cropped land use data for Huizhou, Zhuhai, and Weihai.

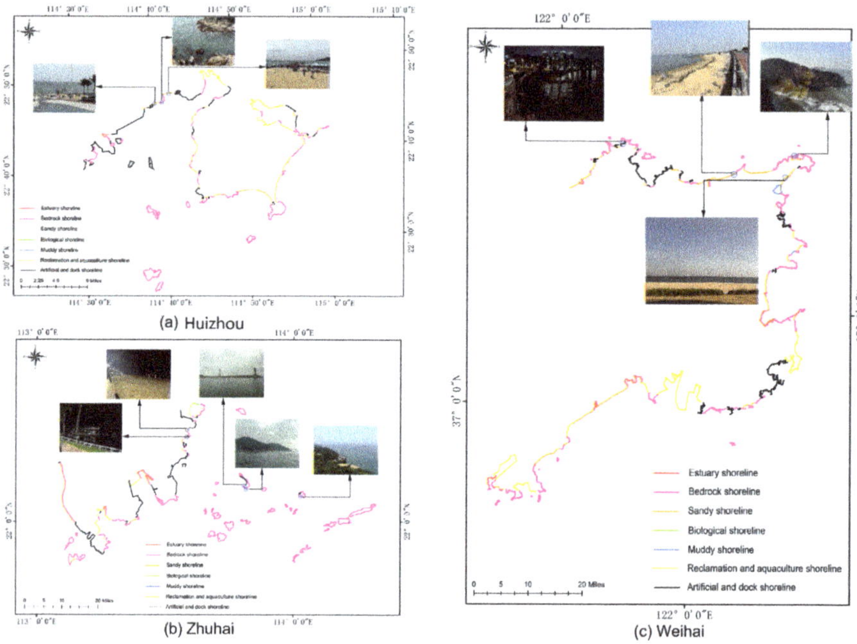

Figure 3. (**a**) Analysis of shoreline types in Huizhou and comparison with on-site research; (**b**) analysis of shoreline types in Weihai and comparison with on-site research; (**c**) analysis of shoreline types in Zhuhai and comparison with on-site research.

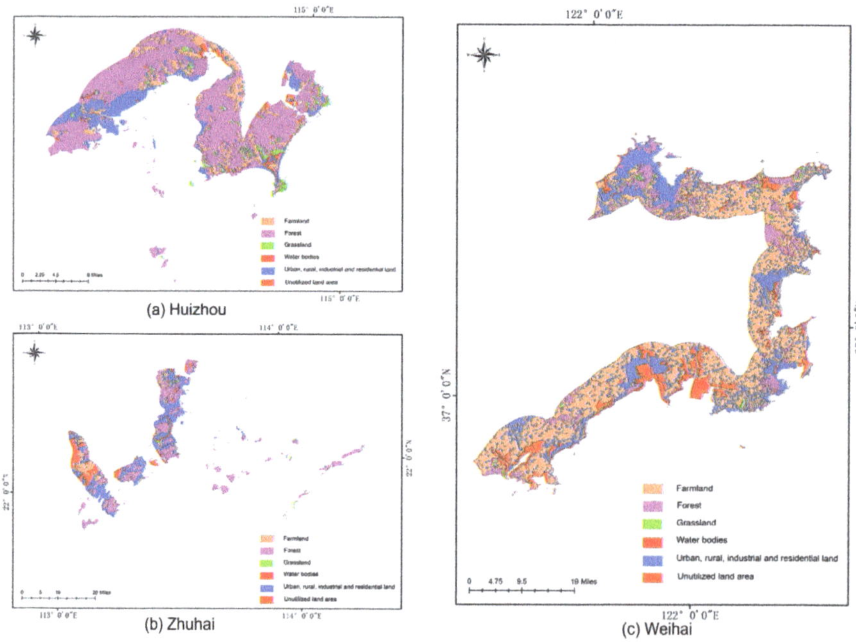

Figure 4. (**a**) Schematic diagram of coastal land use types in Huizhou; (**b**) schematic diagram of coastal land use types in Weihai; (**c**) schematic diagram of coastal land use types in Zhuhai.

4. Methods

4.1. Standardization of Original Index Data

The original data collected varied widely in terms of scale, and the scales vary greatly from one indicator to another. Therefore, it is necessary to standardize the original data before carrying out data analysis. The intention of this standardization is to group the evaluation indicators into a value between 0 and 1, and to eliminate the variability in scale between indicators and their influence on the screening indicators.

Equations (1) and (2) are the formulas used to normalize positive and negative indicators, respectively.

$$x_{ij} = \frac{v_{ij} - \max\limits_{1\leq i\leq m}(v_j)}{\max\limits_{1\leq i\leq m}(v_j) - \min\limits_{1\leq i\leq m}(v_j)} \quad (1)$$

$$x_{ij} = \frac{\max\limits_{1\leq i\leq m}(v_j) - v_{ij}}{\max\limits_{1\leq i\leq m}(v_j) - \min\limits_{1\leq i\leq m}(v_j)} \quad (2)$$

$\max(x_j)$ is the maximum value of all evaluation ranges in the jth indicator;
$\min(x_j)$ is the minimum value of all evaluation ranges in the jth indicator;
m is the total number of indicators (m = 30);
v_{ij} is the value of the jth indicator in the ith region;
x_{ij} is the standardized value of the indicator for the jth indicator in the ith region.

4.2. Entropy Weighted TOPSIS Method

In this study, the entropy weighted TOPSIS method was employed to analyze and evaluate the dimensions of disaster-bearing bodies and disaster-causing factors.

4.2.1. Determining Indicator Weights Using the Entropy Weight Method

(a) After data preprocessing, calculate the proportion "p_{ij}" of index value of project i under index j: "p_{ij}" is calculated as in Equation (3):

$$p_{ij} = \frac{x_{ij}}{\sqrt{\sum\limits_{i=1}^{n} x_{ij}}} \quad (3)$$

p_{ij} is the proportion of the ith sample value under the jth indicator.

(b) Calculate the entropy of each indicator, the formula is shown in Equation (4):

$$e_j = -k\sum_{i=1}^{n} p_{ij}In(p_{ij}) \quad (4)$$

for which $k = 1/\ln(n) > 0$, and $e_j \geq 0$.

e_j is the information entropy of the jth indicator.

(c) Define the degree of difference for the jth indicator as follows:

$$d_j = 1 - e_j \quad (5)$$

(d) Calculate the entropy weight, which is calculated as in Equation (5):

$$w_j = \frac{d_j}{\sum\limits_{j=1}^{m} d_j} \quad (6)$$

w_j is the weight of the jth indicator.

The greater the weight, the more information the indicator reflects, and the more important the indicator is in the comprehensive evaluation.

4.2.2. Using TOPSIS Method to Determine Comprehensive Scores

(a) Build the standardized matrix of weight: Calculate the standardized value "z_{ij}" of weight and build the standardized matrix of weight. The formula is shown in Equation (6):

$$z_{ij} = w_j x_{ij} \quad (7)$$

(b) Determine the ideal solution and the anti-ideal solution, the formulas are shown in Equations (7) and (8):

$$z_j^+ = \{\max Z_{ij}\} \quad (8)$$

$$z_j^- = \{\min Z_{ij}\} \quad (9)$$

Z_i^+ is the ideal solution;
Z_i^- is the anti-ideal solution.

(c) Calculate distance scale: Distance scale calculated by Euclidean distance is the distance between each objective and the ideal solution or the anti-ideal solution. The calculation formulas are shown in Equations (9) and (10).

$$D_i^+ = \sqrt{\sum_{j=1}^{m} (Z_{ij} - Z_j^+)^2} \quad (10)$$

$$D_i^- = \sqrt{\sum_{j=1}^{m} (Z_{ij} - Z_j^-)^2} \quad (11)$$

D_i^+ is the distance between the objective and the ideal solution Z^+;
D_i^- is the distance between the objective and the anti-ideal solution Z^-.

(d) Calculate the closeness degree of the ideal solution. The calculation formula is shown in Equation (11):

$$C_i = \frac{D_i^-}{D_i^- + D_i^+} \quad (12)$$

Evaluation subjects are ranked based on the size of the C_i value, and proximity is used to determine the composite score and to provide the evaluation results. In this study, the entropy-weighted TOPSIS method is employed to assess the vulnerability of the disaster-bearing body and the risk of disaster-causing factors. The higher the value of C_i, the better the result. Therefore, according to the definition, the higher the C_i value, the lower the vulnerability of the disaster-bearing body, and the lower the risk of disaster-causing factors.

4.3. *Weight Rank Sum Ratio (WRSR)*

The rank sum ratio (RSR) is a commonly used evaluation model. It differs from other evaluation models in that it incorporates secondary correction during the calculation process, resulting in improved reliability in its practical applications. In the comprehensive evaluation, the value of the rank sum ratio can contain information about all evaluation indicators, and a larger RSR value indicates a better evaluation result. The advantages of this evaluation method are mainly that it is insensitive to outliers and that it can be graded for evaluation objects. The WRSR method was employed to analyze the stability of the disaster-incubating environment. The calculation steps and formulas are shown in Equations (12)–(14).

$$RSR_i = \frac{1}{mn}\sum_{j=1}^{m} R_{ij} \tag{13}$$

$$WRSR_i = \frac{1}{n}\sum_{j=1}^{m} w_j R_{ij} \tag{14}$$

$$\sum_{j=1}^{m} w_j = 1 \tag{15}$$

$i = 1, 2, \cdots, n$
$j = 1, 2, \cdots, m$
n is the number of evaluated objects
m is the number of indicators
R_{ij} is the rank of the ith row and jth column elements
w_j is the weight of the jth evaluation indicator

The rank of each indicator becomes particularly important when weighting the results of the RSR. According to the "Guideline for risk assessment and zoning of SSD", the evaluation method for the stability of SSDs is divided into four levels of land use stability based on SSDs, with four levels being the highest stability (rank = 4), as shown in Table 4 [62]. Meanwhile, according to Gornitz and Hammar-Klose and Thierer, the stability of natural environment coastlines is divided into five levels, ranging from very low stability (rank = 1) to very high stability (rank = 5), as shown in Table 5 [63,64]. In this study, the disaster-incubating environment indicator weights will be determined based on their ranks.

Table 4. Stability levels of different land use types.

Land Use Type	Stability Level
Farmland	1
Woodland	1
Grassland	1
Water bodies	1
Urban, rural, industrial, and residential land	4
Unutilized land area	1

Table 5. Stability levels of different shoreline types.

Shoreline Type	Stability Level
Estuary shoreline	3
Bedrock shoreline	5
Sandy shoreline	2
Biological shoreline	4
Muddy shoreline	2
The reclamation and aquaculture shoreline	2
The docks, artificial shorelines, and protective embankments	4

The two methods were not combined arbitrarily but selected based on the characteristics of the indicator datasets. The entropy weighted TOPSIS method was applied to the dimensions of disaster-bearing bodies and disaster-causing factors because their associated indicators are numerical and heterogeneous. The weight of indicators needs to be obtained through the entropy weight method, and then the ranking results can be obtained through

TOPSIS. By contrast, for the disaster-incubating environment dimension, indicators were based on the length of different shoreline types and the area of different land use categories. The weights of these indicators are obtained through the authoritative stability scores in the existing literature, which are not suitable for obtaining using the entropy weight method. Furthermore, shoreline and land use types vary significantly across cities, and the evaluation in this dimension emphasizes spatial composition rather than numerical magnitude. Therefore, a more suitable WRSR method was adopted, and ideal-type cities serve as benchmarks to evaluate relative stability across coastal cities.

5. Results

Firstly, the stability of the disaster-incubating environment (S), the risk of disaster-causing factors (R), and the vulnerability of the disaster-bearing bodies (V) are analyzed, and the results are obtained separately. Subsequently, employing the H-E-V framework commonly utilized in flood risk assessments, the formula Flood Risk = Hazard (H) × Exposure (E) × Vulnerability (V) is adapted to compute the capability for responding to SSDs. To facilitate the analysis and observation, the analytical results concerning the disaster-incubating environment, the disaster-bearing body, and the disaster-causing factor were multiplied by 10, respectively. This adjustment facilitates the generation and interpretation shown in Figure 5, which visually depicts the computed disaster response capability. Moreover, the weights of the indicators obtained through the entropy weighting method are displayed in Table 6, and the scores of the evaluation of the response capability to SSDs in coastal cities of China are shown in Table 7. The visualization maps of vulnerability of disaster-bearing bodies, risk of disaster-causing factors, and stability of disaster-prone environments levels based on storm surges are displayed in Figures 6a, 6b and 6c, respectively. The visualization map of storm surge response capability levels for coastal cities in China is displayed in Figure 6d.

Table 6. Entropy-Based Weights of Disaster Response Capability Indicators.

Indicators of the Vulnerability of Disaster-Bearing Bodies.	Weights	Indicators of the Risk of Disaster-Causing Factors	Weights
Number of tidal stations	5.42%	Annual precipitation	0.9789
Mobile phone subscribers	6.38%	Maximum wind speed	0.9912
Expenditure on disaster prevention and emergency management	9.63%	Days with wind speeds exceeding 10.8 m/s	0.9934
Per capita expenditure on education	3.54%	The annual average of the highest tide height	0.9780
Coastline coefficient	1.56%	Maximum tidal height	0.9717
Relative population density (within 10 km of the coastline)	13.10%	Average altitude	0.8942
Proportion of population aged 0~14	3.38%	Average slope	0.9382
Proportion of population aged 65 and above	2.00%		
Area of mariculture	2.95%		
Road mileage per unit land area	8.17%		
Length of drainage pipelines	4.74%		
Fishery output value	11.75%		
Proportion of the primary industry at the district (county) level including the coastline	6.42%		
Proportion of tourist receipts	3.18%		
Number of hospitals per 10,000 people	5.67%		
Number of beds per 10,000 people	1.29%		
Number of doctors per 10,000 people	2.15%		
Total highway passenger traffic volume	1.20%		
Total highway freight traffic volume	0.31%		
Number of shelters	0.46%		

Table 6. *Cont.*

Indicators of the Vulnerability of Disaster-Bearing Bodies.	Weights	Indicators of the Risk of Disaster-Causing Factors	Weights
Proportion of employees covered by basic medical insurance	1.54%		
Per capita disposable income of urban permanent residents	1.64%		
Per capita disposable income of rural permanent residents	0.40%		
Proportion of secondary industry	1.66%		
Proportion of tertiary industry	0.83%		
Insurance penetration	0.65%		

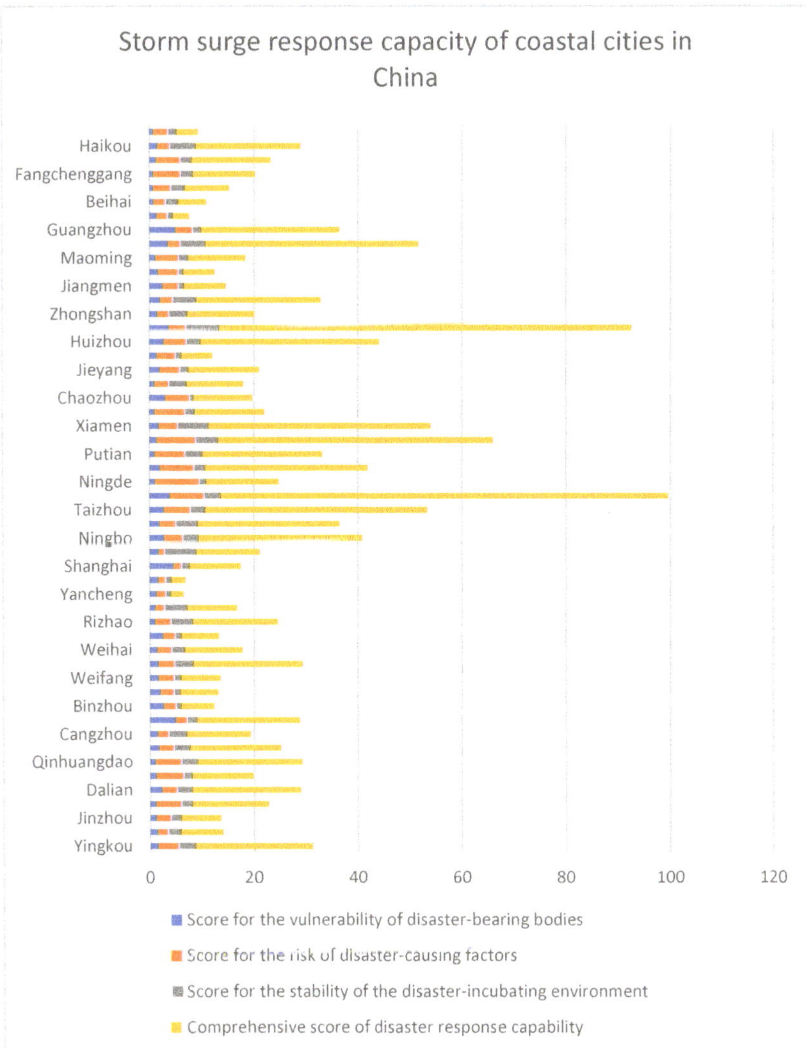

Figure 5. Storm surge response capabilities of coastal cities in China.

Table 7. Score of the response capability evaluation to SSDs in Chinese coastal cities.

City	Vulnerability Scores for Disaster-Bearing Bodies	Score for the Risk of Disaster-Causing Factors	Score for Stability of Disaster-Incubating Environment	Comprehensive Score of DRC
Yingkou	1.83	3.87	3.20	22.64
Panjin	1.71	1.96	2.46	8.24
Jinzhou	1.48	2.72	1.95	7.86
Huludao	1.43	4.77	2.15	14.68
Dalian	2.64	2.77	2.87	20.97
Dandong	1.46	5.23	1.57	11.98
Qinhuangdao	1.37	4.87	3.04	20.28
Tangshan	2.12	2.65	3.12	17.52
Cangzhou	1.78	1.98	3.53	12.44
Tianjin	5.28	2.03	1.86	19.91
Binzhou	2.97	2.21	1.00	6.57
Dongying	2.33	2.45	1.30	7.40
Weifang	1.98	2.84	1.38	7.74
Yantai	1.95	3.05	3.54	21.07
Weihai	1.7	2.68	2.49	11.35
Qingdao	2.84	2.23	1.17	7.39
Rizhao	1.31	2.95	4.22	16.32
Lianyungang	1.4	1.6	4.38	9.82
Yancheng	1.62	1.66	0.97	2.62
Nantong	1.91	1.32	1.15	2.89
Shanghai	4.91	1.37	1.50	10.10
Jiaxing	2.02	1.02	6.01	12.38
Ningbo	3.08	3.45	2.98	31.64
Zhoushan	2.18	3.01	4.19	27.47
Taizhou	3.03	4.92	2.88	42.86
Wenzhou	4.26	6.35	3.18	86.13
Ningde	1.36	8.38	1.23	14.07
Fuzhou	2.31	6.33	2.15	31.44
Putian	1.23	5.74	3.29	23.20
Quanzhou	1.67	7.3	4.34	52.96
Xiamen	2.09	3.48	5.89	42.86
Zhangzhou	1.28	5.69	1.85	13.48
Chaozhou	3.37	4.47	0.76	11.49
Shantou	1.24	2.63	3.41	11.12
Jieyang	2.26	3.75	1.61	13.69
Shanwei	1.65	3.48	1.08	6.22
Huizhou	3.06	4.12	2.74	34.57
Shenzhen	4.01	3.05	6.49	79.41
Zhongshan	1.78	2.12	3.45	13.03
Zhuhai	2.36	2.27	4.49	24.05
Jiangmen	2.79	2.9	1.02	8.29
Yangjiang	1.89	3.8	0.88	6.29
Maoming	1.44	4.35	1.78	11.17
Dongguan	3.81	2.25	4.79	41.07
Guangzhou	5.27	3.14	1.62	26.73
Zhanjiang	1.67	1.94	1.02	3.31
Beihai	1.06	2.15	2.45	5.58
Qinzhou	1	3.23	2.71	8.74
Fangchenggang	1.01	5.09	2.35	12.08
Sanya	1.56	4.48	2.19	15.29
Haikou	1.8	2.28	4.93	20.25
Danzhou	1	2.69	1.61	4.34

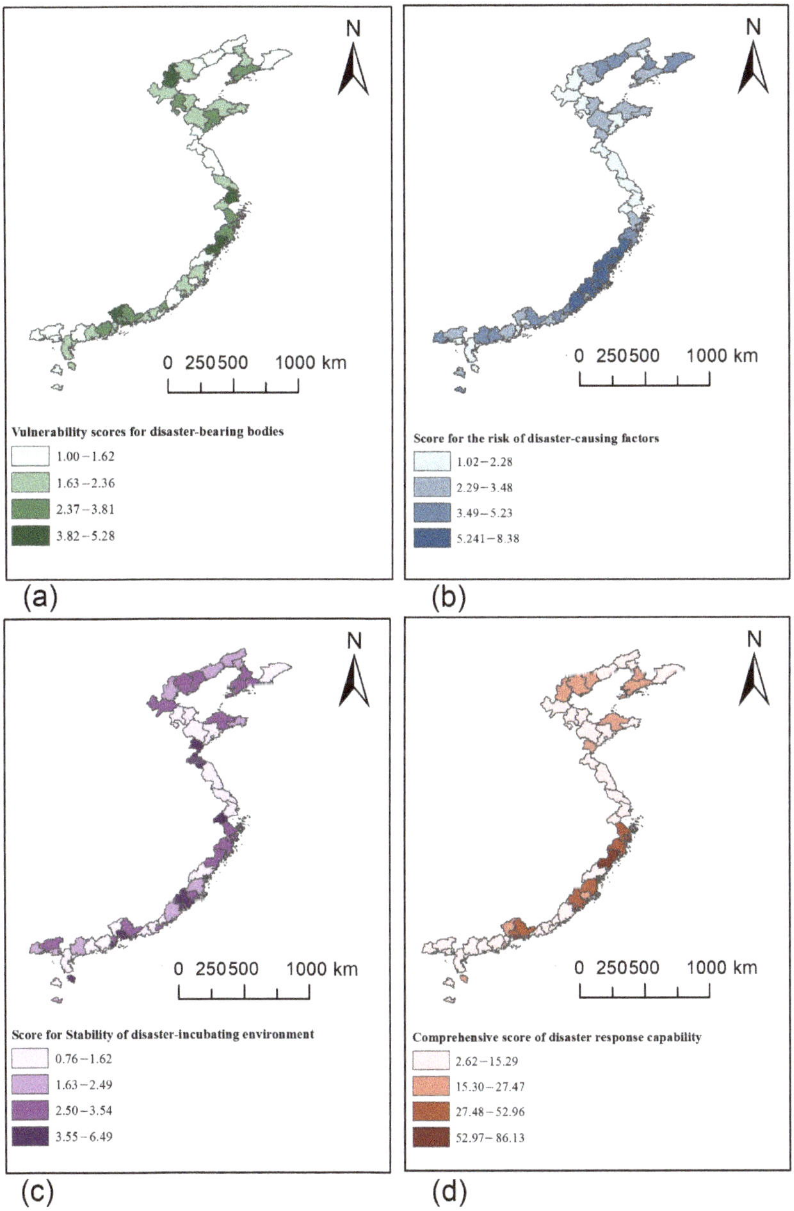

Figure 6. (**a**) Thermal map of vulnerability score of disaster bearing body; (**b**) thermal map of risk score of disaster causing factors; (**c**) thermal map of stability score of disaster- incubating environment; (**d**) thermal map of comprehensive score for DRC.

6. Discussion

6.1. Interpretation of Results

The present analysis focuses on the evaluation of SSD risk and response capabilities across various cities. Scores for the vulnerability of disaster-bearing bodies, the risk of disaster-causing factors, the stability of the disaster-incubating environment, and the overall DRC were obtained. By examining the comprehensive DRC scores for 52 Chinese coastal cities, the various strengths and weaknesses of the DRC of different coastal cities can be more easily identified, which is of great significance for SSD preparedness planning in coastal cities.

According to the study results, Chaozhou scored the lowest in assessing the stability of the disaster-incubating environment, with a score of 0.76. By contrast, Shenzhen scored the highest in the same assessment, scoring 6.49. Shenzhen is a highly urbanized and economically developed prefecture-level city whose coastline consists mainly of rocky and artificial shorelines. Shanghai, an economically developed city, does not have a very high disaster-incubating environmental stability score because a portion of its coastal area is used for agricultural development. Lower scores in this category suggest an environment prone to conditions that can foster disasters, such as unstable geological conditions. For these cities, effective urban planning and environmental management are critical. Storm surge defense strategies for these cities should concentrate on coastal and land use planning while also increasing the resilience of the disaster-bearing bodies to minimize vulnerability. Strategies should include sustainable land use practices, regular environmental assessments, and proactive measures to address and to mitigate potential environmental hazards. Cities with higher scores, indicating a stable environment, should stay active. Continued efforts in environmental monitoring and sustainable practices are essential to maintain stability and to prevent the incubation of disaster-prone conditions.

Based on the scores in the vulnerability component, it is understood that Qinzhou and Danzhou tied for the lowest score of 1, while Guangzhou had the highest score of 5.27. Tianjin, Guangzhou, Shanghai, Wenzhou, and Shenzhen ranked in the top five scores, indicating that these five cities have low vulnerability. Putian, Beihai, Fangchenggang, Qinzhou, and Danzhou are highly vulnerable. This variation indicates a pronounced disparity in how different cities are equipped to handle disasters. Cities with lower scores in this category will likely face more significant challenges due to weaker infrastructure, higher population densities, or insufficient preparedness measures. These cities must prioritize enhancing their resilience through targeted investments in infrastructure fortification, stringent building codes, and comprehensive disaster preparedness programs.

Conversely, cities with higher scores demonstrate a higher degree of resilience. Although better prepared, these cities should continue to innovate and strengthen their existing measures to maintain and improve their preparedness levels. Cities with low vulnerability to hazards are economically developed, while cities with high vulnerability are those with smaller economies and populations along China's coasts. The degree of vulnerability of the disaster-bearers is reflected in the link between the degree of development of the cities. Indeed, the reality is that developed cities are more resilient to disasters.

The risk of disaster-causing factors also shows considerable variation, with scores ranging from 1.02 to 8.38. This wide range underscores the differing levels of exposure to natural or artificial hazards in the cities. Jiaxing has the lowest score of 1.02, while Ningde has the highest score of 8.38. Ningde, Qinzhou, Wenzhou, Fuzhou, and Putian had the top five scores, indicating that their riskiness of disaster-causing factors was low. Meanwhile, according to the results, Yancheng, Lianyungang, Shanghai, Nantong, and Jiaxing have a higher risk of disaster-causing factors. High-risk cities are likely to be more prone to events such as geological instability, extreme weather conditions, or industrial accidents. For these cities, the focus should be on enhancing early warning systems, implementing robust risk mitigation strategies, and maintaining continuous monitoring of potential hazards. On the other hand, cities with lower risk scores must maintain their current safety measures while staying vigilant to new or evolving threats that could impact their risk profiles.

The comprehensive DRC score ranges from 2.62 to 86.12. Yancheng received the lowest score, and Wenzhou received the highest score. The top five cities were Wenzhou, Shenzhen, Quanzhou, Taizhou, and Xiamen, while the bottom five were Shanwei, Beihai, Danzhou, Zhanjiang, Nantong, and Yancheng. Notably, Shanghai scored only 10.09, indicating that developed cities do not necessarily possess more excellent capability to withstand storm surges. A high composite score in environment and climate for coastal cities complements the city's disaster resilience conditions. For example, Shanghai is a developed city. The geographical location, climatic conditions, natural environment, and urban layout are not conducive to resisting storm surges. Although Shanghai has a well-developed disaster prevention infrastructure and adequate staffing and funding for commissioners, the economic losses will be more severe than in other cities.

In summary, analyzing the maximum and minimum values across the four categories provides critical insights into the disaster preparedness and response capabilities of different cities. The disparities highlighted in the vulnerability, risk, environmental stability, and response capability underscore the need for tailored strategies to address the specific challenges faced by each city. For cities with high vulnerability and risk scores, prioritizing resilience and mitigation measures is essential. By contrast, cities with stable environments and robust response capabilities should focus on maintaining and enhancing their existing measures. This comprehensive approach ensures that all cities, regardless of their current scores, are better prepared to handle future disasters and to protect their populations effectively.

6.2. Uncertainty Analysis

While the model demonstrates significant potential in its current form, and the multi-indicator integrated assessment method combining entropy weighted TOPSIS and WRSR was adopted, it also has uncertainties that warrant discussion and refinement in future applications.

Firstly, a primary limitation of the evaluation results is its reliance on data from the year 2020. This temporal constraint may limit the model's ability to capture changes. The reason for using the 2023 remote sensing data is that field surveys were conducted in the same year, making it the most accurate and consistent choice for shoreline classification. Consequently, the results derived from the 2020 dataset might not fully reflect the present-day status or emerging trends in coastal city resilience. This temporal mismatch highlights the need for regular updates to the underlying data to ensure that the model remains reflective of contemporary realities. Although this temporal mismatch may limit the model's ability to fully reflect real-time conditions or recent developments in coastal resilience, the analytical framework itself remains valid and adaptable. As more consistent and updated datasets become available, the model can be recalibrated accordingly. This flexibility allows for its continued application in monitoring and assessing storm surge response capacity over time.

Secondly, although the entropy weighting method was used to improve objectivity, the selection and weighting of indicators still involve subjectivity. The entropy method focuses on the variability of data but may not fully reflect the actual importance in disaster response. The WRSR method helps reduce the effect of extreme values, but its ranking of environmental stability is based on empirical classification, which may oversimplify complex natural processes. In addition, the method assumes that indicators are independent. In reality, urban density, infrastructure, and economic power are often closely related.

This may lead to redundancy and affect the accurate assessment of the vulnerability of the disaster-bearing body.

Finally, the current models are based on cross-sectional analysis and do not reflect changes in storm surge risk and disaster response capacity over time. As climate trends, urban development, and governance continue to change, the results of this study should be seen as a static assessment at a given time rather than a precise forecast of future conditions.

Despite this limitation, the model is inherently designed with flexibility to integrate updated datasets as they become available. This design feature enables the model to remain dynamic and adaptable, allowing for the ongoing evaluation of coastal cities' resilience under changing conditions. Such updates would also provide opportunities for longitudinal analyses to track resilience improvements or deteriorations over time. Methodologically, the model employs a robust framework that facilitates seamless incorporation of new datasets without requiring substantial modifications to its structure. By recalibrating input parameters based on updated information, the model can maintain its validity and reliability in assessing resilience across different time periods. This capability underscores its potential to support evidence-based policymaking and adaptive planning in coastal cities.

In conclusion, although there are current challenges of data temporality, methodological subjectivity, and dynamic adaptation, the storm surge resilience evaluation model's design allows for iterative improvements through data updates. This adaptability ensures its long-term utility in evaluating and enhancing the resilience of coastal cities, thereby supporting sustainable urban development and disaster risk reduction efforts.

7. Conclusions

From the perspective of the disaster response capability related to SSDs, this paper constructed an evaluation indicator system for the storm surge response capacities of 52 coastal cities in China through the H-E-V framework, a functional system for disaster systems, and PPRR theory. The system used in this study consists of three dimensions, which are stability of the disaster-incubating environment (S), the risk of disaster-causing factors (R), and the vulnerability of the disaster-bearing bodies (V). Meanwhile, through government surveys and consultation of statistical yearbooks, a comprehensive evaluation of these cities was conducted using the entropy weighted TOPSIS method and weight rank sum ratio, ultimately producing scores that display their respective SSD response capability levels.

This study has filled a gap in the research on the urban level of DRC and has discussed many aspects of the SSD resilience among coastal cities. In the evaluation of the disaster response capabilities of storm surge, Tianjin, Ningde, and Shenzhen demonstrated strong performance across three critical dimensions: the vulnerability of disaster-bearing entities, the risk associated with disaster-causing factors, and the stability of environments prone to disaster incubation. Each city showed significant strengths in mitigating potential threats and maintaining stability in the face of potential hazards. Conversely, Dandong, which shares the same ranking as Qinzhou, Jiaxing, and Chaozhou, exhibited weaker performance in these crucial areas. These cities faced challenges in minimizing vulnerabilities, managing disaster risks, and ensuring environmental stability, which may necessitate targeted interventions to enhance their disaster resilience. Moreover, Yancheng has the weakest storm surge response capability, while Wenzhou has the strongest. The results not only enable coastal cities to compare and learn from each other more easily, from each other's strengths and weaknesses when it comes to disaster response, but they provide a reference and basis for improving the disaster response capacities and planning mechanisms of coastal cities.

Author Contributions: All authors contributed to the study conception and design. Material preparation, data collection, and analysis were performed by S.C. and L.Z. The first draft of the manuscript was written by S.C., and all authors commented on previous versions of the manuscript. All authors have read and agreed to the published version of the manuscript.

Funding: The authors declare that this study received funding from "Research and Application of Key Technologies for Campus Zero Carbonization Transformation of New Energy Systems" (Project number is 6122020002 and Contract number is Z612302004) by China Yangtze Power Co., Ltd. (CYPC) and Three Gorges Electric Energy Co., Ltd. The funder was not involved in the study design, collection, analysis, interpretation of data, the writing of this article or the decision to submit it for publication.

Data Availability Statement: The raw data supporting the conclusions of this article will be made available by the authors on request.

Acknowledgments: The authors are grateful for the data support provided by the APSEC Center of Tianjin University. At the same time, the financial support for the "Research and Application of Key Technologies for Campus Zero Carbonization Transformation of New Energy Systems" project is appreciated.

Conflicts of Interest: The authors have no relevant financial or non-financial interests to disclose.

References

1. Xin, L.; Yinyin, W. Study on disaster of storm surges and prevention measures in China. In Proceedings of the 2010 2nd IEEE International Conference on Information and Financial Engineering, Chongqing, China, 17–19 September 2010; IEEE: Piscataway, NJ, USA, 2010. [CrossRef]
2. Lin, J.; Xu, Y.; Hou, Y.; Xue, X. Spatio-temporal distribution, composition and influencing factors of economic losses from storm surge disasters: An empirical study from China (2007–2016). *Int. J. Disaster Risk Reduct.* **2023**, *90*, 103669. [CrossRef]
3. Shi, X.; Zhang, Y.; Liu, S.; Yang, L.; Yu, L.; Zhang, Y.; Jia, N.; Tian, Z. Mapping Storm Surge Risk at County Level in Coastal Areas of China. *J. Mar. Sci. Eng.* **2025**, *13*, 427. [CrossRef]
4. Tan, Y.; Zhang, W.; Feng, X.; Guo, Y.; Hoitink, A.J.F. Storm surge variability and prediction from ENSO and tropical cyclones. *Environ. Res. Lett.* **2023**, *18*, 024016. [CrossRef]
5. Meng, D.; Liu, Y.; Wang, Z.; Yang, X.; Liu, X.; Zhang, J.; Gao, K. Decreasing Vulnerability of Storm Surge Disasters in Coastal Cities of China over the Past 30 Years. *J. Mar. Sci. Eng.* **2023**, *11*, 128. [CrossRef]
6. Yi, X.; Sheng, K.; Wang, Y.; Wang, S. Can economic development alleviate storm surge disaster losses in coastal areas of China? *Mar. Policy* **2021**, *129*, 104531. [CrossRef]
7. Xianwu, S.; Jufei, Q.; Bingrui, C.; Xiaojie, Z.; Haoshuang, G.; Jun, W.; Zhuyuan, B. Storm surge risk assessment method for a coastal county in China: Case study of Jinshan District, Shanghai. *Stoch. Environ. Res. Risk Assess.* **2020**, *34*, 627–640. [CrossRef]
8. Huang, X.; Jin, H.; Bai, H. Vulnerability assessment of China's coastal cities based on DEA cross-efficiency model. *Int. J. Disaster Risk Reduct.* **2019**, *36*, 101091. [CrossRef]
9. Blankespoor, B.; Dasgupta, S.; Lange, G.-M. Mangroves as a protection from storm surges in a changing climate. *Ambio* **2016**, *46*, 478–491. [CrossRef] [PubMed]
10. Nguyen, X.H.; Nguyen, X.T.; Nguyen, H.H.; Tran, T.T.; Le, D.Q. Assessment of storm surge risk in aquaculture in the Northern coastal area of Vietnam. *Vietnam. J. Sci. Technol. Eng.* **2018**, *60*, 89–94. [CrossRef]
11. Xu, H.; Xu, K.; Bin, L.; Lian, J.; Ma, C. Joint Risk of Rainfall and Storm Surges during Typhoons in a Coastal City of Haidian Island, China. *Int. J. Environ. Res. Public Health* **2018**, *15*, 1377. [CrossRef]
12. Peng, Z.; Zhang, A.; Bingrui, S.; Wang, Z.; Bai, Y. Design Principles and Construction Process of a Comprehensive Evaluation Index System. *Sci. Res. Manag.* **2020**, *38*, 209–215. (In Chinese)
13. Lu, X.; Han, Z. Emergency management in China: Towards a comprehensive model? *J. Risk Res.* **2018**, *22*, 1425–1442. [CrossRef]
14. Meyer, V.; Scheuer, S.; Haase, D. A multicriteria approach for flood risk mapping exemplified at the Mulde river, Germany. *Nat. Hazards* **2009**, *48*, 17–39. [CrossRef]
15. Haynes, H.; Haynes, R.; Pender, G. Integrating socio-economic analysis into decision-support methodology for flood risk management at the development scale (Scotland). *Water Environ. J.* **2008**, *22*, 117–124. [CrossRef]
16. Giupponi, C.; Mojtahed, V.; Gain, A.K.; Biscaro, C.; Balbi, S. Integrated Risk Assessment of Water-Related Disasters. In *Hydro-Meteorological Hazards, Risks and Disasters*; Elsevier: Amsterdam, The Netherlands; pp. 163–200. [CrossRef]

17. Pachauri, R.; Meyer, L.; Hallegatte, S.; Hegerl, G.; Van Ypersele, J.P.; Brinkman, S. Climate Change 2014 Synthesis Report [Internet]. IPCC. Gian-Kasper Plattner. 2014. Available online: https://www.ipcc.ch/site/assets/uploads/2018/02/SYR_AR5_FINAL_full.pdf (accessed on 29 June 2025).
18. Shi, P. On the Theory and Practice of Disaster Research. *J. Nanjing Univ.* **1991**, *11*, 37–42. (In Chinese)
19. Shi, P. Theory and Practice on Disaster System Research in a Fourth Time. *J. Nat. Disasters* **2005**, *06*, 1–7. (In Chinese)
20. Mileti, D. *Disasters by Design a Reassessment of Natural Hazards in the United States*; J.H. Press: Washington, DC, USA, 1999. [CrossRef]
21. Wisner, B.; Blaikie, P.; Cannon, T.; Davis, I. *At Risk: Natural Hazards, People's Vulnerability and Disasters*, 2nd ed.; Routledge: London, UK, 2004.
22. De Smet, H.; Schreurs, B.; Leysen, J. The Response Phase of the Disaster Management Life Cycle Revisited Within the Context of "Disasters Out of the Box. *J. Homel. Secur. Emerg. Manag.* **2015**, *12*, 319–350. [CrossRef]
23. Anand, A.; Buhagiar, K.; Kozachenko, E.; Parameswar, N. Exploring the role of knowledge management in contexts of crisis: A synthesis and way forward. *Int. J. Organ. Anal.* **2022**, *31*, 2953–2978. [CrossRef]
24. Li, X. Study on Assessment of Urban Composite Disaster-Bearing Capacity. *J. Xi'an Univ. Archit. Technol.* **2012**, *4*, 489–494. (In Chinese)
25. Wang, J.; Gong, Q.; Yu, Y.; Yuan, S.; Chen, J.; Huang, G. Evaluation of Comprehensive Disaster-Bearing Capacity of Urban Natural Disasters in the Guangdong-Hong Kong-Macao Greater Bay Area. *Geogr. Res.* **2020**, *39*, 2189–2199. (In Chinese)
26. Zerger, A.; Smith, D.I.; Hunter, G.J.; Jones, S.D. Riding the storm: A comparison of uncertainty modelling techniques for storm surge risk management. *Appl. Geogr.* **2002**, *22*, 307–330. [CrossRef]
27. Torres-Vera, M.A.; Antonio Canas, J. A lifeline vulnerability study in Barcelona, Spain. *Reliab. Eng. Syst. Saf.* **2003**, *80*, 205–210. [CrossRef]
28. Zhang, M.; Liu, Y.; Yuan, Y. Research on the Assessment of Urban Comprehensive Disaster-Bearing Capacity Based on Variable Fuzzy Clustering. *J. Catastrophology* **2012**, *27*, 135–138. (In Chinese)
29. Mahamoud, A.; Gzam, M.; Mohamed, N.A.; Soulé, H.H.; Montacer, M. A preliminary assessment of coastal vulnerability for Ngazidja Island, Comoros Archipelago, Western Indian Ocean. *Environ. Earth Sci.* **2022**, *81*, 44. [CrossRef]
30. Bormann, H.; Kebschull, J.; Gaslikova, L.; Weisse, R. Model-based assessment of climate change impact on inland flood risk at the German North Sea coast caused by compounding storm tide and precipitation events. *Nat. Hazards Earth Syst. Sci.* **2024**, *24*, 2559–2576. [CrossRef]
31. Qian, X.; Hwang, S.; Son, S. A Study on Key Determinants in Enhancing Storm Surges Along the Coast: Interplay Between Tropical Cyclone Motion and Coastal Geometry. *J. Geophys. Research. Ocean.* **2024**, *129*, e2023JC020400. [CrossRef]
32. Chen, B.; He, J.; He, Z.; Li, L.; Chen, Q.; Li, F.; Chu, D.; Cao, Z.; Yang, X. Potential impacts of storm surge-induced flooding based on refined exposure estimation: A case study in Zhoushan island, China. *Geomat. Nat. Hazards Risk* **2023**, *14*, 2232080. [CrossRef]
33. Liu, X.; Liu, Y.; Wang, Z.; Yang, X.; Zeng, X.; Meng, D. Comprehensive Assessment of Vulnerability to Storm Surges in Coastal China: Towards a Prefecture-Level Cities Perspective. *Remote Sens.* **2023**, *15*, 4828. [CrossRef]
34. Islam, M.F.; Bhattacharya, B.; Popescu, I. Flood risk assessment due to cyclone-induced dike breaching in coastal areas of Bangladesh. *Nat. Hazards Earth Syst. Sci.* **2019**, *19*, 353–368. [CrossRef]
35. Liu, Y.; Lu, C.; Yang, X.; Wang, Z.; Liu, B. Fine-Scale Coastal Storm Surge Disaster Vulnerability and Risk Assessment Model: A Case Study of Laizhou Bay, China. *Remote Sens.* **2020**, *12*, 1301. [CrossRef]
36. Mohanty, M.P.; Sherly, M.A.; Ghosh, S.; Karmakar, S. Tide-rainfall flood quotient: An incisive measure of comprehending a region's response to storm-tide and pluvial flooding. *Environ. Res. Lett.* **2020**, *15*, 064029. [CrossRef]
37. Yu, H.; Shen, Y.; Kelly, R.M.; Qi, X.; Wu, K.; Li, S.; Yu, H.; Bao, X. Trends in social vulnerability to storm surges in Shenzhen, China. *Nat. Hazards Earth Syst. Sci.* **2020**, *20*, 2447–2462. [CrossRef]
38. Choudhury, M.-U.-I.; Haque, C.E.; Hostetler, G. Transformative learning and community resilience to cyclones and storm surges: The case of coastal communities in Bangladesh. *Int. J. Disaster Risk Reduct.* **2021**, *55*, 102063. [CrossRef]
39. Kantamaneni, K.; Panneer, S.; Sudha Rani, N.N.V.; Palaniswamy, U.; Bhat, L.D.; Jimenez-Bescos, C.; Rice, L. Impact of Coastal Disasters on Women in Urban Slums: A New Index. *Sustainability* **2022**, *14*, 3472. [CrossRef]
40. Bernard, A.; Long, N.; Becker, M.; Khan, J.; Fanchette, S. Bangladesh's vulnerability to cyclonic coastal flooding. *Nat. Hazards Earth Syst. Sci.* **2022**, *22*, 729–751. [CrossRef]
41. Ghosh, S.; Mistri, B. Cyclone-induced coastal vulnerability, livelihood challenges and mitigation measures of Matla–Bidya inter-estuarine area, Indian Sundarban. *Nat. Hazards* **2023**, *116*, 3857–3878. [CrossRef] [PubMed]
42. Zhao, Y.; Liu, Q.; Huang, R.; Pan, H.; Xu, M. Recent Evolution of Coastal Tidal Flats and the Impacts of Intensified Human Activities in the Modern Radial Sand Ridges, East China. *Int. J. Environ. Res. Public Health* **2020**, *17*, 3191. [CrossRef]

43. Cui, Y.; Yan, F.; He, B.; Ju, C.; Su, F. Characteristics of shoreline changes around the South China Sea from 1980 to 2020. *Front. Mar. Sci.* **2022**, *9*, 1005284. [CrossRef]
44. Gao, Y.; Wang, H.; Liu, G.M.; Sun, X.Y.; Fei, X.Y.; Wang, P.T.; Lv, T.T.; Xue, Z.S.; He, Y.W. Risk assessment of tropical storm surges for coastal regions of China. *J. Geophys. Res. Atmos.* **2014**, *119*, 5364–5374. [CrossRef]
45. Zhang, C.; Yin, K.; Shi, X.; Yan, X. Risk assessment for typhoon storm surges using geospatial techniques for the coastal areas of Guangdong, China. *Ocean. Coast. Manag.* **2021**, *213*, 105880. [CrossRef]
46. Wang, Y.; Chen, X.; Wang, L. Differential Semi-Quantitative Urban Risk Assessment of Storm Surge Inundation. *ISPRS Ann. Photogramm. Remote Sens. Spat. Inf. Sci.* **2022**, *10*, 177–185. [CrossRef]
47. Li, K.; Li, G.S. Vulnerability assessment of storm surges in the coastal area of Guangdong Province. *Nat. Hazards Earth Syst. Sci.* **2011**, *11*, 2003–2010. [CrossRef]
48. Li, K.; Li, G.S. Risk assessment on storm surges in the coastal area of Guangdong Province. *Nat. Hazards* **2013**, *68*, 1129–1139. [CrossRef]
49. Hatzikyriakou, A.; Lin, N. Impact of performance interdependencies on structural vulnerability: A systems perspective of storm surge risk to coastal residential communities. *Reliab. Eng. Syst. Saf.* **2016**, *158*, 106–116. [CrossRef]
50. Abdrabo, K.I.; Kantoush, S.A.; Esmaiel, A.; Saber, M.; Sumi, T.; Almamari, M.; Elboshy, B.; Ghoniem, S. An integrated indicator-based approach for constructing an urban flood vulnerability index as an urban decision-making tool using the PCA and AHP techniques: A case study of Alexandria, Egypt. *Urban Clim.* **2023**, *48*, 101426. [CrossRef]
51. Bathi, J.; Das, H. Vulnerability of Coastal Communities from Storm Surge and Flood Disasters. *Int. J. Environ. Res. Public Health* **2016**, *13*, 239. [CrossRef] [PubMed]
52. Pastor-Escuredo, D.; Morales-Guzmán, A.; Torres-Fernández, Y.; Bauer, J.-M.; Wadhwa, A.; Castro-Correa, C.; Romanoff, L.; Lee, J.G.; Rutherford, A.; Frias-Martinez, V.; et al. Flooding through the lens of mobile phone activity. In Proceedings of the IEEE Global Humanitarian Technology Conference (GHTC 2014), San Jose, CA, USA, 10–13 October 2014; IEEE: Piscataway, NJ, USA, 2014. [CrossRef]
53. Toya, H.; Skidmore, M. Cellular Telephones and Natural Disaster Vulnerability. *Sustainability* **2018**, *10*, 2970. [CrossRef]
54. Guo, J.; Song, W.; Zheng, H. Assessment of Comprehensive Fishery Loss due to Storm Surge Based on Modified Order Relation Weighting Model. *J. Coast. Res.* **2020**, *104*, 541–545. [CrossRef]
55. Wang, L.; Hu, S.; Han, J.-C.; Hu, P.; Yu, X. Mitigation measures of storm surge inundation at an onshore aquaculture farm. *J. Hydrol.* **2024**, *638*, 131443. [CrossRef]
56. Phillips, B.F.; Pérez-Ramírez, M. (Eds.) *Climate Change Impacts on Fisheries and Aquaculture*; John Wiley & Sons, Ltd: Hoboken, NJ, USA, 2017. [CrossRef]
57. United Nations Office for Disaster Risk Reduction. Sendai Framework for Disaster Risk Reduction 2015–2030. *UNDRR*. 2015. Available online: https://www.undrr.org/publication/sendai-framework-disaster-risk-reduction-2015-2030 (accessed on 29 June 2025).
58. Disaster Resilience Scorecard for Cities | Making Cities Resilient. *Mcr2030.Undrr.org*. Available online: https://mcr2030.undrr.org/disaster-resilience-scorecard-cities (accessed on 25 August 2021).
59. APEC Integrated Urban Planning Report–Combining Disaster Resilience with Sustainability | APEC. *APEC*. 2025. Available online: https://www.apec.org/Publications/2021/03/APEC-Integrated-Urban-Planning-Report (accessed on 29 June 2025).
60. Urban Socio-Economic Survey Department of the National Bureau of Statistics. In *China Urban Statistical Yearbook 2021*; Xu, L., Ed.; China Statistics Press: Beijing, China, 2022. (In Chinese)
61. National Marine Data and Information Service. In *Tide Table*; Lu, Y., Ed.; China Ocean Press: Beijing, China, 2019; Volume 1–3. (In Chinese)
62. China Standardization Committee. *Guideline for Risk Assessment and Zoning of Storm Surge Disaster*; China Quality Inspection Press: Beijing, China, 2020; pp. 22–24. Available online: https://max.book118.com/html/2018/0830/7053131031001145.shtm (accessed on 29 June 2025). (In Chinese)
63. Gornitz, V. Global coastal hazards from future sea level rise. *Glob. Planet. Chang.* **1991**, *89*, 379–398. [CrossRef]
64. Hammar-Klose, E.S.; Thieler, E.R. *Coastal Vulnerability to Sea-Level Rise: A Preliminary Database for the U.S. Atlantic, Pacific, and Gulf of Mexico Coasts*; Data Series; U.S. Geological Survey: Reston, VA, USA, 2001. [CrossRef]

water

MDPI

Article

Seasonal Circulation Characteristics of Oceanic System in the Beibu Gulf Based on Observations and Numerical Simulations

Gongpeng Liu [1], Na Zhang [2], Yuping Yang [2] and Chenghao Wang [2,*]

1 POWERCHINA Zhongnan Engineering Corporation Limited, Changsha 410014, China; liugp01@msdi.cn
2 Tianjin Research Institute for Water Transport Engineering, Ministry of Transport, Tianjin 300456, China; sunrain79516@163.com (N.Z.); 12031081@mail.sustech.edu.cn (Y.Y.)
* Correspondence: oucwch@126.com

Abstract

The Beibu Gulf's ocean circulation system regulates regional marine ecosystems, sediment transport, and coastal geomorphology while also supporting vital economic activities. This study integrates one-year current observations from four in-situ current observation stations (B1−B4) with simulations using the Regional Ocean Modeling System (ROMS) to characterize circulation dynamics in the gulf. Observations show persistent northward subtidal currents west of Hainan Island year-round, primarily sustained by tidal-induced residual currents. These currents briefly reverse southward during strong northerly wind events, whereas subtidal currents in the northern Beibu Gulf are more wind-dependent, showing pronounced seasonal variations. Numerical results confirm that winter circulation is dominated by a basin-wide cyclonic gyre driven by northeasterly monsoons. In summer, circulation in the northern gulf is cyclonic under southeasterly winds, but turns anticyclonic when southwesterly winds prevail, indicating strong sensitivity to summer monsoon wind direction. By combining multi-station observations and numerical simulations, this study provides a systematic characterization of the seasonal circulation of the oceanic system in the Beibu Gulf, offering new insights into its dynamic mechanisms.

Keywords: Beibu Gulf; ocean circulation; monsoonal influence; seasonal variation

1. Introduction

The Beibu Gulf is a semi-enclosed shelf sea in the northwestern South China Sea (17°–22° N, 105°–110° E) spanning approximately 130,000 km^2 with a mean depth of 50 m, bordered by China and Vietnam (Figure 1). This region holds strategic importance due to its rich oil, gas, and biological resources. Despite decades of study, its complex circulation patterns remain controversial. Early efforts, such as the 1964 China–Vietnam Joint Oceanic Survey, outlined the Gulf's key hydrodynamic features, but sparse observations and conflicting model results have hindered a consensus on its circulation mechanisms [1,2].

During winter (Dec−Feb), the Beibu Gulf is dominated by strong northeasterly monsoons, forming a basin-scale cyclonic circulation. High-salinity water (>34 PSU) from the open South China Sea enters the Gulf through the central and eastern bay mouth, then flows northward along the eastern coast of Hainan Island under the influence of northeasterly winds. Near ~20° N, this flow is deflected westward by the southward western boundary current and westward flow from the Qiongzhou Strait, eventually merging into the western coastal current that continues southward along the Vietnam coast [3–5]. While this forms the framework of winter cyclonic circulation, the relative influence of tidal mixing versus wind forcing remains debated [6]. Spring (Mar–May) circulation may exhibit a gulf-scale cyclonic pattern, with an enclosed cyclonic gyre in the north and an open cyclonic gyre in the south [5]. Recent model results further suggest that the northern and southern Gulf are governed by an enclosed and an open cyclonic gyre, respectively [7]. However, a major limitation of previous research has been the lack of long-term, multi-station oceanic current observations. Most studies have relied on short-term or single-season datasets, leading to inconsistent conclusions. For instance, the Qiongzhou Strait (QS), a key exchange pathway between the South China Sea (SCS) and Beibu Gulf, was traditionally thought to flow eastward in summer and westward in winter [5,6,8], but recent current measurements and models show predominantly westward flow year-round, with brief eastward reversals in spring [7–9]. The underlying mechanisms, such as tide-induced residual current, sea-level gradients, and monsoon winds, remain poorly understood.

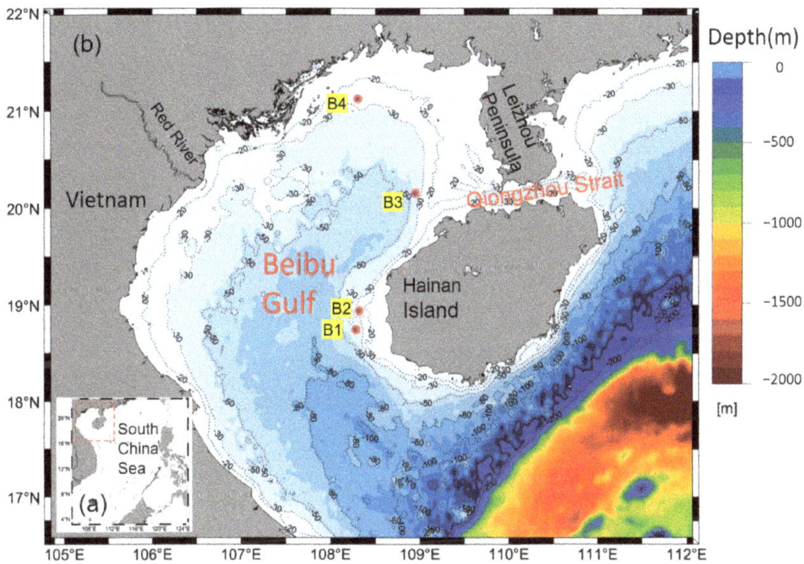

Figure 1. (**a**) Geographical location map of the Beibu Gulf; (**b**) Bathymetry and observation stations in the Beibu Gulf. B1 to B4 are four in-situ current observation stations.

Compared with winter, the circulation structure of the Beibu Gulf in summer (Jun–Aug) is more complex. Although southwesterly winds dominate, their weak intensity and variability contribute to diverse flow patterns. On the one hand, high-salinity water from the northwestern South China Sea continues to enter through the central and eastern parts of the bay mouth, but its path and extent vary under the influence of the southwesterly monsoon. On the other hand, waters along the western coast experience high temperatures and low salinity (<33 PSU) due to summer heating and freshwater runoff, forming a coastal current that flows southward from the northwestern coast [10,11].

There remains academic debate over the specific pattern of summer circulation in the Beibu Gulf. Some studies describe an anticyclonic surface circulation driven by the dominant southwesterly monsoon and enhanced river runoff along the Vietnamese coast [1,12], while others suggest a persistent cyclonic pattern shaped by wind-driven and density-driven circulation [13–15]. In recent years, researchers have employed more advanced numerical methods combined with dynamic field analyses to investigate the Gulf's circulation mechanisms. These studies reveal that circulation is affected by multiple factors, including topography, waves, typhoons, open-sea water exchange, seawater density distribution, and river discharge [16–18]. Findings also suggest that in summer, the southern Gulf is dominated by an anticyclonic eddy, while the northern Gulf features a cyclonic gyre [9].

In summary, the Beibu Gulf exhibits complex and seasonally variable circulation influenced by multiple factors. The scarcity of long-term current velocity observations has limited understanding of its circulation structure and its seasonal dynamics. This study addresses these gaps by using year-long current observations from four ADCP stations (B1–B4) along with high-resolution Regional Ocean Modeling System (ROMS) simulations.

2. Materials and Methods

2.1. Field Observations and Data Processing

We deployed four ADCP (Acoustic Doppler Current Profiler) stations (B1–B4) to investigate the characteristics of tidal and subtidal currents in the Beibu Gulf. Stations B1 and B2 were located off the western coast of Hainan Island, Station B3 lay west of the Qiongzhou Strait, and Station B4 was in the northern Beibu Gulf (Figure 1). The water depths of the four stations ranged between 30 and 40 m. Detailed information, including geographical coordinates, water depths, observation periods, and measured parameters, is summarized in Table 1. Currents were recorded every 30 min, with each value representing the average of continuous sampling over at least 100 s. The current profile cell size was set to 1 or 2 m. Observations spanned multiple seasons to capture seasonal variations in oceanic currents.

Table 1. Observation information and parameters.

Station	Longitude (° E)	Latitude (° N)	Water Depth (m)	Observation Period	Measured Parameters
B1	108.318	18.940	38	1 November 2021–13 November 2022	Water Level Current
B2	108.280	18.744	27	1 November 2022–6 February 2023	
B3	108.945	20.154	41	2 November 2021–31 October 2022	
B4	108.298	21.126	29	15 July 2021–30 June 2022	

Based on continuous ocean current observations, a filtering method was applied to separate astronomical tidal currents and low-frequency currents from the measured data. The measured currents were first decomposed into eastward and northward components, and each component was then filtered to remove near-inertial oscillations and tidal currents specific to the South China Sea, yielding the subtidal currents.

To isolate subtidal flows from the observed current data (which includes both astronomical tides and low-frequency currents), a low-pass filtering approach was employed. Given that the observation obtained was annual continuous data, the specific steps are as follows:

First, the observed currents were decomposed into eastward and northward components. Then, low-pass filtering was applied to each component [19]. The filter form is expressed as the following formula:

$$X_F = \mathrm{pl66tn}\,(x, dt, T) \tag{1}$$

where x represents the time-series data, dt is the sampling interval, T is the cut-off period of the filter (40 h), and X_F is the filtered data. The choice of the 40-h cut-off period was closely related to the regional inertial periods and tidal characteristics. The near-inertial period at the latitude of the current observation station was calculated using the formula for inertial periods ($T_i = 2\pi/f$, where $f = 2\Omega\sin\varphi$ is the Coriolis parameter, Ω is the angular velocity of the Earth's rotation, and φ is the latitude). For the study area, the computed near-inertial period was approximately 37.2 h. Moreover, Diurnal tides typically had periods around 24 h (e.g., M_2 tide had a period of about 12.42 h, and diurnal tides like K_1 had periods near 23.93 h). By setting a 40-h cut-off period, the filter could suppress signals with periods shorter than 40 h, including diurnal tides and near-inertial oscillations (37.2 h). This ensured that the filtered data (X_F) primarily retained the low-frequency subtidal flow components, preventing contamination from these higher-frequency motions.

In addition to in-situ observations, hourly wind data from the European Centre for Medium-Range Weather Forecasts (ECMWF) ERA5 reanalysis dataset with a spatial resolution of 0.25° × 0.25° were used to analyze wind–current interactions. The data corresponded to the observation period and were obtained from the Copernicus Climate Data Store [20].

2.2. Model Setup and Configuration

This study employed the Regional Ocean Modeling System (ROMS) numerical model to simulate the dynamic environment of the Beibu Gulf [21]. A two-layer nested grid scheme was adopted to capture the circulation of the broader South China Sea. The outer grid (first layer) spanned 3°–27° N and 105°–125° E (Figure 1a), covering most of the South China Sea, with a horizontal resolution of approximately 5 km using a curvilinear orthogonal grid. The inner grid (second layer) covered 16.5°–22° N and 105.6°–121° E, including the Beibu Gulf and the deep-water area southeast of Hainan Island, with a resolution of approximately 1.5 km (Figure 1b). Model bathymetry was based on the global depth data from the General Bathymetric Chart of the Oceans (GEBCO), and the vertical domain was divided into 20 sigma layers. Atmospheric forcing was provided by ERA5 reanalysis dataset, including wind fields, sea surface air temperature and pressure, shortwave and longwave radiation, relative humidity, precipitation, and other variables at a temporal resolution of 1 h and spatial resolution of 0.25°×0.25°. Momentum, heat, and salt fluxes across the air–sea interface were calculated using the bulk formula in ROMS. The initial temperature–salinity field and boundary conditions for the first-layer grid were derived from the global HYCOM (Hybrid Coordinate Ocean Model) reanalysis (1/12°resolution), providing ocean forecast and reanalysis data for temperature, salinity, sea level, and momentum at model initialization [22]. The open-boundary data had a time step of one day and provided the same parameters. For the second-layer grid, boundary conditions such as sea level, temperature, salinity, and currents were inherited from the first-layer grid. In addition, tidal forcing was sourced from the TPXO global tidal model, which supplied ROMS with 13 tidal constituents, including semi-diurnal, diurnal, and shallow-water tides: MM, MF, Q_1, O_1, P_1, K_1, N_2, M_2, S_2, K_2, MN_4, M_4, and MS, with a spatial resolution of 1/30° [23].

Model performance was assessed using the Pearson correlation coefficient (R), root mean square error (RMSE), and model skill score (Skill), defined as the following formulae:

$$R = \frac{\sum_{n=1}^{N}\left(M_n - \overline{M_n}\right)\left(O_n - \overline{O_n}\right)}{\sqrt{\left[\sum_{n=1}^{N}\left(M_n - \overline{M_n}\right)^2\right]\left[\sum_{n=1}^{N}\left(O_n - \overline{O_n}\right)^2\right]}} \quad (2)$$

$$RMSE = \left[\frac{1}{N}\sum_{n=1}^{N}(O_n - M_n)^2\right]^{1/2} \quad (3)$$

$$Skill = 1 - \frac{\sum_{n=1}^{N}|M_n - O_n|^2}{\sum_{n=1}^{N}\left(|M_n - \overline{O_n}|^2 + |O_n - \overline{O_n}|^2\right)} \quad (4)$$

where O_n and M_n are observations and model results at the same time and location, the overbar is the mean of the data, and N is the total number of data points.

Table 2 summarizes the key validation metrics of the ROMS model at the four stations, including correlation (R), root mean square error (RMSE), and model skill (Skill) for water level and current velocity (eastward and northward components). High correlation values (R > 0.85 for most cases) and reasonable RMSE, along with high Skill scores, indicated that the ROMS model could effectively reproduce the observed water level and current velocity patterns, providing a reliable basis for subsequent analyses. Part of the model validation figures can be found in the Supplementary Materials.

Table 2. Model verification results.

Station	Parameter	Component	Correlation (R)	RMSE	Skill
B1	Water Level	-	0.96	0.15 m	0.95
B1	Current Velocity	Eastward (u)	0.85	0.09 m/s	0.86
B1	Current Velocity	Northward (v)	0.87	0.12 m/s	0.88
B2	Water Level	-	0.94	0.18 m	0.94
B2	Current Velocity	Eastward (u)	0.89	0.10 m/s	0.90
B2	Current Velocity	Northward (v)	0.88	0.11 m/s	0.89
B3	Water Level	-	0.92	0.10 m	0.93
B3	Current Velocity	Eastward (u)	0.91	0.12 m/s	0.90
B3	Current Velocity	Northward (v)	0.87	0.13 m/s	0.86
B4	Water Level	-	0.94	0.13 m	0.95
B4	Current Velocity	Eastward (u)	0.87	0.10 m/s	0.85
B4	Current Velocity	Northward (v)	0.88	0.12 m/s	0.86

3. Results

3.1. Current off the Western Coast of Hainan Island

In the Beibu Gulf, circulation is seasonally driven by the northeast monsoon in winter and the southwest monsoon in summer. Accordingly, the period from October to March is defined as the winter half-year (northeast monsoon period), and from April to September as the summer half-year (southwest monsoon period) [24].

3.1.1. Winter Half-Year

During the winter half-year, currents at both Stations B1 and B2 were predominantly northward (Figure 2b–g), opposite to the prevailing northerly/northeasterly monsoon (i.e., winds blowing from north to south) (Figure 2a). Subtidal currents, however, flowed from south to north. At Station B1, the vertical distribution of mean current speeds across different layers (surface, 0.2H, 0.4H, 0.6H, 0.8H, bottom, and the vertically average, the "H" refers to the water depth at the observation station) was relatively uniform in some months (e.g., November 2021: 12.8 cm/s at the surface, 13.3 cm/s at 0.2H, and 11.6 cm/s at 0.4H). Although the vertical variation at Station B2 was more complex, the overall northward-directed reverse flow pattern remained (Figure 3).

Maximum current speeds fluctuated at both stations. At B1, speeds were relatively low in January (19.9 cm/s at the surface) and increased significantly in March (up to 31.0 cm/s). At B2, surface speeds peaked at 33.4 cm/s in December 2022 and dropped to 14.9 cm/s in January 2023. Minimum speeds generally fell below 5 cm/s, indicating periods of weak flow.

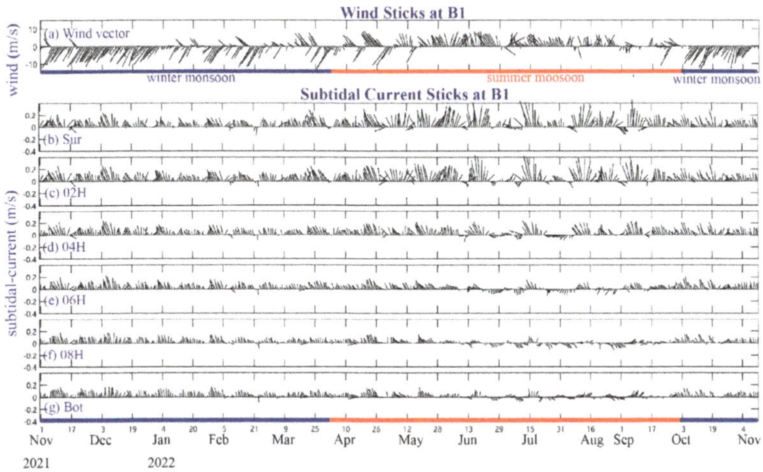

Figure 2. (**a**) Daily-averaged wind vectors (ERA5 dataset) at Station B1; (**b**–**g**) Observed daily-averaged subtidal current vectors at Station B1, showing profiles at the surface, 0.2H, 0.4H, 0.6H, 0.8H, and bottom layers, respectively.

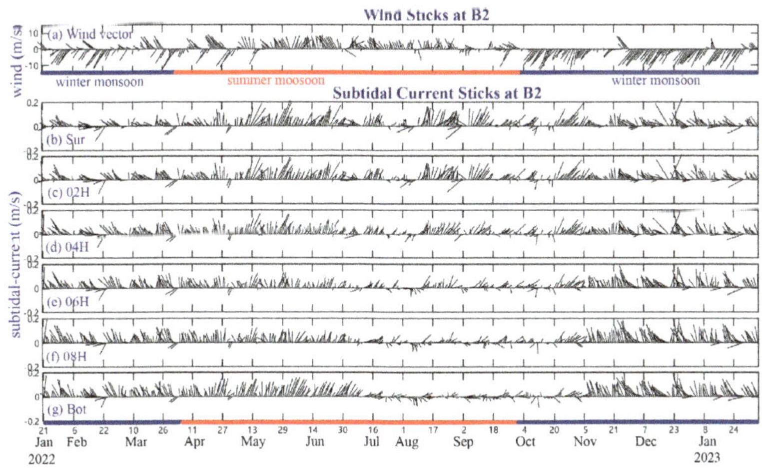

Figure 3. (**a**) Daily-averaged wind vectors (ERA5 dataset) at Station B2; (**b–g**) Observed daily-averaged subtidal current vectors at Station B2, showing profiles at the surface, 0.2H, 0.4H, 0.6H, 0.8H, and bottom layers, respectively.

Figure 4 presents rose diagrams of observed subtidal current vectors at Stations B1 and B2 for both the winter and summer half-years. During the winter half-year, subtidal currents at Station B1 (Figure 4a) predominantly flowed northwards (NNW, N, NNE), with the highest frequency in the N direction and current speeds mainly ranging from 0.05 to 0.2 m/s. At Station B2 (Figure 4c), currents also favored northern directions (NNW, N, NNE) but were more dispersed compared to B1, with the NNE direction most frequent and speeds mostly between 0 and 0.15 m/s.

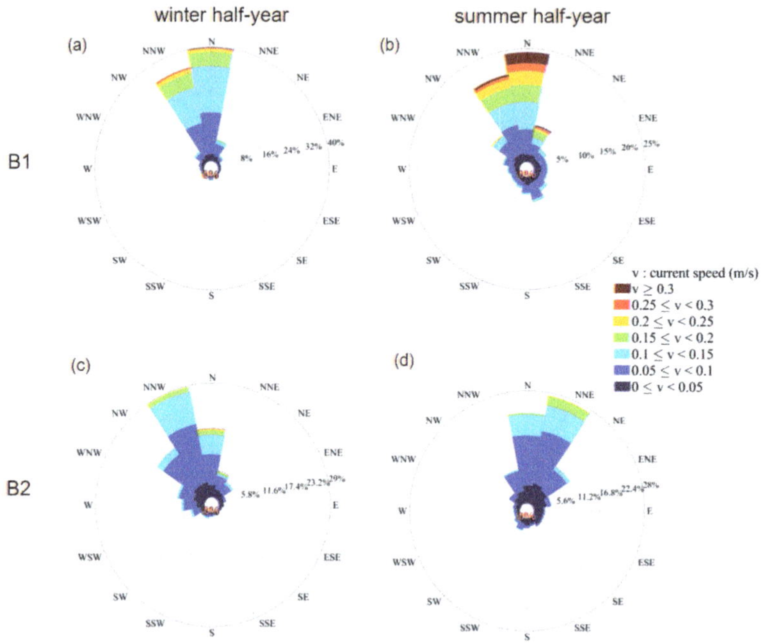

Figure 4. Rose diagram of observed subtidal current vectors at (**a**) Station B1 during the winter half-year; (**b**) Station B1 during the summer half-year; (**c**) Station B2 during the winter half-year; (**d**) Station B2 during the summer half-year.

3.1.2. Summer Half-Year

Subtidal currents off the western coast of Hainan Island exhibited more complex characteristics in summer than in winter.

Both Stations B1 and B2 maintained a dominant northward flow, but current speeds and vertical distributions were more variable. At B1, maximum current speeds increased significantly, up to 47.3 cm/s in July and 49.1 cm/s in September. Vertical gradients became evident in most months (e.g., in June, surface mean speed reached 24.0 cm/s, while the bottom was only 5.8 cm/s), indicating strong vertical shear.

At B2, maximum current speeds were also relatively high in some months (e.g., 18.0 cm/s at the surface in August 2022 and 23.6 cm/s in October 2022). Mean current speeds showed significant vertical shear, and, although the flow remained northward, current directions were more dispersed compared to the winter half-year.

In the summer half-year, subtidal currents at Station B1 (Figure 4b) continued to flow predominantly northward (N, NNW, NNE), though with some changes in frequency distribution. The N direction remained the most frequent, and current speeds covered a wider range, including higher values (e.g., 0.15–0.3 m/s). At Station B2 (Figure 4d), currents were also mainly oriented northward and northeastward (N, NNE, NE), with NNE dominant. Similar to B1, current speeds at B2 spanned a broader range than in the winter half-year, reflecting more dynamic summer conditions. Overall, both stations exhibited a persistent northward flow tendency across seasons, though the directional frequencies and current speed ranges differed between winter and summer. In summary, subtidal currents off the western coast of Hainan Island at Stations B1 and B2 were consistently northward throughout the year, with no significant seasonal reversal in the main flow direction.

3.2. Current Structure in the Northern Beibu Gulf

This section focuses on the observation stations in the northern Beibu Gulf (see Figure 1). We analyzed the seasonal variation patterns of subtidal currents using multi-layer subtidal current vector data from Stations B3 and B4 (see Figures 5 and 6) and subtidal current rose diagrams for the winter and summer half-years (see Figure 7), aiming to characterize the regional circulation structure.

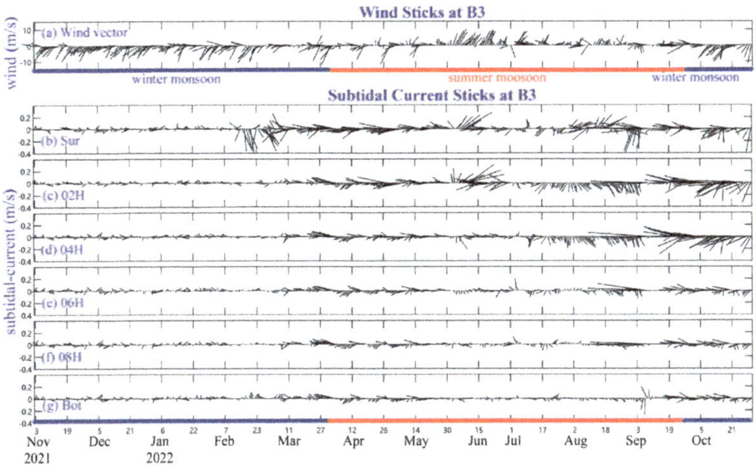

Figure 5. (**a**) Daily-averaged wind vectors (ERA5 dataset) at Station B3; (**b**–**g**) Observed daily-averaged subtidal current vectors at Station B3, showing profiles at the surface, 0.2H, 0.4H, 0.6H, 0.8H, and bottom layers, respectively.

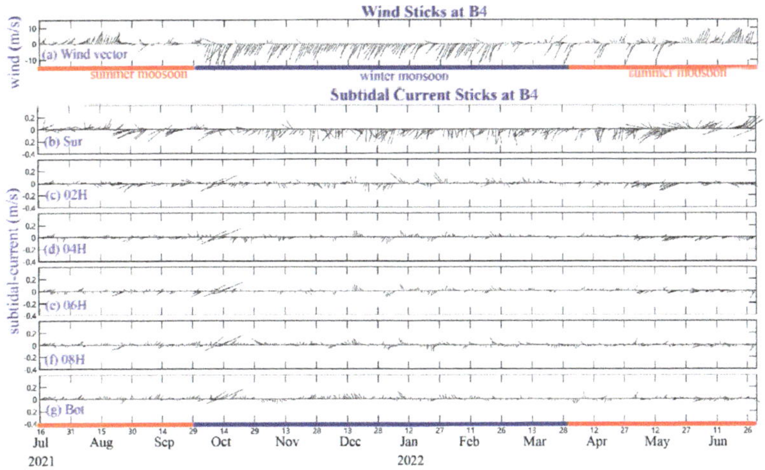

Figure 6. (**a**) Daily-averaged wind vectors (ERA5 dataset) at Station B4; (**b**–**g**) Observed daily-averaged subtidal current vectors at Station B4, showing profiles at the surface, 0.2H, 0.4H, 0.6H, 0.8H, and bottom layers, respectively.

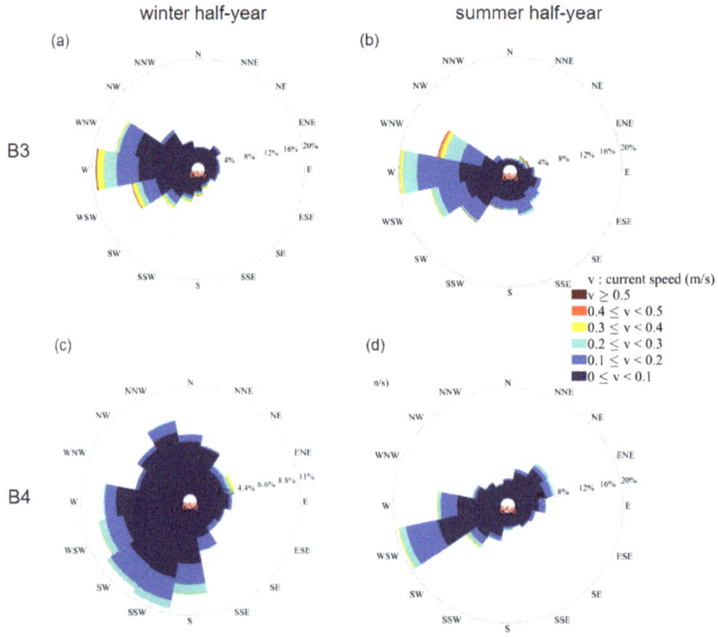

Figure 7. Rose diagram of observed subtidal current vectors at (**a**) Station B3 during the winter half-year; (**b**) Station B3 during the summer half-year; (**c**) Station B4 during the winter half-year; (**d**) Station B4 during the summer half-year.

3.2.1. Winter Half-Year

The winter subtidal current rose diagram at Station B3 (Figure 7) shows dominant flow directions concentrated in the west-northwest sector (WNW, W, WSW), with the highest frequency towards the west (W). Current speeds primarily ranged from 0.1 to 0.3 m/s. According to the subtidal current vector time series (Figure 5), under the influence of the winter northerly monsoon, subtidal currents flowed westward or northwestward. Vertically, flow directions from the surface to the upper-middle layers (0.2H–0.6H) were highly consistent. The bottom layer showed a slight southwestward deflection due to seabed friction, but the overall flow remained west-northwestward. Velocity differences between surface and bottom layers were minor: surface velocities peaked at 35 cm/s (mean 8.2 cm/s), while bottom velocities reached up to 25 cm/s (mean 6.5 cm/s).

At Station B4, the winter current rose diagram (Figure 7) shows a wide dispersion of flow directions, with dominant sectors from south (S) to west (W) and north-northwest (NNW) sectors. Current speeds were primarily between 0.1 and 0.3 m/s. Combined with the vector time series (Figure 6), the results indicate that the Beibu Gulf is dominated by northerly winds. Observations revealed that the surface subtidal current at Station B4 generally aligned with the winter wind direction, showing a southward tendency. In contrast, the middle and bottom layers displayed markedly different velocity directions, characterized by northward transport against the prevailing wind. This vertical reversal is likely caused by "compensation currents" in the Guangxi coastal region, where northward flow in the deeper layers replenishes the water removed by the southward surface flow.

3.2.2. Summer Half-Year

In the summer half-year, the subtidal current rose diagram at Station B3 (Figure 7b) showed similarly dispersed flow directions as in winter, with high-frequency flows in the WSW-W-WNW sector, indicating dominant westward transport. Although NNE and NE flows were less frequent, occasional strong events in these directions suggest transient dynamic disturbances within the summer circulation.

The subtidal current vector time series (Figure 5) revealed a distinct phased circulation pattern, which was caused by the weak summer monsoon in 2022. From April to May, the westward flow pattern persisted from the winter season. Around June 15, intensifying southerly winds triggered a strong northeastward current in the surface and 0.2H layers, with peak velocities reaching 0.38 m/s. This wind-induced current showed vertical attenuation. The surface and subsurface layers responded quickly. In contrast, changes in flow direction and velocity in the middle (0.4H–0.6H) and bottom (0.8H) layers were delayed.

At Station B4, the summer rose diagram (Figure 7d) reveals a significantly decreased dispersion in subtidal current directions, presenting a bimodal concentration in the southwest-northeast direction. Currents in the W-WSW and SW directions accounted for approximately 40%, while NE and ENE flows represented 15%. Overall, the southwestward currents dominated. Velocity ranged from 0.1 to 0.4 m/s with low-velocity flows (0–0.2 m/s) comprising more than 60%, reflecting a seasonal pattern dominated by weak currents.

An analysis of the subtidal current vector time series (Figure 6) reveals clear phased responses in the circulation system. From July to September 2021, the summer monsoon was weak. This resulted in highly dispersed subtidal current directions, with flow patterns fluctuating across layers and no dominant trends emerging. During April–May 2022, weak southerly winds and occasional strong northerly events drove southwestward transport at Station B4, though with poor vertical consistency. In June, as the summer monsoon intensified, surface subtidal currents responded rapidly, showing a pronounced northeastward deflection (maximum velocity of 0.35 m/s). However, currents below the surface showed limited response and occasionally moved opposite to the surface flow. The middle and lower layers (0.4H to bottom) were constrained by dynamic factors such as water mass inertia and seabed friction, resulting in significant deviations from surface flow. Vertical flow reversal occurred in approximately 40% of the observation period, highlighting strong stratification between wind-driven surface circulation and inertia-dominated deeper-layer flow in the summer circulation regime.

3.3. *Circulation Structure of the Beibu Gulf Under Different Monsoon Conditions*

The circulation structure of the Beibu Gulf is significantly influenced by seasonal monsoons, with distinct patterns during the winter and summer monsoon periods. Based on both observational and simulation results, this section analyzes the circulation characteristics under each monsoon regime.

During the winter monsoon, northeasterly winds dominated the Beibu Gulf's dynamics. Wind vectors at Station B1 (Figure 8a) indicated a persistent northeasterly wind from February 19 to 27, which drove the ocean circulation's response and adjustment. Layered observations of the north-south subtidal current at Station B1 (surface, 0.6H middle layer, bottom, and vertical average) showed the following:

From February 15 to 22, all layers displayed a stable northward flow with velocities up to 15 cm/s. From February 22 to 25, under intensified northeasterly winds, currents temporarily reversed southward, though velocities remained at 15 cm/s. After the wind weakened, currents resumed their northward direction, directly reflecting the monsoonal influence on current direction (Figure 8b).

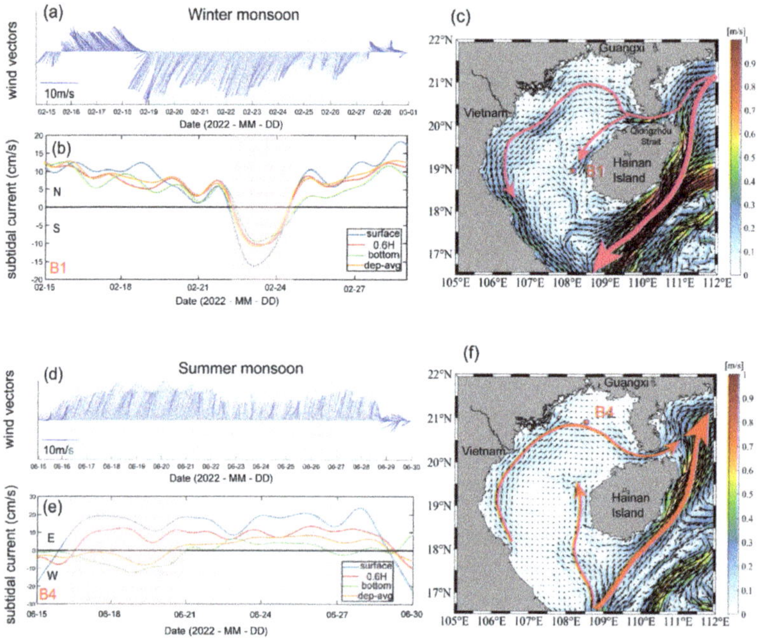

Figure 8. (**a**) Wind vectors during winter northerly wind events; (**b**) Subtidal currents at Station B1 (surface, 0.6H, bottom, and depth-averaged layers) during the winter northerly wind events. Positive values indicate northward flow, while negative values indicate southward flow; (**c**) Vertically averaged flow field in the Beibu Gulf during the northerly wind period. (**d**) Wind vectors during summer southerly wind events; (**e**) Subtidal current at Station B4 (surface, 0.6H, bottom, and depth-averaged layers) during the summer southerly wind events. Positive values indicate northward flow, while negative values indicate southward flow; (**f**) Vertically averaged flow field in the Beibu Gulf during the southerly wind period.

The vertically averaged current field from February 22 to 25 (Figure 8c) shows a strong westward flow through the Qiongzhou Strait, driven by the coastal current of the South China Sea. On the western side of the Qiongzhou Strait, the current splits into two branches. One branch extends northwest. It forms a cyclonic circulation in the northern Beibu Gulf. Then, it turns southward along Vietnam's coast. The other branch flows southward along the west coast of Hainan Island. This caused the local current at Station B1 to shift from northward to southward. These findings confirm that winter monsoonal northerly wind events significantly shape the circulation structure. The Beibu Gulf circulation responds strongly to weather-scale northerly wind events, reflecting dynamic adjustments driven by short-term wind forcing.

During the summer monsoon, southwesterly winds dominated. Wind vectors at Station B4 (Figure 8d) showed persistent southerly winds from June 15 to 29, resulting in layered and stage-wise circulation patterns:

From June 15 to 21, subtidal currents showed vertical stratification: the surface and mid-depth (0.6H) layers flowed eastward (up to 30 cm/s and ~10 cm/s, respectively), while the bottom layer flowed westward. After June 21, all layers shifted to uniform eastward flow, reflecting the summer monsoon's effect on flow stratification (Figure 8e).

Simulation results revealed that southwesterly winds drove a anticyclonic circulation in the northern Beibu Gulf (Figure 8f). The western coastal current flowed northward along the Vietnam coast, turned east near Station B4, and extended into the Qiongzhou Strait, demonstrating how the summer monsoon dynamically shapes the regional anticyclonic circulation.

Figure 9c,d show the numerically simulated, vertically averaged circulation patterns in the Beibu Gulf for winter and summer half-years. In winter, the gulf exhibited a distinct cyclonic circulation structure (Figure 9c). Driven by prevailing northeasterly winds (Figure 9a), a strong South China Sea western boundary current developed east of Hainan Island, flowing southwestward. Upon entering the Beibu Gulf, a small branch of this current turned northward west of Hainan Island, following the direction of the tidal subtidal currents. In the area about north of 19° N, this branch split into two parts. One part flowed northwestward through the central Beibu Gulf. It formed a cyclonic circulation together with the Vietnam coastal current. The other part continued northward along the western coast of Hainan Island. Then, it turned northwest near the Qiongzhou Strait. It formed another cyclonic gyre together with the Guangxi coastal current. The simulated winter circulation aligns with previous studies, confirming it as a typical winter pattern in which subtidal currents west of Hainan Island play a key role.

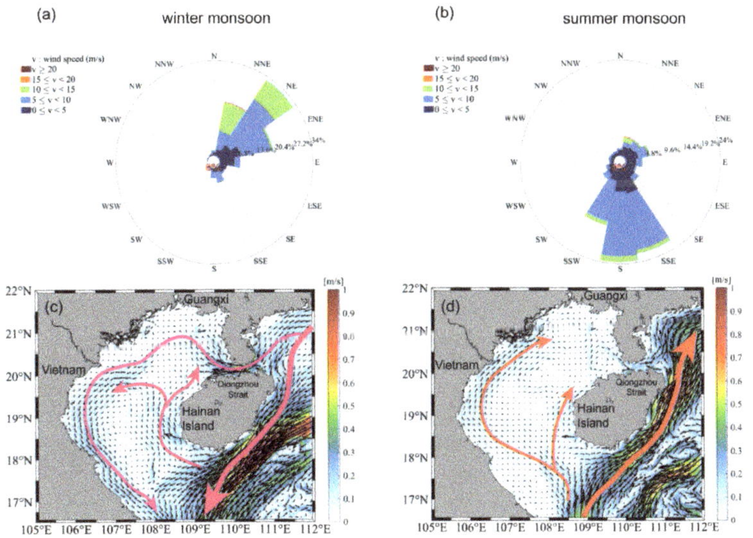

Figure 9. (**a**) Wind rose diagram of winter monsoon in the Beibu Gulf (winter half-year of 2022); (**b**) Wind rose diagram of summer monsoon in the Beibu Gulf (summer half-year of 2022); (**c**) Vertically averaged circulation in the Beibu Gulf in winter half-year; (**d**) Vertically averaged circulation in the Beibu Gulf in summer half-year.

Figure 8c depicts the circulation in the Beibu Gulf during a strong winter northerly wind event, while Figure 9c illustrates the mean circulation structure under the winter monsoon. A comparison reveals notable similarities: both exhibited east-to-west flow through the Qiongzhou Strait and a cyclonic circulation pattern in the northern and western gulf. The key difference lies in the coastal current west of Hainan Island: during the strong northerly wind event, the current flowed southward, whereas in the winter-half-year mean flow field, it flowed northward.

Figure 8f illustrates the circulation structure of the Beibu Gulf under southwest monsoons, while Figure 9d depicts the summer half-year averaged flow field. The two circulation patterns differed significantly. Under southwest winds, the Beibu Gulf circulation generally exhibited a basin-scale anticyclonic circulation (Figure 8f), which only closed in the northern gulf due to the northward coastal current west of Hainan Island. In contrast, the summer-averaged flow field (Figure 9d) indicated a less stable circulation. Driven by prevailing southerly winds (Figure 9b), the South China Sea western boundary current east of Hainan Island flowed northeastward from the southwest. A branch entered the central Beibu Gulf, moved northward along its western margin, and weakened off the Guangxi coast. Although subtidal currents west of Hainan Island were weaker than in winter, they still maintained a northward direction. Seasonal mean model results further showed northward subtidal currents on both the western and eastern sides of the gulf, albeit reduced velocities compared to winter.

4. Discussion

4.1. The Mechanism of the Northward Flow off the Western Coast of Hainan Island

The direction of subtidal currents west of Hainan Island plays a critical role in shaping the overall circulation structure of the Beibu Gulf. Year-round observations at Stations B1 and B2 show that subtidal currents in the offshore waters west of Hainan generally flowed northward throughout the year. Even under winter northerly winds, the subtidal currents at both stations generally flowed northward, counter to wind direction, only reversing southward during strong northerly wind events (e.g., in February 2022). In summer, the influence of southerly winds further reinforced this northward trend. The northward coastal current contributes directly to the formation of the wintertime cyclonic circulation in the Beibu Gulf. Therefore, further research is needed to understand the mechanisms sustaining this northward flow west of Hainan Island.

Figure 10 presents the time-series curves of tidal and subtidal currents averaged along the north-south vertical transects at Stations B1 and B2. The subtidal current intensity shows a strong correlation with the spring and neap tides. Specifically, at Station B2, enhanced northward subtidal currents coincided with spring tides, while during the neap tides, the currents weakened or even reversed. This pattern suggests that the subtidal current may be a tide-induced flow driven by nonlinear effects during tidal wave propagation. Wavelet power spectrum analysis, used to examine the energy distribution of signals in the time-frequency domain, was applied in this study to analyze the power spectra of the observed north-south tidal and subtidal currents, as shown in Figure 11. At both stations, the tidal currents exhibited a significant period of approximately 1 day, consistent with the diurnal tidal regime in this region. The power spectrum of the subtidal currents showed an obvious ~14-day period, corresponding closely to the spring–neap tidal cycle. Although subtidal currents in the winter half-year are driven northward against the prevailing southward winter monsoon winds, their vertical structure and speed fluctuations are significantly influenced by factors such as the tidal cycle. This indicates a distinct dynamic response to the combined effects of the monsoon and tides.

Figure 12a,b depict the tidal-induced residual currents averaged vertically during spring tide and neap tide, respectively. During spring tide, the tidal-induced residual currents showed relatively stronger intensities and more complex spatial patterns, especially in the Qiongzhou Strait and around the Hainan Island. The direction of the tidal-induced residual currents during spring tide is characterized by a notable northward flow along the western coast of Hainan Island, accompanied by a weak cyclonic circulation in the northern part of the gulf. In contrast, the neap tide residual currents are generally weaker and the flow structures are more subdued.

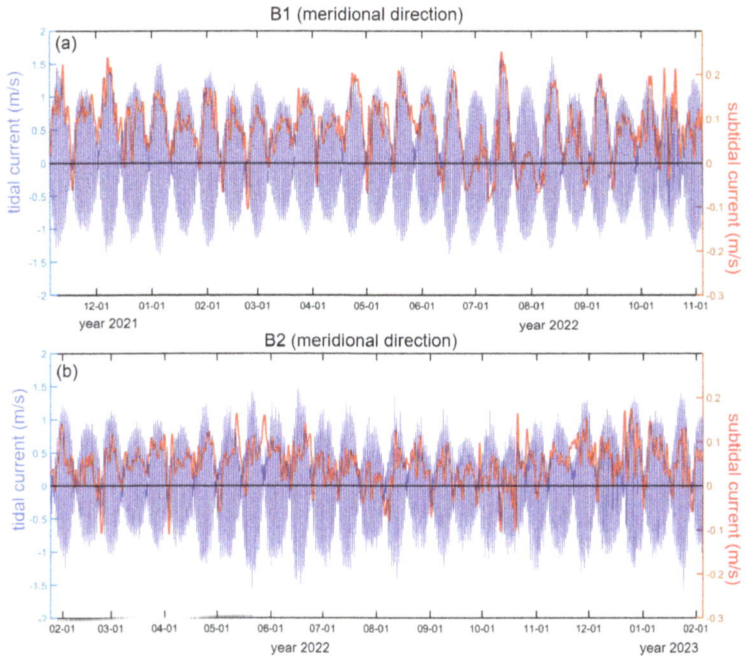

Figure 10. Depth-averaged tidal current and subtidal currents in the meridional (north-south) direction at (**a**) Station B1 and (**b**) Station B2.

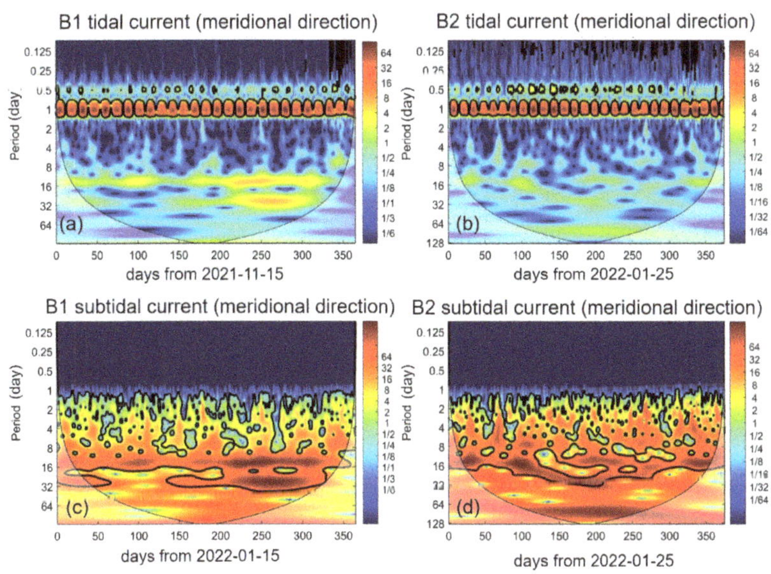

Figure 11. Wavelet power spectra of tidal currents (**a**,**b**) and subtidal currents (**c**,**d**) in the north-south direction at Station B1 and Station B2. Color bar numbers represent normalized power spectral density.

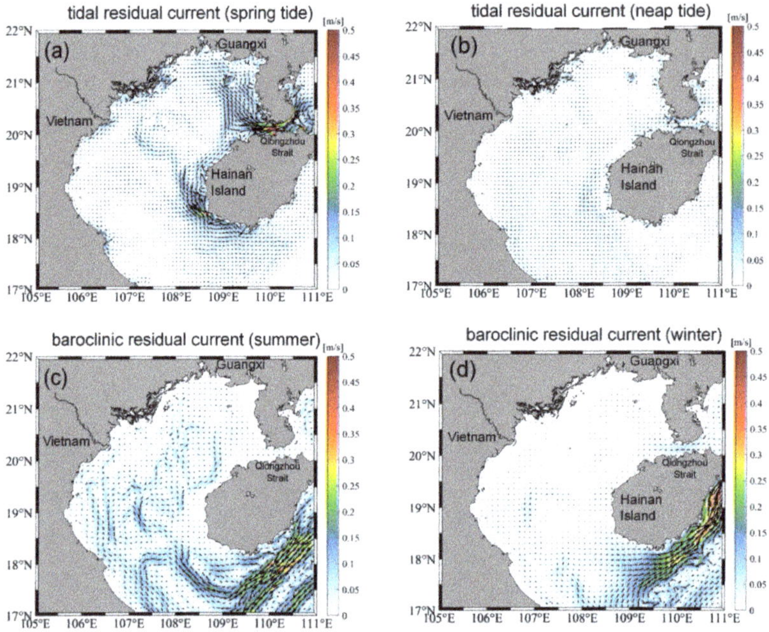

Figure 12. (**a**) Vertical-averaged tidal-induced residual current during spring tide; (**b**) Vertical-averaged tidal-induced residual current during neap tide; (**c**) Vertical-averaged baroclinic residual current in summer; (**d**) Vertical-averaged baroclinic residual current in winter.

Figure 12c,d illustrate the baroclinic residual currents averaged vertically in summer and winter. A notable feature is that in the outer area of the Beibu Gulf, the baroclinic currents on the southeast side of Hainan Island show opposite directions in winter and summer. Within the Beibu Gulf, the summer baroclinic circulation is slightly stronger than that in winter. However, the baroclinic-induced circulation is very weak in the coastal area of Hainan Island (with water depth less than 30 m) and in the Qiongzhou Strait. This further confirms that the baroclinic effect is not the dominant driver of the northward residual current west of Hainan Island. Instead, the above analysis indicates that the northward subtidal current in this region is primarily modulated by tidal processes.

4.2. Diverse Circulation Structures of the Beibu Gulf in the Summer Half-Year

The summer circulation structure of the Beibu Gulf has been a subject of ongoing debate, with three prevailing views: a basin-scale anticyclonic circulation, a basin-scale cyclonic circulation, and a two-gyre system featuring a cyclonic gyre in the north and an anticyclonic gyre in the south [11]. Different studies support different structures, primarily due to the lack of long-term in-situ current observations.

Using long-term current measurements from Station B3 (west of the Qiongzhou Strait) and Station B4 (northern Beibu Gulf), this study shows that the summer circulation structure in the northern Beibu Gulf is predominantly cyclonic. Statistical results from Figure 7b,d show that, during the summer half-year, subtidal currents at Station B3 were primarily westward (20%), west-southwest (14%), and west-northwest (12%), with westerly components accounting for about 46%. At Station B4, dominant directions included west-southwest (18%), west (12%), and southwest (8%), with southwestward components totaling about

38%. These subtidal current directions, combined with the stations'geographic locations, indicate a cyclonic circulation structure in the Beibu Gulf during summer.

However, both observations and numerical simulations at Station B4 under typical summer southwesterly winds reveal an anticyclonic circulation structure, suggesting the possible coexistence of cyclonic and anticyclonic structures during summer. Figure 13a,c depict southeasterly summer monsoon wind fields, while Figure 13b,d show the corresponding current fields. Notably, the northern Beibu Gulf exhibits a distinct cyclonic circulation under southeasterly winds, whereas southwesterly winds induce an anticyclonic pattern (Figure 9d). This indicates that different summer monsoon directions (southwesterly vs. southeasterly) lead to opposite circulation patterns. Therefore, discrepancies in simulation timeframes may account for the contrasting circulation structures reported in previous studies, underscoring the critical role of monsoon direction in shaping summer circulation in the Beibu Gulf.

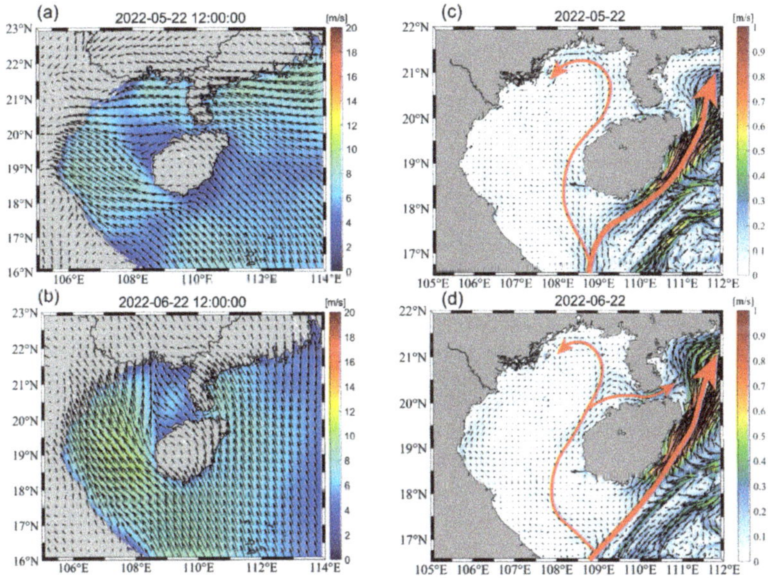

Figure 13. Cyclonic circulation structure of the Beibu Gulf in summer. (**a**) Wind field over the Beibu Gulf on 22 May 2022; (**b**) Vertically averaged circulation on 22 May 2022; (**c**) Wind field over the Beibu Gulf on 22 June 2022; (**d**) Vertically averaged circulation on 22 June 2022.

5. Conclusions

This study investigates the circulation characteristics of the Beibu Gulf by integrating long-term current observations from four ADCP stations (B1–B4) with high-resolution simulations using the Regional Ocean Modeling System (ROMS).

Observations revealed persistent year-round northward subtidal residual currents west of Hainan Island, primarily driven by tidal dynamics, as indicated by their 14-day periodicity aligned with the spring neap tide cycles. These currents only briefly shift southward during intense winter northerly wind events (e.g., February 2022). In the northern Beibu Gulf, subtidal currents show pronounced seasonal variations: winter is characterized by westward-dominated flows (0.1–0.3 m/s) at Station B3 and vertical flow reversals at Station B4 (surface southward, mid-bottom northward), while summer exhibits bimodal southwest-northeast circulation, with wind-driven eastward surface currents (up to 0.38 m/s) and inertia-maintained westward subsurface flows.

Numerical simulations confirmed that winter circulation forms a basin-scale cyclonic gyre driven by northeasterly monsoons, with westward flows through the Qiongzhou Strait interacting with a northern cyclonic cell. These results align with observations and highlight the combined influence of monsoon forcing and tide-induced residual currents. Summer circulation varies with wind direction: southeasterly winds generate a northern cyclonic gyre, while southwesterly winds produce an anticyclonic structure (e.g., June 2022 simulation). This resolves long-standing debates by identifying monsoon wind direction as the primary driver of summer circulation polarity.

By integrating multi-station observations and numerical modeling, this study addresses gaps in long-term circulation data, clarifying the mechanism of tide-driven year-round northward currents west of Hainan Island and the summer circulation's response to monsoon winds. The findings offer new insights into shelf–sea circulation dynamics and support ecological management efforts in the ecologically sensitive Beibu Gulf, such as nutrient transport and pollutant dispersion forecasting.

Supplementary Materials: The following supporting information can be downloaded at: https://www.mdpi.com/article/10.3390/w17162365/s1, Figure S1: Water Level Validation for Station B1 from November 2021 to April 2022; Figure S2: Water Level Validation for Station B1 from April 2022 to September 2022; Figure S3: Water Level Validation for Station B2 from April 2022 to September 2022; Figure S4: Water Level Validation for Station B3 from June 2022 to March 2023; Figure S5: Verification of Depth-Averaged Flow Velocity and Direction at Station B1 in March 2022; Figure S6: Verification of Depth-Averaged Flow Velocity and Direction at Station B1 in February 2022.

Author Contributions: Conceptualization, G.L. and N.Z.; methodology, G.L.; software, G.L.; validation, G.L., N.Z., and Y.Y.; formal analysis, G.L.; investigation, G.L.; resources, G.L.; data curation, G.L.; writing—original draft preparation, G.L.; writing—review and editing, N.Z., Y.Y., and C.W.; visualization, G.L.; supervision, C.W.; project administration, C.W.; funding acquisition, C.W. All authors have read and agreed to the published version of the manuscript.

Funding: The works presented in this paper is financially sponsored by the Research Innovation Fund of Tianjin Research Institute for Water Transport Engineering, Ministry of Transport, China, grant number TKS20230108; and the National Key Research and Development Program of China, grant number 2023YFB2603803.

Data Availability Statement: Data are contained within the article.

Acknowledgments: We sincerely thank the three reviewers and editors for their valuable comments and dedicated efforts that have significantly improved this manuscript.

Conflicts of Interest: Author Gongpeng Liu was employed by the company POWERCHINA Zhongnan Engineering Corporation Limited. The remaining authors declare that the research was conducted in the absence of any commercial or financial relationships that could be construed as a potential conflict of interest.

References

1. Science and Technology Committee of the People's Republic of China (STCPRC). *Reports of Sino—Vietnamese Joint Comprehensive Marine Survey in the Beibu Gulf*; Science and Technology Committee of the People's Republic of China: Beijing, China, 1964.
2. Tan, G.H. Preliminary analysis of hydrologic structure and hydrologic feature in the sea region of the Beibu Gulf. *Trans. Oceanol. Limnol.* **1987**, *4*, 7–15. (In Chinese with English Abstract)
3. Gao, J.; Shi, M.; Chen, B.; Guo, P.; Zhao, D. Responses of the circulation and water mass in the Beibu Gulf to the seasonal forcing regimes. *Acta Oceanol. Sin.* **2014**, *33*, 1–11. [CrossRef]

4. Chen, Z.H.; Qiao, F.L.; Xia, C.S.; Wang, G. The numerical investigation of seasonal variation of the cold water mass in the Beibu Gulf and its mechanisms. *Acta Oceanol. Sin.* **2015**, *34*, 44–54. [CrossRef]
5. Zu, T.T. Analysis of the Current and Its Mechanism in the Gulf of Beibu. Master's Dissertation, Ocean University of China, Qingdao, China, 2005. (In Chinese with English Abstract).
6. Zhang, G.R.; Pang, W.R.; Lan, J. The characteristics of the volume transport in spring and winter in the northern and eastern Beibu Gulf. In *The Essay Collection of the Researches on the Ocean Science in the Beibu Gulf-PT II*; Ocean Press of China: Beijing, China, 2009; pp. 127–138. (In Chinese with English Abstract)
7. Gao, J.; Mo, L.; Lu, H.; Meng, X.; Wu, G.; Wang, D.; Nguyen, K.-C.; Hu, B.; Tran, A.T. Spring circulation characteristics and formation mechanism in the Beibu Gulf. *Front. Mar. Sci.* **2024**, *11*, 1398702. [CrossRef]
8. Zhu, X.-H.; Ma, Y.-L.; Guo, X.; Fan, X.; Long, Y.; Yuan, Y.; Xuan, J.-L.; Huang, D. Tidal and residual currents in the Qiongzhou Strait estimated from shipboard ADCP data using a modified tidal harmonic analysis method. *J. Geophys. Res. Oceans* **2014**, *119*, 8039–8060. [CrossRef]
9. Gao, J.S.; Chen, B.; Shi, M.C. Summer circulation structure and formation mechanism in the Beibu Gulf. *Sci. China Ser. D-Earth Sci.* **2015**, *58*, 286–299. [CrossRef]
10. Chen, B.; Shi, M. Research Progress on Marine Circulation in the Beibu Gulf. *Guangxi Sci.* **2019**, *26*, 595–603.
11. Gao, J.; Wu, G.; Ya, H. Review of the circulation in the Beibu Gulf, South China Sea. *Cont. Shelf Res.* **2017**, *138*, 106–119. [CrossRef]
12. Wang, D.R. Study of the Dynamic-Thermodynamic Mechanic of Beibu Bay Cool Water Masses. Doctoral Dissertation, Ocean University of China, Qingdao, China, 1998.
13. Xia, H.Y.; Li, S.H.; Shi, M.C. Three-D numerical simulation of wind driven current and density current in the Beibu Gulf. *Acta Oceanol. Sin.* **2001**, *20*, 455–472.
14. Ding, Y.; Chen, C.; Beardsley, R.C.; Bao, X.; Shi, M.; Zhang, Y.; Lai, Z.; Li, R.; Lin, H.; Viet, N.T. Observational and model studies of the circulation in the Beibu Gulf, South China Sea. *J. Geophys. Res. Oceans* **2013**, *118*, 6495–6510. [CrossRef]
15. Wu, D.; Wang, Y.; Lin, X.; Yang, J. On the mechanism of the cyclonic circulation in the Gulf of Tonkin in the summer. *J. Geophys. Res.* **2008**, *113*, C09029. [CrossRef]
16. Sun, H.L.; Huang, W.M.; Zhao, J.S. Three-dimensional numerical simulation of tide induced, wind-driven and thermohaline residual currents in the Beibu Bay. *Ocean. Limnol. Sin.* **2001**, *32*, 561–568. (In Chinese with English Abstract)
17. Gao, J.; Xue, H.; Chai, F.; Shi, M. Modeling the circulation in the Gulf of Tonkin, South China Sea. *Ocean. Dyn.* **2013**, *63*, 979–993. [CrossRef]
18. Yang, J.L.; Jiang, S.C.; Wu, J.S.; Xie, L.L.; Zhang, S.W.; Bai, P. Effects of wave-current interaction on the waves, cold-water mass and transport of diluted water in the Beibu Gulf. *Acta Oceanol. Sin.* **2020**, *39*, 25–40. [CrossRef]
19. Beardsley, R.C.; Limeburner, R.; Rosenfeld, L.K. *CODE-1 Moored Array and Large-Scale Data Report*; Woods Hole Oceanographic Institution: Woods Hole, MA, USA, 1983; pp. 1–2.
20. Hersbach, H.; Bell, B.; Berrisford, P.; Hirahara, S.; Horányi, A.; Muñoz-Sabater, J.; Nicolas, J.; Peubey, C.; Radu, R.; Schepers, D.; et al. The ERA5 global reanalysis. *Q. J. R. Meteorol. Soc.* **2020**, *146*, 1999–2049. [CrossRef]
21. Shchepetkin, A.F.; McWilliams, J.C. The regional ocean modeling system (ROMS): A split-explicit, free-surface, terrain-following coordinate ocean model. *Ocean. Model.* **2005**, *9*, 347–365. [CrossRef]
22. HYCOM Consortium. The Hybrid Coordinate Ocean Model (HYCOM) system: Overview and applications. *Ocean. Sci. Discuss.* **2019**, *16*, 1–36.
23. Egbert, G.D.; Erofeeva, S.Y. Efficient inverse modeling of barotropic ocean tides. *J. Atmos. Ocean. Technol.* **2002**, *19*, 183–204. [CrossRef]
24. Cao, Z.-Y.; Bao, M.; Guan, W.-B.; Chen, Q. Water-mass evolution and the seasonal change in northeast of the Beibu Gulf, China. *Oceanol. Limnol. Sin.* **2019**, *50*, 532–542.

water

MDPI

Article

Distribution of Excess Pore Water Pressure in Layered Seabed Induced by Internal Solitary Waves

Hao Tian [1,2], Lei Jia [3], Jingtao Zhao [1,2,*], Libo Wang [1,2], Jing Kan [1,2], Fuyu Wu [1,2] and Zhuangcai Tian [3,4,*]

1 Qingdao Institute of Marine Geology, China Geological Survey, Qingdao 266237, China
2 Laboratory for Marine Mineral Resources, Qingdao Marine Science and Technology Center, Qingdao 266237, China
3 State Key Laboratory of Intelligent Construction and Healthy Operation and Maintenance of Deep Underground Engineering, China University of Mining and Technology, Xuzhou 221116, China
4 Research Center for Deep Ocean Science and Underwater Engineering, China University of Mining and Technology, Xuzhou 221008, China
* Correspondence: zhaojingtao113@163.com (J.Z.); zhuangcaitian@cumt.edu.cn (Z.T.)

Abstract

The research focuses on the complex stratification of the seabed, which is consistent with real marine geological conditions. This article presents the effects of four parameters—seabed shear modulus, permeability coefficient, porosity, and saturation—on the distribution of excess pore water pressure within the seabed by internal solitary waves (ISWs). Additionally, the changes in excess pore water pressure distribution in layered seabed are analyzed. The findings indicate that increases in saturation and permeability coefficient lead to deeper penetration of excess pore water pressure into the seabed by ISWs. Conversely, the effects of shear modulus and porosity are relatively minor and inversely related to the depth of influence of excess pore water pressure. When stratification occurs in the permeability coefficient and saturation of the seabed, significant alterations are observed in the downward propagation of excess pore water pressure. Saturation stratification exhibits similar effects, with soil layers exhibiting higher saturation levels being more conducive to the transmission of excess pore water pressure by ISWs. These research findings hold substantial implications for assessing seabed liquefaction and surface erosion processes induced by ISWs.

Keywords: excess pore water pressure; layered seabed; internal solitary waves; saturation; permeability

1. Introduction

The complex hydrodynamic environment of the ocean significantly impacts the seabed sedimentary environment, submarine safety, marine engineering structures, and seabed stability. This environment is influenced by various factors, including waves, storm surges, tides, ocean currents, and internal waves [1–3]. Internal solitary wave (ISW) is a specific type of nonlinear internal wave characterized by their solitary wave morphology, occurring in stably stratified seawater. ISWs are characterized by short periods, large amplitudes, high flow velocities, and rapid propagation speeds, primarily affecting deep-water environments [4,5]. ISWs can induce changes in water pressure at the seabed surface, which subsequently alters the excess pore water pressure within the seabed. Variations in excess pore water pressure are crucial indicators of the dynamic response of the seabed [6,7].

Due to the challenges and high costs associated with in situ observations of seabed responses, the existing research predominantly relies on laboratory flume experiments and numerical simulations.

Numerical simulation studies have primarily concentrated on the dynamic responses within a horizontal and uniform seabed. Qiao et al. (2016) developed a model of an infinite-depth horizontal seabed to calculate the forces exerted by ISWs and analyzed the excess pore water pressure generated in the seabed [8]. Rivera Rosario et al. (2017) utilized the Dubreil–Jacotin–Long equation to solve for the velocity and density fields produced by the propagation of ISWs over a horizontal seabed, discovering that the excess pore water pressure in the seabed can also be negative [9]. They noted that the gradient of excess pore water pressure would drive upward flow of excess pore water, potentially leading to seabed instability [9]. Tian et al. (2019) assessed the excess pore water pressure in sandy silt and clayey silt seabed and evaluated the depth of ISW influence on the seabed [10]. Additionally, Tian et al. (2023) analyzed the spatial and temporal variations in the dynamic response of the seabed induced by ISWs [11]. Collectively, these numerical studies provide a comprehensive analysis of the dynamic effects of ISWs on a horizontal seabed, as well as the distribution characteristics of excess pore water pressure within the seabed.

The dynamic effects of ISWs on a sloping seabed have primarily been investigated through laboratory flume experiments. Qiao et al. (2018) observed that both the pressure exerted by ISWs and the excess pore water pressure within the seabed exhibited negative pressure changes at the slope and at the summit of the slope [12]. They also noted that the amplitude of ISW pressure increased as the slope angle was raised from 0.071 to 0.16. In a subsequent study, Li et al. (2021) [13] varied the slope angles to 3°, 6°, and 9° and found that the excess pore water pressure in the seabed consistently increased or decreased at the base, mid-section, and top of the slope, revealing differences in the intensity of dynamic responses between these locations [13,14]. Tian et al. (2023) employed large eddy simulation to investigate four mechanisms of ISW breaking and, for the first time, analyzed the pressures and excess pore water pressures induced by ISWs on a sloping seabed [15]. The pressure generated by ISWs underwent a polarity reversal, with the rate of this reversal increasing as the slope angle increased [16–18]. Compared to a horizontal seabed, the changes in excess pore water pressure resulting from ISW propagation on a sloping seabed exhibited significant differences, indicating that the dynamic response on a sloping seabed is more complex. However, excess pore water pressure distribution in a layered seabed is ignored.

ISWs predominantly influence deep-water environments and pose a risk to seabed stability [16–21]. Current research has not adequately addressed the complex stratification of the seabed, which diverges from real marine geological conditions. In the context of intricate geological formations and marine sedimentation, the seabed typically displays heterogeneous and stratified characteristics. The physical and mechanical properties of various soil layers within the seabed are not uniform, which subsequently influences the distribution of excess pore water pressure among these layers [22,23]. Zhang et al. (2024) demonstrated that under wave–current interactions, the permeability coefficient significantly affects the distribution of excess pore water pressure in a stratified seabed [23]. When an ISW interacts with a layered seabed, the resulting seabed response may differ from that observed in a homogeneous seabed. However, current research has yet to address the complexities associated with seabed stratification, which diverges from actual marine geological conditions. This article aims to analyze the distribution of excess pore water pressure in a layered seabed under the influence of ISWs through numerical simulations.

2. Numerical Model of ISW and Layered Seabed

Utilizing the Navier–Stokes equations, the Biot consolidation equation, the volume of fluid method for tracking layered fluid interfaces, and the method for generating ISWs, numerical models of a flume and seabed were developed. These models were subsequently coupled under conditions of continuous water pressure at the seabed surface to facilitate the numerical simulation of the interaction between ISWs and a sloping seabed. A detailed description of this process can be found in reference [15].

2.1. Seabed Model

In Biot consolidation theory, the dissipation of excess pore water pressure and the displacement of the soil skeleton in the porous media are coupled. We adopted the governing equations of Biot consolidation theory in references [10,15] to calculate the excess pore water pressure in the seabed induced by ISWs. The seepage continuous equation of excess pore water is

$$K_x \frac{\partial^2 p}{\partial x^2} + K_Y \frac{\partial^2 p}{\partial z^2} - \gamma_\omega n \beta \frac{\partial p}{\partial t} = \gamma_\omega \frac{\partial \varepsilon}{\partial t}$$

where p is the excess pore water pressure. Kx and K_Y are the permeability coefficients of sediment in the X-direction and Y-direction, respectively. In this study, the permeability of the seabed is isotropic. The γ_w is the bulk density of excess pore water, n is the porosity of sediment, β is the compression coefficient of pore fluid, and ε is the volumetric strain of water.

The boundary conditions of the governing equation are as follows. The normal stress σ and shear stress τ on the seabed surface are generally negligible when the water viscosity and friction are ignored. The excess pore water pressure p on the seabed surface equals the wave-induced pressure ΔP [15]. Namely, on the seabed surface, we have

$$\sigma = 0,\ \ \tau = 0,\ \ p = \Delta P$$

The bottom of the seabed is considered impermeable, and the soil displacement at this section is zero. Therefore,

$$u = 0,\ \ w = 0,\ \ \partial p/\partial y = 0$$

where u is the horizontal displacement of soil, w is the vertical displacement of soil, and y is the Cartesian coordinate.

Assuming that the horizontal displacement on the left and right sides of the seabed boundary is 0, and the normal flow of excess pore water is 0, we have

$$u = 0,\ \ \partial p/\partial x = 0$$

where x is the Cartesian coordinate.

Tian et al. (2023) reported that the excess pore water pressure in the seabed induced by ISWs propagates to a depth of 0.15 m [15]. Beyond this depth, the excess pore water pressure remains unaffected. According to the findings of Rivera Rosario et al. (2017) [9], it is established that the excess pore water pressure in the seabed due to ISWs is confined to a specific depth range. When the depth exceeds this range, the excess pore water pressure remains unchanged. Consequently, the stratification in the deep seabed exerts less influence on the seabed response compared to that in the shallow seabed. For this study, the seabed was divided into two layers, with the specific dimensions of the model illustrated in Figure 1. The upper layer of the seabed, depicted in purple, has a thickness of 0.1 m, while the lower layer, shown in gray, represents the remainder of the seabed. The slope of the seabed is set at 9°, and the ISWs are modeled as breaking waves in the form of rolling waves. The parameters associated with the ISWs in the model are detailed in Table 1.

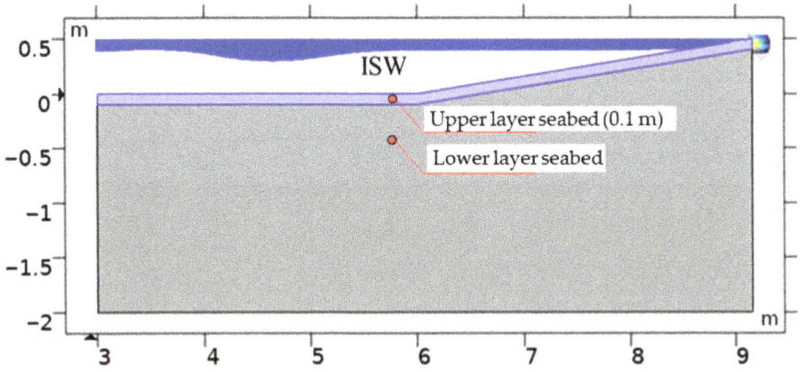

Figure 1. Schematic diagram of model for the interaction between ISW and layered seabed.

Table 1. The parameters of the ISW.

Upper Water Depth (m)	Density of Upper Water Depth (kg/m^3)	Lower Water Depth (m)	Density of Lower Water Depth (kg/m^3)	Amplitude (m)	Wave Length (m)
0.1	998	0.4	1025	0.112	1.232

2.2. ISW Model

In this paper, fluid was regarded as an ideal incompressible fluid. We investigated the change in flow field disturbed by ISWs by Large Eddy Simulations (LESs). The governing Navier–Stokes and density transport equations were solved. A standard lock-release setup was used to generate ISWs. The interface between the stratified fluids was captured by the volume of fluid (VOF) method. At the top of the flume, the effect of surface waves was ignored, and the rigid cover assumption was adopted; therefore, the symmetry boundary condition was selected. Wall boundary conditions were adopted on both sides and at the bottom of the flume and the inclined terrain.

This study established a total of eight experimental cases, with consistent physical model dimensions across all conditions. An identical ISW load was applied to each case, and the relevant parameters of the ISW are detailed in Table 1. The eight cases were categorized into four groups. The grouping and seabed parameters for each case are summarized in Table 2.

In Group 1, the permeability coefficient, porosity, saturation, and Poisson's ratio of the seabed remain unchanged across layers, while only the shear modulus varies. In case 1, the shear modulus of the upper soil is relatively low (1×10^6), while the shear modulus of the lower soil is significantly higher (1×10^7). Conversely, in case 2, the results are compared to illustrate the influence of the shear modulus parameter in seabed layering.

In Group 2, only the permeability coefficient varies with depth, while all other parameters are held constant. In case 3, the permeability coefficient of the upper soil is relatively low (1×10^{-7}), whereas the lower soil exhibits a higher permeability coefficient (1×10^{-4}). The opposite scenario is observed in case 4.

In Group 3, the porosity of the seabed varies with depth, while all other parameters remain constant. In case 5, the porosity of the upper layer of the seabed is 0.4, and that of the lower layer is 0.8. In case 6, the porosity values are reversed.

In Group 4, we examine variations in saturation with depth, while all other parameters are unchanged. In case 7, the saturation of the upper layer of the seabed is relatively low (0.9), while the lower layer is fully saturated (saturation = 1). This relationship is inverted in case 8.

We calculated Biot consolidation equations using COMSOL Multiphysics 5.6. The generation, propagation, and breaking of ISWs were simulated by LES. The numerical model aligns with the model of reference [15] and was validated by the paper; further validation is not necessary in this manuscript. For additional details, please refer to reference [15].

Table 2. The parameters of the layered seabed.

Group	Case	Water Layer	Shear Modulus	Permeability Coefficient	Porosity	Saturation	Poisson's Ratio
Group 1	Case 1	Upper Lower	1×10^6 1×10^7	1.33×10^{-4}	0.717	0.98	0.31
	Case 2	Upper Lower	1×10^7 1×10^6	1.33×10^{-4}	0.717	0.98	0.31
Group 2	Case 3	Upper Lower	1×10^7	1×10^{-7} 1×10^{-4}	0.717	0.98	0.31
	Case 4	Upper Lower	1×10^7	1×10^{-4} 1×10^{-7}	0.717	0.98	0.31
Group 3	Case 5	Upper Lower	1×10^7	1.33×10^{-4}	0.4 0.8	0.98	0.31
	Case 6	Upper Lower	1×10^7	1.33×10^{-4}	0.8 0.4	0.98	0.31
Group 4	Case 7	Upper Lower	1×10^7	1.33×10^{-4}	0.717	0.9 1	0.31
	Case 8	Upper Lower	1×10^7	1.33×10^{-4}	0.717	1 0.9	0.31

3. Results and Discussion

3.1. Distribution of Excess Excess Pore Water Pressure of Homogeneous Seabed

Under the influence of wave action, the variation in excess pore water pressure distribution within the seabed, attributed to stratification, is primarily a result of changes in the relevant physical and mechanical parameters among the seabed layers. Consequently, prior to investigating the impact of seabed stratification on excess pore water pressure distribution, this section first examines the effects of four key parameters—shear modulus, permeability coefficient, porosity, and saturation—on excess pore water pressure distribution within a homogeneous seabed.

Figure 2 illustrates a schematic representation of the numerical model utilized for parameter analysis, where the dimensions of the seabed and the parameters of ISW are consistent with case 3 of reference [15]. The analysis involves modifying the relevant parameters of the seabed soil to facilitate comparison of the results. The analysis is conducted at a time point of t = 27 s, during which the entire ISW interacts with the horizontal seabed. A vertical intercept is established at a cross-section located at x = 4.5 m on the seabed to monitor the distribution characteristics of excess pore water pressure along the intercept, with a length of 1 m.

The permeability coefficient of the seabed soil on the northern continental slope of the South China Sea ranges approximately from 10^{-5} to 10^{-7} m/s. The shear modulus typically varies around 10^7 Pa. Porosity values are generally high, predominantly ranging between 0.5 and 0.7, with extreme cases reaching approximately 0.8. The saturation level is close to complete saturation; however, some gas-bearing seabeds may not be fully saturated, yet their saturation remains above 0.9. To comprehensively assess the impact of these parameters without introducing excessive distortion, the range of parameter values is appropriately expanded. The permeability coefficient is varied to include medium permeability values of 10^{-4}, 10^{-5}, 10^{-6}, and 10^{-7} m/s. Considering the presence of a gas-bearing seabed, saturation values of 0.90, 0.93, 0.96, and 1.00 are employed. Porosity values are set at 0.5, 0.6, 0.7, and 0.8, respectively. The shear modulus is varied around 10^7 Pa, with values of $1 \times 10^6, 5 \times 10^6, 1 \times 10^7$, and 5×10^7 Pa sequentially applied.

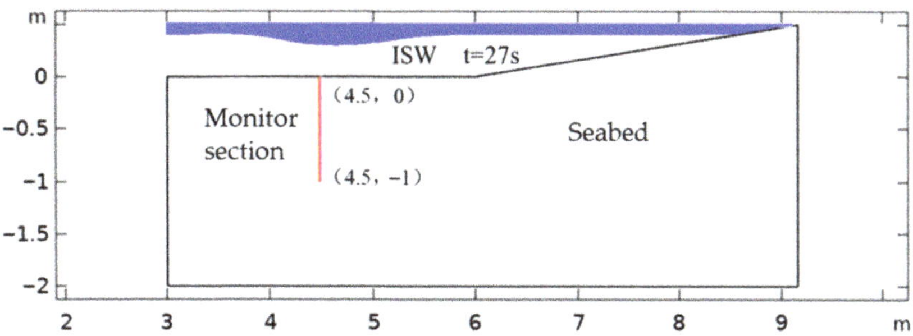

Figure 2. Schematic diagram of parameter analysis numerical model and monitoring cut-off position.

Figure 3 illustrates the distribution of excess pore water pressure with depth by ISW, within a 1 m range on the seabed. The distribution of excess pore water pressure demonstrates consistent characteristics. At the surface of the seabed, the continuous condition of water pressure being equal to the pressure of ISWs, along with the negative force exerted by ISWs, results in a negative excess pore water pressure. As depth increases, the excess pore water pressure gradually rises at varying rates, influenced by differences in parameters, and approaches zero. Specific cases may exhibit unique conditions, which will be elaborated upon in subsequent sections. The absolute value of excess pore water pressure decreases with depth, indicating that the influence of ISWs on the seabed diminishes as depth increases. These results are consistent with reference [10].

Figure 3a compares the effects of varying shear modulus on excess pore water pressure in the seabed. Within the depth range of 0 to 0.2 m, the results across the four cases show minimal differences. Below a depth of 0.2 m, the results for shear moduli G2, G3, and G4 become more similar and approach zero at a depth of 0.5 m (Figure 3a). In contrast, the excess pore water pressure under shear modulus G1 tends towards −0.3 Pa at 0.5 m. Overall, at any given depth, a smaller seabed shear modulus correlates with a greater change in excess pore water pressure induced by ISWs.

Figure 3b examines the impact of differing permeability coefficients on excess pore water pressure in the seabed. Variations in permeability coefficients significantly influence the distribution of excess pore water pressure (Figure 3b). Under the permeability coefficient K1, the excess pore water pressure approaches zero at a depth of 0.4 m. When the permeability coefficient is K2, this pressure reaches zero at a depth of 0.2 m. For permeability coefficient K3, the rate of change in excess pore water pressure with depth accelerates, reaching zero at approximately 0.5 m. In the case of permeability coefficient K4, the excess pore water pressure caused by ISWs is limited to a depth of about 0.2 m at the seabed surface, with no further influence detected below this depth. Comparative analysis indicates that permeability coefficients significantly affect the distribution of excess pore water pressure in the seabed. Higher permeability coefficients allow ISWs to exert influence at greater depths, while lower coefficients result in a more rapid decrease in excess pore water pressure with depth, thereby reducing the depth of ISW influence on the seabed.

Figure 3c illustrates the relationship between porosity and excess pore water pressure in the seabed. At a constant depth, an increase in porosity corresponds to a reduction in the absolute value of excess pore water pressure, indicating a diminished impact of ISWs on the seabed (Figure 3c). Conversely, the depth of ISW influence on the seabed is inversely related to porosity.

Figure 3d examines the effect of saturation on excess pore water pressure within the seabed. Variations in saturation lead to notable differences in the distribution of excess pore water pressure (Figure 3d). In cases where the seabed is not fully saturated (S1, S2, S3), excess pore water pressure approaches zero at a specific depth, which increases with rising saturation levels. In contrast, when the seabed is fully saturated (saturation S4), the change in excess pore water pressure is gradual within a depth range of 1 m and does not stabilize. At a depth of 1.0 m, the excess pore water pressure is approximately −2.8 Pa.

From the analysis of these four parameters, it can be concluded that the saturation and permeability coefficient of the seabed significantly affect excess pore water pressure by ISW, while the effects of shear modulus and porosity are comparatively minor. The seabed's response to ISWs is directly proportional to both saturation and permeability coefficient and inversely proportional to shear modulus and porosity.

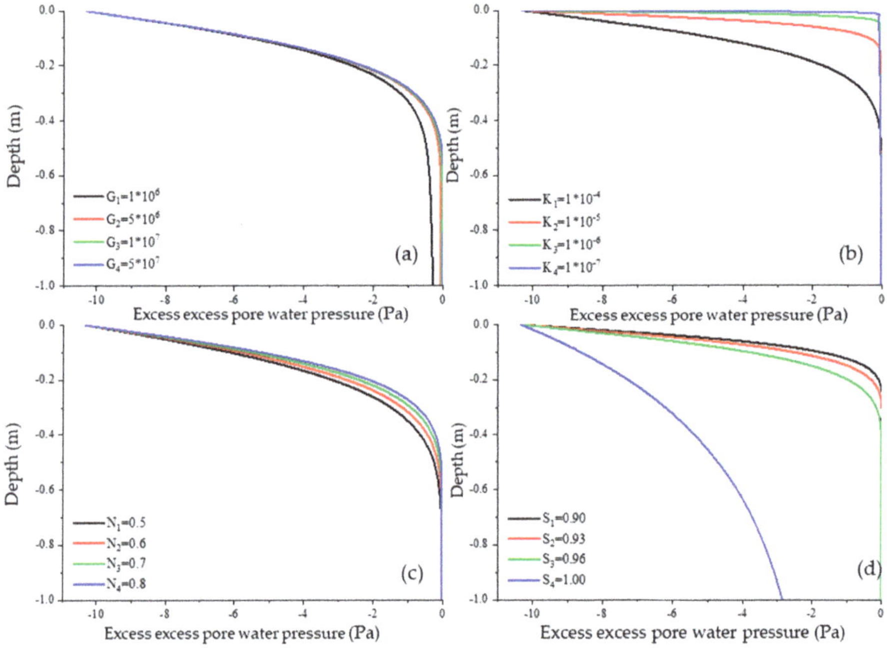

Figure 3. The influence of seabed parameters on the distribution of excess pore water pressure. (**a**) varying shear modulus, (**b**) varying permeability coefficients, (**c**) varying porosity, (**d**) varying saturation.

3.2. *Influence of Shear Modulus Stratification*

In cases 1 and 2, while all other parameters were maintained constant, only the shear modulus of the seabed was varied, with its specific layered distribution being depicted in Figure 4. The response of excess pore water pressure in the seabed induced by ISWs is illustrated in Figure 4. This figure highlights several critical moments, including the propagation of ISWs over the horizontal seabed, their interaction with the slope at the foot of the slope, the point of breaking upon contact with the slope, and the subsequent phase after breaking. The dynamics of shoaling ISWs and the associated changes in excess pore water pressure within the seabed have been previously discussed by reference [15].

Upon comparison, no significant differences were observed in the results between the two cases, indicating that the interlayer was excessively smooth. Furthermore, no considerable impact of shear modulus stratification on the distribution of excess pore water pressure was detected (Figure 5). Consequently, it can be concluded that variations in shear modulus do not play a critical role in influencing the seabed response in a layered seabed (see Figure 5).

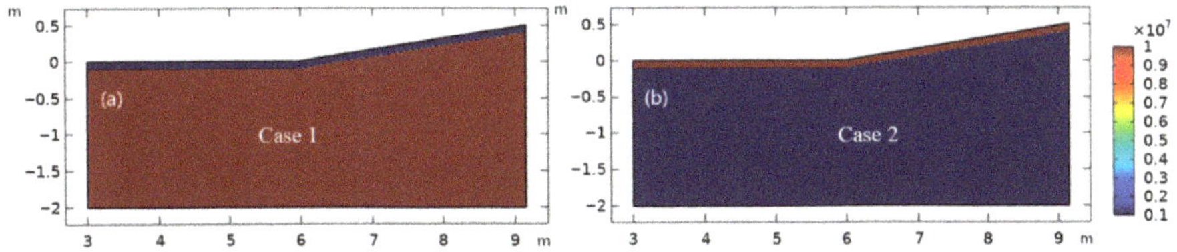

Figure 4. (**a**,**b**) The distribution of shear modulus in the seabed. See Table 2 for more information in cases 1 and 2.

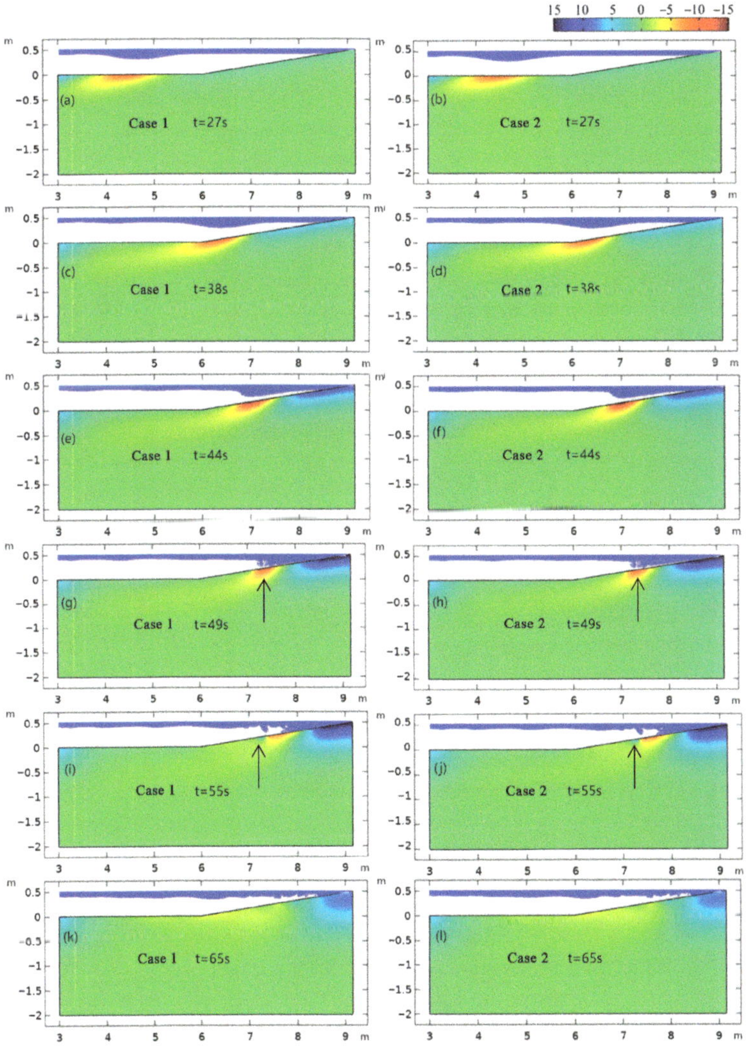

Figure 5. Excess pore water pressure induced by ISW in stratified seabed (cases 1 and 2). Black arrow indicates obvious changes. (**a**,**c**,**e**,**g**,**i**,**k**) are case 1, (**b**,**d**,**f**,**h**,**j**,**l**) are case 2.

3.3. Influence of Permeability Coefficient Stratification

In cases 3 and 4, the permeability coefficient of the seabed varies by layer. In case 3, the upper layer of the seabed consists of soil with a low permeability coefficient, while the lower layer exhibits a high permeability coefficient. Conversely, the characteristics are reversed in case 4, as illustrated in Figure 6.

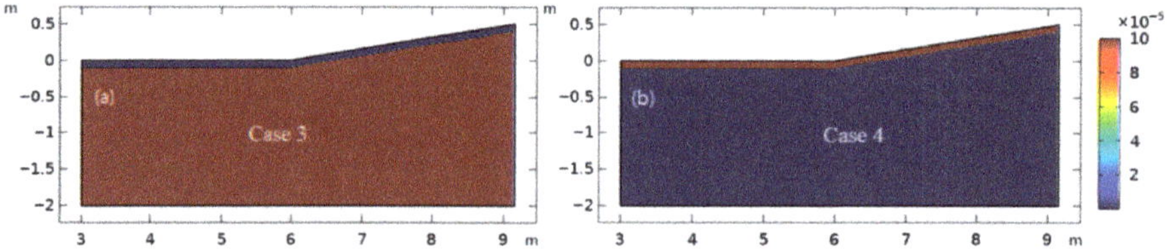

Figure 6. (**a**,**b**) The distribution of the permeability coefficient in the seabed. See Table 2 for more information in cases 3 and 4.

Figure 7 depicts the location of ISW action and the distribution of excess pore water pressure within the seabed. The layered differences in permeability coefficients result in significant variations in excess pore water pressure distribution induced by ISWs. In case 3, throughout the duration of ISW action, no excess pore water pressure was generated within the seabed; instead, excess pore water pressure was confined to the shallow surface of the seabed, rendering it difficult to discern in the cloud map (Figure 7). This observation indicates that when the permeability of the surface soil is poor, the effects of ISWs are limited to a shallow depth range, without propagating into the deeper layers of the seabed. Consequently, the excess pore water pressure within the seabed remains unchanged, and the original equilibrium state is preserved.

When the seabed soil is homogeneous and exhibits uniform permeability, the maximum (absolute value) of excess pore water pressure is observed in the surface layer of the seabed. And as it extends deeper, the pressure exhibits a gradual decline, ultimately approaching zero, with no further influence at greater depths. In case 4 (Figure 7b), the upper soil layer demonstrates strong permeability; thus, under ISW action, excess pore water pressure develops within this layer. However, due to the poor permeability of the lower soil layer, which behaves similarly to an impermeable layer, the transmission of excess pore water pressure is completely halted at the interface between the seabed layers. This results in the excess pore water pressure exhibiting minimal variation within the upper soil layer, where permeability is greater, leading to a more uniform distribution that abruptly reaches zero upon encountering the "impermeable layer."

A comparison of cases 3 and 4 reveals that soil with a low permeability coefficient and poor permeability effectively functions as an "impermeable layer," significantly influencing the transmission of excess pore water pressure induced by ISWs into the deeper strata of the seabed. Therefore, in real seabed environments, it is crucial to consider the permeability stratification of seabed soils.

3.4. Influence of Porosity Stratification

In cases 5 and 6, all seabed parameters remain consistent and do not vary, with the exception of porosity. The stratification of porosity in these two cases is illustrated in Figure 8. In case 5, the porosity of the upper layer of the seabed soil is relatively low, at 0.4, while the porosity of the lower layer is comparatively high, at 0.8. In contrast, case 6 exhibits the opposite porosity distribution.

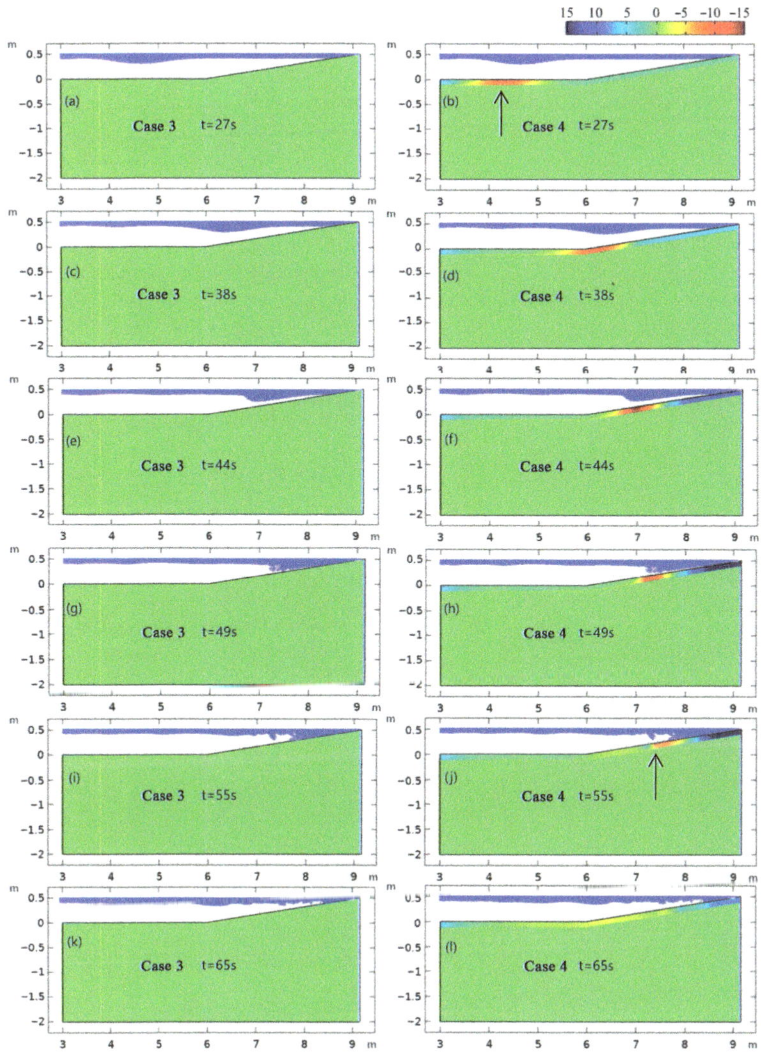

Figure 7. Excess pore water pressure induced by an ISW in a stratified seabed (cases 3 and 4). Black arrow indicates obvious changes. (**a**,**c**,**e**,**g**,**i**,**k**) are case 3, (**b**,**d**,**f**,**h**,**j**,**l**) are case 4.

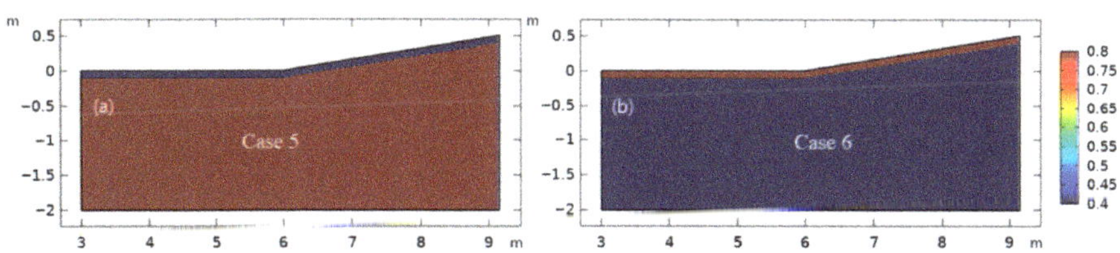

Figure 8. (**a**,**b**) The distribution of porosity in the seabed. See Table 2 for more information in cases 5 and 6.

Figure 9 presents the cases characterized by porosity stratification. In cases 5 and 6, the variation in excess pore water pressure induced by ISWs at the stratification interface is relatively gradual, lacking the abrupt changes observed in case 4. The difference in porosity of the upper soil layers between cases 5 and 6 results in a slight variation in the extent of influence exerted by excess pore water pressure (Figure 9). Specifically, in case 6, the excess pore water pressure at the seabed is transmitted to greater depths compared to case 5. This phenomenon is evident throughout the entire duration of ISW activity, including instances of negative excess pore water pressure during valley action and positive excess pore water pressure near the slope crest during the ISW breaking phase.

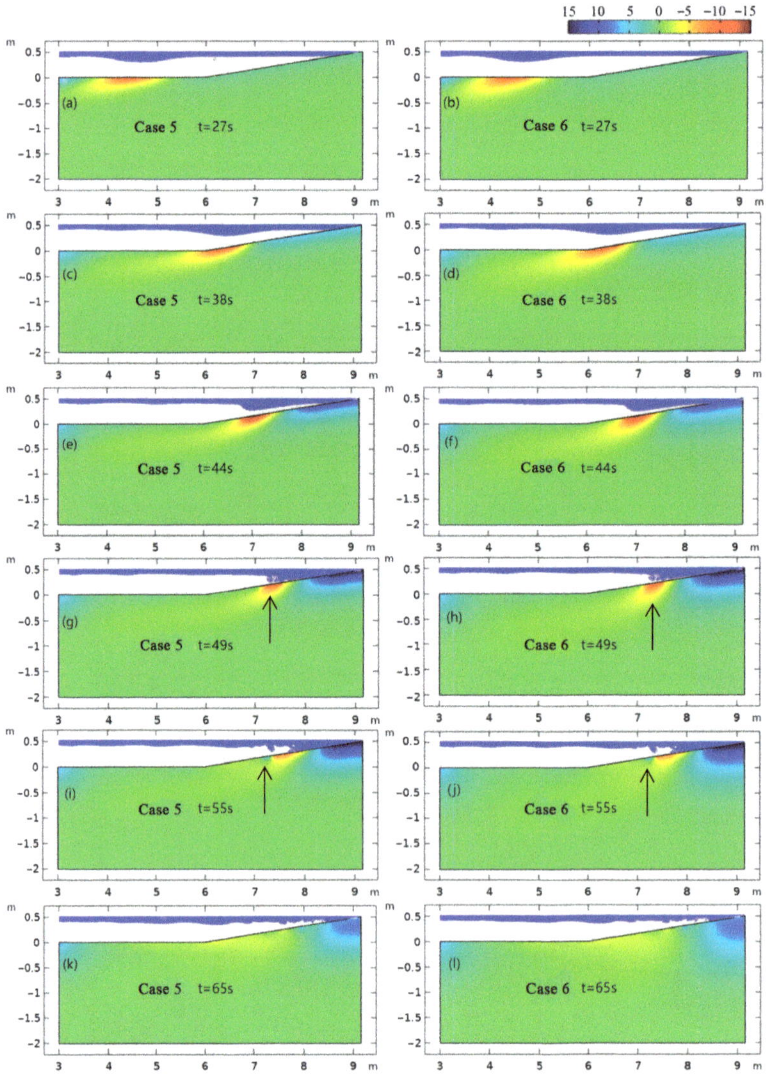

Figure 9. Excess pore water pressure induced by an ISW in a stratified seabed (cases 5 and 6). Black arrow indicates obvious changes. (**a**,**c**,**e**,**g**,**i**,**k**) are case 5, (**b**,**d**,**f**,**h**,**j**,**l**) are case 6.

3.5. Impact of Saturation Stratification

We compare the response of ISWs to excess pore water pressure in the seabed during saturation stratification by controlling variables for cases 7 and 8, as illustrated in Figure 10.

Figure 11 demonstrates that the stratification of saturation significantly influences the distribution of excess pore pressure.

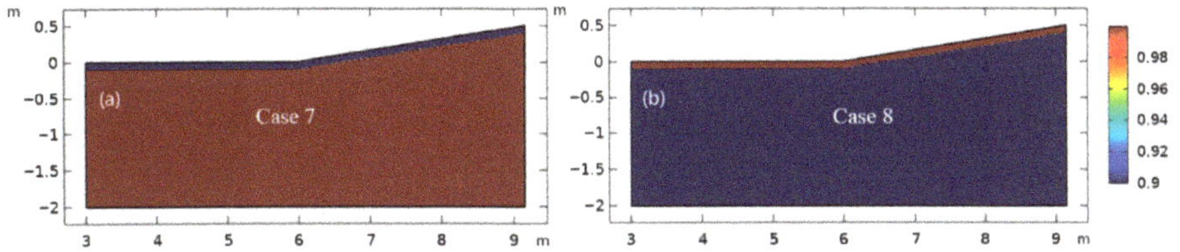

Figure 10. (**a**,**b**) The distribution of saturation in the seabed. See Table 2 for more information in cases 7 and 8.

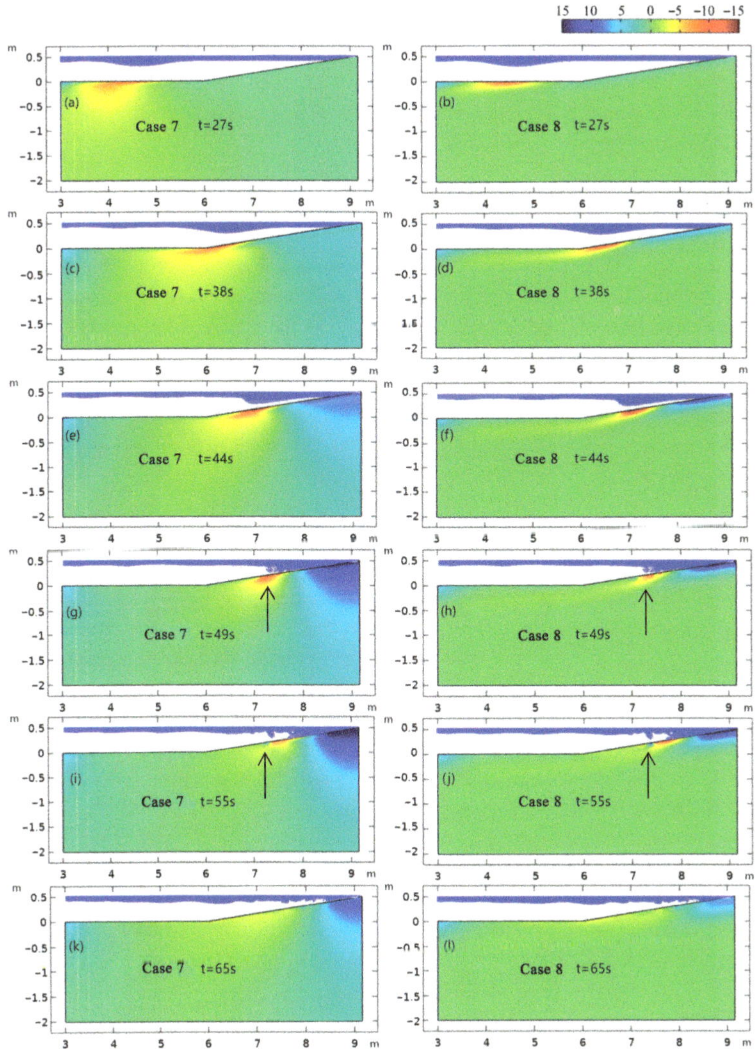

Figure 11. Excess pore water pressure induced by an ISW in a stratified seabed (cases 7 and 8). Black arrow indicates obvious changes. (**a**,**c**,**e**,**g**,**i**,**k**) are case 7, (**b**,**d**,**f**,**h**,**j**,**l**) are case 8.

In case 7, the saturation level of the upper soil layer is set at 0.9, while the lower soil layer is fully saturated. Under the influence of ISWs, excess pore water pressure generated in the upper soil layer is concentrated beneath the trough of the ISWs. As this pressure propagates downward and encounters the stratification, the increased saturation in the soil leads to a pronounced diffusion of excess pore water pressure to the lower left side (Figure 11). This indicates that higher saturation enhances the diffusion and transmission of excess pore water pressure. As the ISW approaches the foot of the slope, the excess pore water pressure in the seabed also diffuses downward due to the increased saturation of the lower soil layer. Furthermore, the excess pore water pressure (positive values) resulting from the breaking of ISWs near the top of the slope exhibits significant downward propagation (Figure 11).

In contrast, the saturation stratification for case 8 is reversed compared to that of case 7, leading to markedly different results. In this case, seabed saturation decreases from 1 at the surface to 0.9 at depth, which hinders the transmission of excess pore water pressure in the lower soil layer. Consequently, the distribution of excess pore pressure in case 8 is shallower than that in case 7 at the same time. The reduction in soil saturation suppresses the deeper diffusion of pore pressure (Figure 11).

4. Conclusions

The seabed response induced by ISWs and their shoaling in the presence of stratification within the seabed soil is investigated by numerical simulations. The research first examines the effects of four key parameters: seabed shear modulus, permeability coefficient, porosity, and saturation, on the distribution of excess pore water pressure in the seabed. Subsequently, a layered seabed model is developed to analyze the variations in excess pore water pressure distribution resulting from stratified seabed parameters. Our results indicate that the saturation and permeability coefficient of the seabed significantly influence the excess pore water pressure generated by ISWs. As both saturation and permeability increase, the excess pore water pressure induced by ISWs penetrates deeper into the seabed. In contrast, the effects of shear modulus and porosity are relatively minor and inversely related to the depth of influence of excess pore water pressure. When stratification occurs in the permeability coefficient and saturation of the seabed, notable changes arise in the downward propagation of excess pore water pressure. Low-permeability soil layers impede the transmission of excess pore water pressure. If such layers are situated in the upper portion of the seabed, excess pore water pressure is confined to the surface layer. Conversely, if low-permeability layers are located deeper within the seabed, they function as impermeable barriers, obstructing the downward movement of excess pore water pressure. Saturation stratification exhibits similar effects, with soil layers exhibiting higher saturation being more conducive to the transmission of excess pore water pressure.

Ocean engineering is developing rapidly all over the world and is constantly moving towards deep water. This study provides a basis for evaluating seabed liquefaction, erosion processes, stability, and risk assessment by ISWs and contributes to researching seabed topography induced by ISWs [24,25]. The research results provide scientific guidance for the exploitation of submarine resources and the prevention and control of deep-sea engineering geological disasters.

Author Contributions: H.T. wrote the manuscript, assisted by L.J. and Z.T., with comments from all of the authors. L.J. collected data. L.J., Z.T., and J.Z. designed the field experiment. L.W., J.K., and F.W. assisted with cable fault data. All authors have read and agreed to the published version of the manuscript.

Funding: This project was provided by the Qingdao Postdoctoral Funding Project (No. QDBSH20240202079), National Natural Science Foundation of China (No. 41976049), China Marine Geological Survey Project (No. DD20243114), and Natural Science Foundation of Shandong Province (No. ZR2021MD074).

Data Availability Statement: The data within the manuscript is available from the corresponding author upon reasonable request.

Conflicts of Interest: The authors declare no competing interests.

References

1. Huang, X.; Chen, Z.; Zhao, W.; Zhang, Z.; Zhou, C.; Yang, Q.; Tian, J. An extreme internal solitary wave event observed in the northern South China Sea. *Sci. Rep.* **2016**, *6*, 30041. [CrossRef] [PubMed]
2. Zhang, X.; Huang, X.; Zhang, Z.; Zhou, C.; Tian, J.; Zhao, W. Polarity Variations of Internal Solitary Waves over the Continental Shelf of the Northern South China Sea: Impacts of Seasonal Stratification, Mesoscale Eddies, and Internal Tides. *J. Phys. Oceanogr.* **2018**, *48*, 1349–1365. [CrossRef]
3. La Forgia, G.; Tokyay, T.; Adduce, C.; Constantinescu, G. Bed shear stress and sediment entrainment potential for breaking of internal solitary waves. *Adv. Water Resour.* **2020**, *135*, 103475. [CrossRef]
4. Tian, Z.; Guo, X.; Qiao, L.; Jia, Y.; Yu, L. Experimental investigation of slope sediment resuspension characteristics and influencing factors beneath the internal solitary wave-breaking process. *Bull. Eng. Geol. Environ.* **2019**, *78*, 959–967. [CrossRef]
5. Tian, Z.; Huang, J.; Xiang, J.; Zhang, S.; Wu, J.; Liu, X.; Luo, T.; Yue, J. Interaction between internal solitary waves and the seafloor in the deep sea. *Deep. Undergr. Sci. Eng.* **2024**, *3*, 149–162. [CrossRef]
6. Tian, Z.; Jia, Y.; Du, Q.; Zhang, S.; Guo, X.; Tian, W.; Zhang, M.; Song, L. Shearing stress of shoaling internal solitary waves over the slope. *Ocean Eng.* **2021**, *241*, 110046. [CrossRef]
7. Xiang, J.; Tian, Z.; Dong, Y.; Zhao, E.; Song, L.; Yue, J.; Zhou, G.; Liu, X.; Guo, X.; Jia, Y. Dynamic response of seabed in wind power of deep-shallow sea induced by internal solitary waves. *Mar. Georesour. Geotechnol.* **2024**, *43*, 1198–1214. [CrossRef]
8. Qiao, L.; Guo, X.; Tian, Z.; Jia, Y.; Zhang, M. Analysis on internal solitary wave-induced dynamic response characteristics of surface sediments in the Northern South China Sea. *Chin. J. Undergr. Space Eng.* **2016**, *12* (Suppl. S2), 604–611.
9. Rivera-Rosario, G.A.; Diamessis, P.J.; Jenkins, J.T. Bed failure induced by internal solitary waves. *J. Geophys. Res. Oceans* **2017**, *122*, 5468–5485. [CrossRef]
10. Tian, Z.; Chen, T.; Yu, L.; Guo, X.; Jia, Y. Penetration depth of the dynamic response of seabed induced by internal solitary waves. *Appl. Ocean Res.* **2019**, *90*, 101867. [CrossRef]
11. Tian, Z.; Jia, L.; Hu, N.; Wang, S.; Zhang, M.; Zhou, G. Spatial and temporal variation process of seabed dynamic response induced by the internal solitary wave. *Acta Oceanol. Sin.* **2023**, *42*, 142–149. [CrossRef]
12. Qiao, L.; Guo, X.; Tian, Z.; Yu, L. Experimental analysis of pore pressure characteristics of slope sediments by shoaling internal solitary waves. *Haiyang Xuebao* **2018**, *40*, 68–76. [CrossRef]
13. Li, Y.; Liu, L.; Gao, S.; Zhang, Y.; Xiong, X. Experimental study on dynamic response characteristics of continental shelf slope to internal solitarywaves. *Haiyang Xuebao* **2021**, *43*, 126–134. [CrossRef]
14. Li, Y.; Liu, L.; Zhou, Q.; Hui, Y. Experimental study on the dynamic process and characteristics of slope sediments after breaking of internal solitary waves. *Haiyang Xuebao* **2022**, *44*, 42–50. [CrossRef]
15. Tian, Z.; Jia, L.; Xiang, J.; Yuan, G.; Yang, K.; Wei, J.; Zhang, M.; Shen, H.; Yue, J. Excess pore water pressure and seepage in slopes induced by breaking internal solitary waves. *Ocean Eng.* **2023**, *267*, 113281. [CrossRef]
16. Cai, S.; Xie, J.; He, J. An Overview of Internal Solitary Waves in the South China Sea. *Surv. Geophys.* **2012**, *33*, 927–943. [CrossRef]
17. Grimshaw, R. Internal Solitary Waves. In *Environmental Stratified Flows. Topics in Environmental Fluid Mechanics*; Grimshaw, R., Ed.; Springer: Boston, MA, USA, 2003; Volume 3. [CrossRef]
18. Grimshaw, R.; Pelinovsky, E.; Talipova, T. Modelling Internal Solitary Waves in the Coastal Ocean. *Surv. Geophys.* **2007**, *28*, 273–298. [CrossRef]
19. Grimshaw, R.; Pelinovsky, E.; Talipova, T.; Kurkina, O. Internal solitary waves: Propagation, deformation and disintegration, Nonlin. *Process. Geophys.* **2010**, *17*, 633–649. [CrossRef]
20. Miles, J.W. On internal solitary waves. *Tellus* **1979**, *31*, 456–462. [CrossRef]
21. Sutherland, B.R.; Barrett, K.J.; Ivey, G.N. Shoaling internal solitary waves. *J. Geophys. Res. Oceans* **2013**, *118*, 4111–4124. [CrossRef]
22. Tian, Z.; Liu, C.; Ren, Z.; Guo, X.; Zhang, M.; Wang, X.; Song, L.; Jia, Y. Impact of seepage flow on sediment resuspension by internal solitary waves: Parameterization and mechanism. *J. Oceanol. Limnol.* **2022**, *41*, 444–457. [CrossRef]

23. Zhang, Z.; Chen, J.; Zhu, Z.; Pan, Y.T.; Sun, M. Sand liquefaction and seepage pore pressure around shield tunnels in multilayered seabed under action of waves and currents. *Chin. J. Geotech. Eng.* **2024**, *46*, 683–693.
24. Tian, Z.; Liu, C.; Jia, Y.; Song, L.; Zhang, M. Submarine Trenches and Wave-Wave Interactions Enhance the Sediment Resuspension Induced by Internal Solitary Waves. *J. Ocean Univ. China* **2022**, *22*, 983–992. [CrossRef]
25. Fan, C.; Tian, Z.; Cui, K.; Huang, J.; Bian, S.; Yang, L.; Zhang, T. A new discovery of source contribution and transport mechanism of clay minerals in Taiwan Canyon-Manila Trench. *Deep. Sea Res. Part I Oceanogr. Res. Pap.* **2025**, *224*, 104571. [CrossRef]

water

MDPI

Article

Effect of Relative Wavelength on Excess Pore Water Pressure in Silty Seabeds with Different Initial Consolidation Degrees

Hongyi Li [1], Yaqi Zhang [1,*], Aidong Ma [1], Mingzheng Wen [2], Zixi Zhao [1] and Shaotong Zhang [1]

1 Frontiers Science Center for Deep Ocean Multispheres and Earth System, Key Lab of Submarine Geosciences and Prospecting Techniques, MOE, State Key Laboratory of Physical Oceanography, College of Marine Geosciences, Ocean University of China, Qingdao 266100, China; lihongyi01211@163.com (H.L.); mad@stu.ouc.edu.cn (A.M.); zhaozixi@stu.ouc.edu.cn (Z.Z.); shaotong.zhang@ouc.edu.cn (S.Z.)

2 Tianjin Geological Survey Center, China Geological Survey, Tianjin 300170, China; wenmzh@foxmail.com

* Correspondence: zhangyaqi@ouc.edu.cn

Abstract

Wave-induced silty seabed liquefaction is one of the key threats to offshore infrastructure stability. The excess pore pressure (EPP) response is the key parameter for judging seabed liquefaction. Many studies have examined the EPP response to surface waves in initially well-consolidated seabed; few works have explored initially less-consolidated seabed, which is widely distributed in estuaries due to the massive river sediment discharge and, thereafter, rapid accumulation. Here, we investigate the EPP response of silty seabed with various initial consolidation degrees using wave flume experiments. We found that (1) in initially liquefied seabed, the EPP magnitude monotonically increases with wavelength, while in initially consolidated seabed, there is a maximal response wavelength which is inversely related to consolidation degree. (2) Furthermore, we found two opposite EPP responses to cyclic surface wave loading under varying seabed conditions in initially liquefied and consolidated seabeds. That is, under the same waves, the EPP magnitude is inversely related to the consolidation degree in initially liquefied seabed, while the EPP magnitude is positively related to the consolidation degree in initially consolidated seabed. In other words, the influence of initial seabed consolidation degree on EPP magnitude behaves like a "√" shaped curve. Our findings provide some implications for further understandings of wave-induced silty seabed liquefaction.

Keywords: wave-seabed interactions; seabed liquefaction; liquefaction degree; negative excess pore water pressure; Yellow River Delta

1. Introduction

Wave-induced seabed liquefaction may cause catastrophic damage to offshore infrastructure, resulting in huge economic losses. Previous incidents serve as clear evidence of this risk, including the damage to the breakwater in Niigata's West Port in 1976 [1], the repeated ruptures of a buried steel pipeline in Lake Ontario [2], and the breakage of the submarine cable in the Chengdao oilfield in 2003 [3]. Additionally, Puzrin et al. [4] hypothesized the liquefaction of a breakwater caisson foundation in Barcelona under intense wave action. Li et al. [5] used field data and a model to identify multiple liquefaction events in Jiangsu Province, confirming liquefaction as a critical factor in such incidents.

According to Terzaghi's principle of effective stress [6], in saturated seabeds, external wave loading increases pore water pressure, with the excess over static pressure termed

Excess Porewater Pressure (EPP). Liquefaction occurs when EPP surpasses the initial effective stress of the seabed. Two mechanisms of liquefaction are identified: transient liquefaction, common in well-drained sands, and residual liquefaction, more frequent in silty seabeds [7–10]. Experiments by Tzang [11] showed silty seabeds exhibit both oscillations and pore water accumulation due to their higher compressibility and lower permeability, making them more prone to liquefaction [12].

Silty sediments are widely distributed around the world. The modern Yellow River Delta in China is a typical silt-clay coast, with surface sediments primarily composed of silts. Due to the rapid sedimentation of Yellow River derived sediments which are often in an unconsolidated state, they are characterized by high water content, high pore ratio, and lower strength, thus are highly susceptible to soil liquefaction under cyclic loading of waves [13].

Existing studies have shown that the main factors affecting seabed liquefaction include the hydrodynamics (wave period, wave height, water depth, and the history of wave loading, etc.), as well as soil parameters (permeability coefficient, shear modulus, consolidation coefficient, and saturation of the seabed, etc.). The present work selects two of the most representative key indicators: the relative wavelength, defined as the ratio of wavelength (λ) of surface water waves to water depth (h), i.e., λ/h. This dimensionless parameter is used to represent different wave conditions in this study (associated wave frequency and periodicity), and the consolidation degree is defined as the complementary number of the liquefaction degree, where the liquefaction degree is the ratio of excess pore water pressure to initial vertical effective stress, reflecting critical soil properties such as density and moisture content. It has also been demonstrated in previous research that wavelength and wave period significantly influence the EPP response in seabeds [14–22]. Consolidation degree also greatly affects other properties of the seabed [23]. Therefore, we focused on studying the effects of relative wavelength and consolidation degree and investigated the coupled effects of wavelength and consolidation degree on cumulative pore pressure response.

A review of existing research in this area is provided below. Ippen [14] highlighted that in water depths exceeding half the wavelength, the influence of short waves decreases with depth, whereas long waves become more significant. Okusa [15] reported that long-period waves experience less damping and create more noticeable pore pressure delays in silty seabeds than short-period waves. Maeno and Hasegawa [16] discovered that low-frequency waves propagate more effectively through seabeds. The numerical calculations by Liu et al. [17] found that as the wave period increases, the liquefaction depth decreases, indicating that shorter wave periods make the seabed more prone to liquefaction. In recent years, some scholars focus on the seabed response under random wave action. Niu et al. [18] found that random waves generate higher cumulative EPP than regular waves due to significant low-frequency contributions. Zhang et al. [19] found that EPP near a monopile increases with wave period under random waves, similar to regular waves, but random waves contribute more significantly to EPP. Klammler et al. [20] analyzed wave pressure and liquefaction during a storm in Panama City Beach, categorizing waves into infra-gravity (<0.075 Hz), swell (0.075–0.25 Hz), and short-wave (>0.25 Hz) bands. They found swell waves were the primary cause of sediment destabilization, with short waves contributing $\approx$ 20% and infra-gravity waves $\approx$ 10%. Xu et al. [21] noted that seabed liquefaction during storm waves developed slowly, taking about 1.5 h, because early high-frequency wind waves had little effect, whereas later, low-frequency, long-period swells had a much stronger influence on the seabed. Xu et al. [22] conducted flume experiments using spectral analysis to separate random waves into long and short components. They studied EPP responses to these waves at different depths in a silty seabed, finding long waves transmit more easily and contribute significantly to accumulation at certain depths, increasing liquefaction risk.

The sediment consolidation degree also plays a significant role in the EPP response under waves [23–25]. Liu et al. [23] studied Yellow River Delta sediments through experiments and field observations, finding that key geotechnical properties like unit weight and shear strength improved with consolidation time, while water content and void ratio declined. Ren et al. [24] found that under wave loading, highly consolidated silty seabeds initially fail by shear, while low-consolidation seabeds fail by liquefaction. Pore water pressure development also varies with consolidation degrees. Chen et al. [25] microscopically explained the findings from Ren et al. [24] by simulating sediment particle arrangement during consolidation using discrete element methods. They observed that low consolidation results in an open microstructure, while higher consolidation increases particle overlap, significantly reducing permeability due to enhanced sediment cohesion.

In summary, previous studies have only focused on the EPP response under different wave components or consolidation degrees, separately. Few studies have experimentally investigated the EPP response considering both wavelength (wave period, etc.) and consolidation degree (liquefied state or non-liquefied state), simultaneously. Therefore, the present study investigates the EPP responses under different wavelengths for initially liquefied or non-liquefied seabed of varying consolidation degrees through 24 groups of controlled wave flume experiments.

2. Materials and Methods

2.1. Experimental Flume and Instruments

A total of 24 groups of experiments were carried out in a small wave flume; this is because the initial seabed conditions are more controllable in a small tank, and it is also more effective to conduct multiple comparison groups in a small wave flume. As shown in Figure 1, the main structure consisted of two parts: the wave flume and the soil tank. The flume was 2.4 m long, 0.23 m wide, and 0.6 m high, with a wave generation device (a curved slider) on the right side of the flume, which allows different wave conditions to be set by changing the motor's operating frequency. The left side of the wave flume was filled with porous material to eliminate reflected waves, which effectively absorbed wave energy and significantly reduced the impact of wave reflection on the experimental results. The wave flume was made entirely of transparent acrylic panels, which facilitated the observation and recording of the experimental process.

The soil tank was located 0.6 m away from the wave generation device, with a length of 0.6 m and a height of 0.16 m. It featured a detachable design; thus, the seabed could be prepared outside and then placed into the water flume. Two pore pressure sensors were vertically placed at the center of the soil tank, at 5.5 cm and 10.7 cm depth from the seabed surface, respectively, to measure the EPP response at different depths. To prevent sudden changes in wave shape, a slope was set up between the wave generator and the soil tank. A wave gauge was installed on the left side of the tank to measure the wave conditions during the experiment.

The pore pressure sensor used was the DY2012 high-precision digital pore water pressure sensor produced by Chengdu Yiyunda Technology (Chengdu, China). The sensor uses the latest international SOC (System on Chip) as the CPU, combined with high precision and high stability reference source technology, as well as a series of advanced technologies for signal acquisition and processing, communication, and bus development. It features software zeroing, floating-point data processing, bidirectional communication with the host computer, and standardized digital output, allowing it to be used without any additional data acquisition equipment.

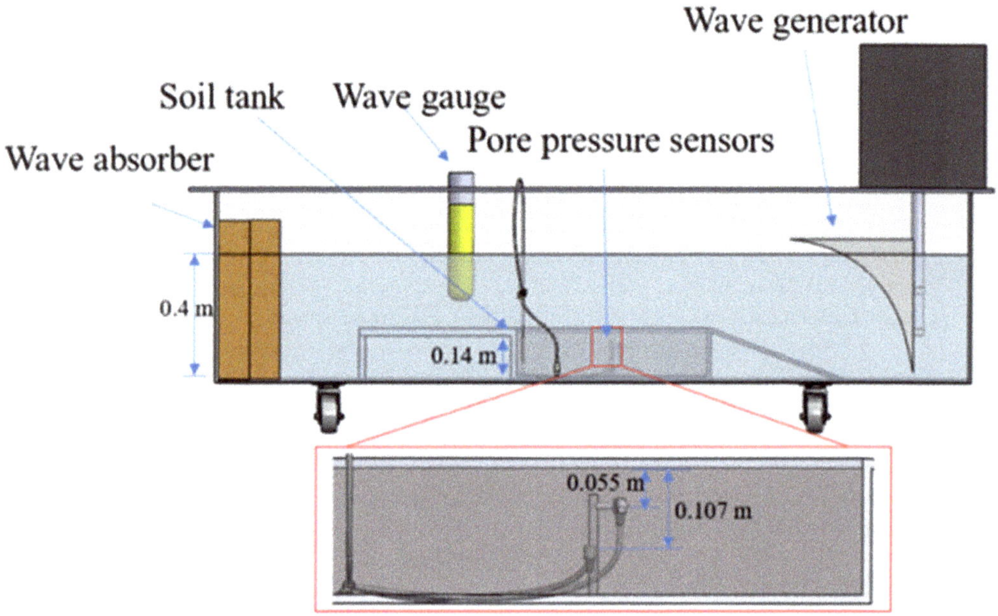

Figure 1. Layout of the experimental flume and instruments.

The wave gauge was the TWAVE101 produced by Beijing Haizhou Saiwei Technology (Beijing, China). This instrument features high-precision pressure and temperature recording capabilities. The parameters of the pore pressure sensor and wave gauge are shown in Table 1.

Table 1. Parameters of the instruments used in the experiments.

Instruments	Model	Sampling Rate	Precision	Range of Measurement
Pore pressure sensor	DY2012	20 Hz	0.1%	0–1 MPa
Wave gauge	TWAVE101	8 Hz	0.03%	/

2.2. Experimental Soil

The experimental soil was taken from the Chengdao area in the northern part of the Yellow River Delta, as marked by the red star in Figure 2. The reason for selecting this sampling location is that, through field experiments, it was found that the soil in this area is highly susceptible to liquefaction under cyclic vibration loads. To analyze the physical and mechanical properties of the liquefiable soil, we collected cylindrical samples with a length of 25 cm from three sites in this area. The cylindrical samples were divided into layers every 5 cm, and tests for water content and particle size analysis were conducted. Particle size analysis was performed using a laser particle size analyzer. The experimental results are shown in Figure 3. The water content of the soil layers was in the range of 21–28%. The clay content was 6–19%, the silt content was 69–86%, the sand content was 2–22%, and the median grain size was 0.03 mm.

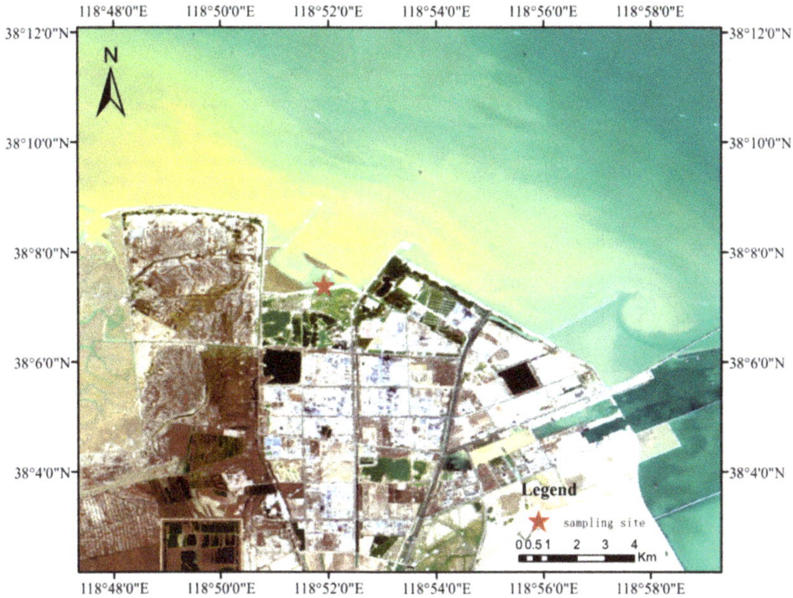

Figure 2. Sampling location of the experimental soil.

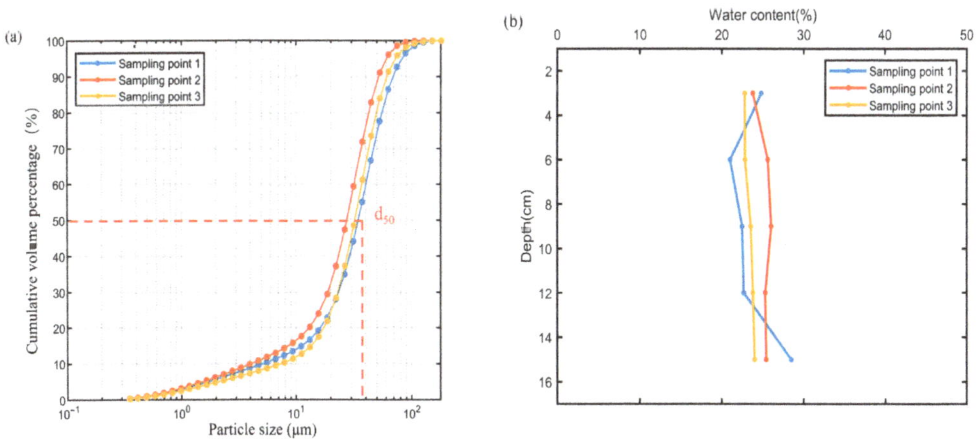

Figure 3. (**a**) Particle size distribution curve of the experimental soil; (**b**) water content of cylindrical samples along the depth.

2.3. Data Processing Methods

Criteria for Seabed Liquefaction:

According to Terzaghi's principle of effective stress, the total stress in normally consolidated seabed is equal to the sum of the effective stress between soil particles and the pore water pressure. Under cyclic loading from waves, the pore water pressure continues to increase, leading to a decrease in the vertical stress between soil particles. When the pore water pressure exceeds the initial effective stress of the soil particles, liquefaction occurs in the seabed. Regarding the moment when seabed liquefaction begins, some researchers have proposed different criteria for identification. Zen and Yamazaki [7] discovered that silty soil begins to liquefy under wave action when pore water pressure rises suddenly. Sumer et al. [26] believe that liquefaction starts in the seabed when the pore water pressure reaches its maximum value, initially occurring near the mudline and gradually developing downward. Some researchers have proposed one-dimensional and three-dimensional criteria for liquefaction identification. Okusa [27], based on the principle of effective stress, argued that if the vertical effective stress exceeds the soil weight, liquefaction will occur in the seabed. He proposed the following criteria for liquefaction of the soil under one-dimensional conditions:

$$-(\gamma_s - \gamma_w)z = \sigma'_{z0} \leq \sigma'_{zu} \tag{1}$$

where γ_s represents unit weight of soil, γ_w represents unit weight of water, z represents the thickness of the soil layer, σ'_{z0} represents the initial vertical effective stress of the soil, and σ'_{zu} represents the critical vertical effective stress.

Tasi [28] built upon this foundation by applying the average effective stress and extending it to three-dimensional conditions:

$$-\frac{1+2k_0}{3}(\gamma_s - \gamma_w)z = \frac{1}{3}\left(\sigma'_{xd} + \sigma'_{yd} + \sigma'_{zd}\right) \tag{2}$$

where k_0 represents soil static lateral pressure coefficient.

Zen and Yamazaki [7] established a one-dimensional liquefaction identification criterion based on the development of pore water pressure:

$$-(\gamma_s - \gamma_w)z = \sigma'_{z0} \leq p - p_b \tag{3}$$

where p represents pore water pressure generated by vibration, and p_b represents static pore water pressure.

Based on Yamazaki's liquefaction identification criterion, Jeng [29] revised it and extended it to three-dimensional conditions:

$$-(\gamma_s - \gamma_w)\left(\frac{1+2k_0}{3}\right)z \leq p - p_b \tag{4}$$

Jia et al. [30] proposed the concept of liquefaction degree to determine whether liquefaction occurs in the soil and the state of liquefaction. To determine whether soil liquefaction has occurred, we refer to the liquefaction degree proposed by Jia et al. [30]. This is defined as the ratio of excess pore water pressure to initial vertical effective stress, which is consistent with the conventional concept of excess pore pressure ratio in soil mechanics. Additionally, in our experimental set up, the seabed was maintained in an initially liquefied state with limited consolidation time, resulting in a high degree of liquefaction. This condition significantly amplified the soil's dynamic response to wave loading, making the wave-induced soil motions visually detectable during experiments. Observations of soil circular motion during experiments were used as supplementary qualitative evidence to further confirm the occurrence of liquefaction. These criteria—liquefaction degree and soil circular motion—were used to assess the occurrence of liquefaction [26,30,31]. When the liquefaction degree is greater than 1, meaning that the excess pore water pressure exceeds the initial vertical effective stress, the soil is considered to be fully liquefied. Due to the short consolidation time of the experimental seabed in this study, the initial excess pore water pressure of the seabed gradually decreased to 0 with the consolidation time before wave loading. Consequently, there exists an initial liquefaction degree. Therefore, in this study, ${f_d}'$ is used to represent the initial liquefaction degree of the seabed.

$${f_d}' = \frac{p_{exc}}{\sigma'_0} \tag{5}$$

where the excess pore pressure (p_{exc}) was calculated by removing the overlying static water pressure (p_w) from the total pressure (p_m) measured at D = 5.5, 10.7 cm subsurface:

$$p_{exc} = p_m - p_w \tag{6}$$

In Equation (5), σ_0' is the initial vertical effective stress of the seabed when it is in a normally consolidated state.

$$\sigma_0' = \gamma' z \tag{7}$$

where z represents the vertical distance from the seabed surface to the pore pressure measurement, and γ' is the buoyant unit weight of normally consolidated soil in the Yellow River Delta, given by $\gamma' = \gamma_s - \gamma_w = 5.46\ \mathrm{kN/m^3}$ [32]. The effective stresses at depths of 5.5 cm and 10.7 cm below the seabed are 0.30 kPa and 0.58 kPa, respectively.

2.4. Experimental Process

It should be emphasized in advance that there are two variables in this experiment, the relative wavelength (λ/d), which is the ratio of wavelength (λ) to water depth (d) and the seabed consolidation degree. Wavelengths were determined through photographic measurement and compared with values derived from the linear wave dispersion relationship using wave gauge data. Although minor discrepancies were observed between the photographically measured and numerically derived wavelengths, the directly measured data were considered more accurate and were therefore selected for defining the wave conditions in this experimental study. In this study, the consolidation degree was used as an indicator to describe the initial state of the seabed soil before wave loading. The initial state of the experimental seabed was set as two types: liquefied and non-liquefied. When the seabed is in a liquefied state, the initial state of the soil includes excess pore water pressure. Therefore, the liquefaction degree (f_d') was used to characterize the different consolidation degrees of the soil in the liquefied soil experiments. In contrast, for normally consolidated seabeds, the initial excess pore water pressure is 0, making it impossible to use the liquefaction degree (f_d') to distinguish the consolidation degrees in the consolidated seabed experiments. Instead, consolidation time was used to differentiate these cases. In this study, the different liquefaction degrees in the liquefied seabed and the consolidation times in the non-liquefied seabed both reflect varying degrees of soil consolidation.

The pore pressure sensor was fixed at 5.5 and 10.7 cm depth using a frame along the center of the soil tank (shown in Figure 1) to ensure the pore water pressure sensor would not move during the experiments. With reference to the water content measured in the experimental soil sampled in the field in Section 2.2, the soil was prepared with a water content of 26% by mixing sediment and water in a ratio of 1:7.5, mixing well so that a small amount of air could be expelled from the soil, and then prepared as a saturated soil, and then backfilled into the soil tank. Song et al. [33] used a sub-bottom profiler to delineate the extent of seabed liquefaction in the Chengdao area of the Yellow River subaqueous delta. They identified liquefied soils based on acoustic reflection characteristics, such as chaotic reflections and poorly developed internal stratification in the imaging. Borehole sampling was conducted in both liquefied and non-liquefied areas, followed by a detailed analysis of the physical and mechanical properties of the soils. Based on their findings, the density of the liquefied surface silt typically ranges from 1.78 to 2.12 ($\mathrm{g/cm^3}$), with an average value of 1.98 ($\mathrm{g/cm^3}$), while the density of the non-liquefied soil ranges from 1.75 to 2.07 ($\mathrm{g/cm^3}$), with an average of 1.97 ($\mathrm{g/cm^3}$). The dry density of the liquefied soil ranges from 1.39 to 1.77 ($\mathrm{g/cm^3}$), with an average of 1.59 ($\mathrm{g/cm^3}$), whereas the dry density of the non-liquefied soil ranges from 1.22 to 1.76 (g/cm^3), with an average of 1.57 (g/cm^3). The void ratio of the liquefied soil varies between 0.516 and 0.889, with an average of 0.67, while the void ratio of the non-liquefied soil ranges from 0.52 to 0.976, with an average of 0.709. The specific gravity of the soil ranges from 2.67 to 2.72 (g/cm^3), with an average of 2.69 (g/cm^3). For each group of experiments, after the soil tank was placed in the wave flume, water was added immediately to a water depth of 0.25 m. In order to avoid disturbing the sediments during the addition of water, we extended the pipe into the bottom of the flume, added water away from the soil tank, and lowered the speed of the water flow when the water surface was gradually elevated close to the surface of the sediment. After adding water, the sediments were allowed to stand underwater for drainage and consolidation.

We use "rounds" to denote the major experimental categories based on different initial soil conditions (liquefied/non-liquefied), comprising six rounds in total. "Groups" are used to distinguish the different wave conditions within each round (four groups per round, totaling 24 groups). This distinction clearly reflects the two variable levels in the experimental design: initial state (rounds) and wave conditions (groups). The seabed for each round was completely remolded. In rounds I, II, and III of the experiments, the initial consolidation time of the soil was 0.25, 0.33, and 0.5 h. Due to the short consolidation time, the initial EPP of the three rounds was not 0 before wave loading, i.e., there was an initial liquefaction degree. With the increase in consolidation time, the initial liquefaction degree at 10.7 cm gradually decreased to, respectively, $f_d' = 0.68$, $f_d' = 0.60$, $f_d' = 0.34$.

The initial consolidation times of experiment rounds IV, V, and VI were 3 h, 12 h, and 23 h, respectively, and the three rounds of seabed soils were in the state of consolidation, i.e., the initial EPP was 0, and the degree of consolidation increased with the consolidation time. Wave loads were applied to the seabeds of the six rounds (I–IV) in four stalls, and the duration of each stall was 600 s.

A total of 24 groups of experiments were conducted, and the thickness of the consolidated sediments in each experiment was reduced from 16 cm at the beginning to 14 cm, which simulated the process of rapid consolidation of the sediments in the estuary of the Yellow River. After each wave loading, each seabed was rested for a certain period of time to ensure that the seabeds maintained the same initial conditions (i.e., the same initial pore water pressure) during each wave loading. The experimental parameters are summarized in Table 2.

Table 2. Experimental wave and seabed conditions.

Experiment Number	H (m)	λ (m)	d (m)	λ/d	T (h)	$f_d'(D = 10.7$ cm)	Initial Seabed State Before Wave Loading
I-1	0.035	0.454	0.25	1.816			Liquefied
I-2	0.037	0.435	0.25	1.74	0.25 h	$f_d' = 0.68$	Liquefied
I-3	0.044	0.305	0.25	1.22			Liquefied
I-4	0.046	0.275	0.25	1.1			Liquefied
II-1	0.035	0.454	0.25	1.816			Liquefied
II-2	0.037	0.435	0.25	1.74	0.33 h	$f_d' = 0.60$	Liquefied
II-3	0.044	0.305	0.25	1.22			Liquefied
II-4	0.046	0.275	0.25	1.1			Liquefied
III-1	0.035	0.454	0.25	1.816			Liquefied
III-2	0.037	0.435	0.25	1.74	0.5 h	$f_d' = 0.34$	Liquefied
III-3	0.044	0.305	0.25	1.22			Liquefied

Table 2. *Cont.*

Experiment Number	H (m)	λ (m)	d (m)	λ/d	T (h)	$f_d{}'(D = 10.7$ cm)	Initial Seabed State Before Wave Loading
III-4	0.046	0.275	0.25	1.1			Liquefied
IV-1	0.035	0.454	0.25	1.816	3 h	/	Non-Liquefied
IV-2	0.037	0.435	0.25	1.74			Non-Liquefied
IV-3	0.044	0.305	0.25	1.22			Non-Liquefied
IV-4	0.046	0.275	0.25	1.1			Non-Liquefied
V-1	0.035	0.454	0.25	1.816	12 h	/	Non-Liquefied
V-2	0.037	0.435	0.25	1.74			Non-Liquefied
V-3	0.044	0.305	0.25	1.22			Non-Liquefied
V-4	0.046	0.275	0.25	1.1			Non-Liquefied
VI-1	0.035	0.454	0.25	1.816	23 h	/	Non-Liquefied
VI-2	0.037	0.435	0.25	1.74			Non-Liquefied
VI-3	0.044	0.305	0.25	1.22			Non-Liquefied
VI-4	0.046	0.275	0.25	1.1			Non-Liquefied

Note: H is the wave height, λ is the wavelength, d is the water depth, T is the seabed consolidation time, D is the depth of the soil, and $f_d{}'$ is the initial liquefaction degree. In oceanography, waves can usually be decomposed into different types of components. The seabed response to different types of wave action varies. Waves can be classified according to the relative wavelength perspective, such as the ratio of wavelength (λ) to water depth (d), which is used to measure the propagation characteristics of waves. In this paper, λ/d is used to measure the effect of wavelength change on the pore water pressure response.

3. Results and Analysis

Figure 4 and Figure 7 show the measured EPP time series in all the 24 groups of experiments; it can be seen that during the initial stage of wave action, the EPP rapidly accumulates at all depths under any wavelengths. The cumulative EPP magnitude increases with depth, which is consistent with the conclusion by Clukey et al. [12] that the maximum EPP occurs at a certain depth within the seabed rather than at the sediment surface.

The initial EPP of the seabed in rounds I, II, and III of the experiments was not 0 before wave loading, and the soil performed circular motions when exposed to wave action. Therefore, we believe that the soils in the I, II, and III rounds were in a liquified state before wave loading; under wave action, the EPP further accumulated, resulting in an increased degree of liquefaction.

However, in rounds IV, V, and VI of the experiments, the initial EPP before wave loading was 0. Although the EPP also accumulated under wave action, it remained lower than the initial effective stress of the seabed. Therefore, no liquefaction occurred in rounds IV, V, and VI of the experiments.

In this section, we found that the cumulative magnitude of EPP in the silty seabed of the Yellow River Delta under wave action does not monotonically increase with wavelength. The EPP response to relative wavelength differs for initially liquefied or non-liquefied seabed. Therefore, this section discusses the EPP response of liquefied (Section 3.1) and non-liquefied (Section 3.2) soils under different relative wavelengths, separately.

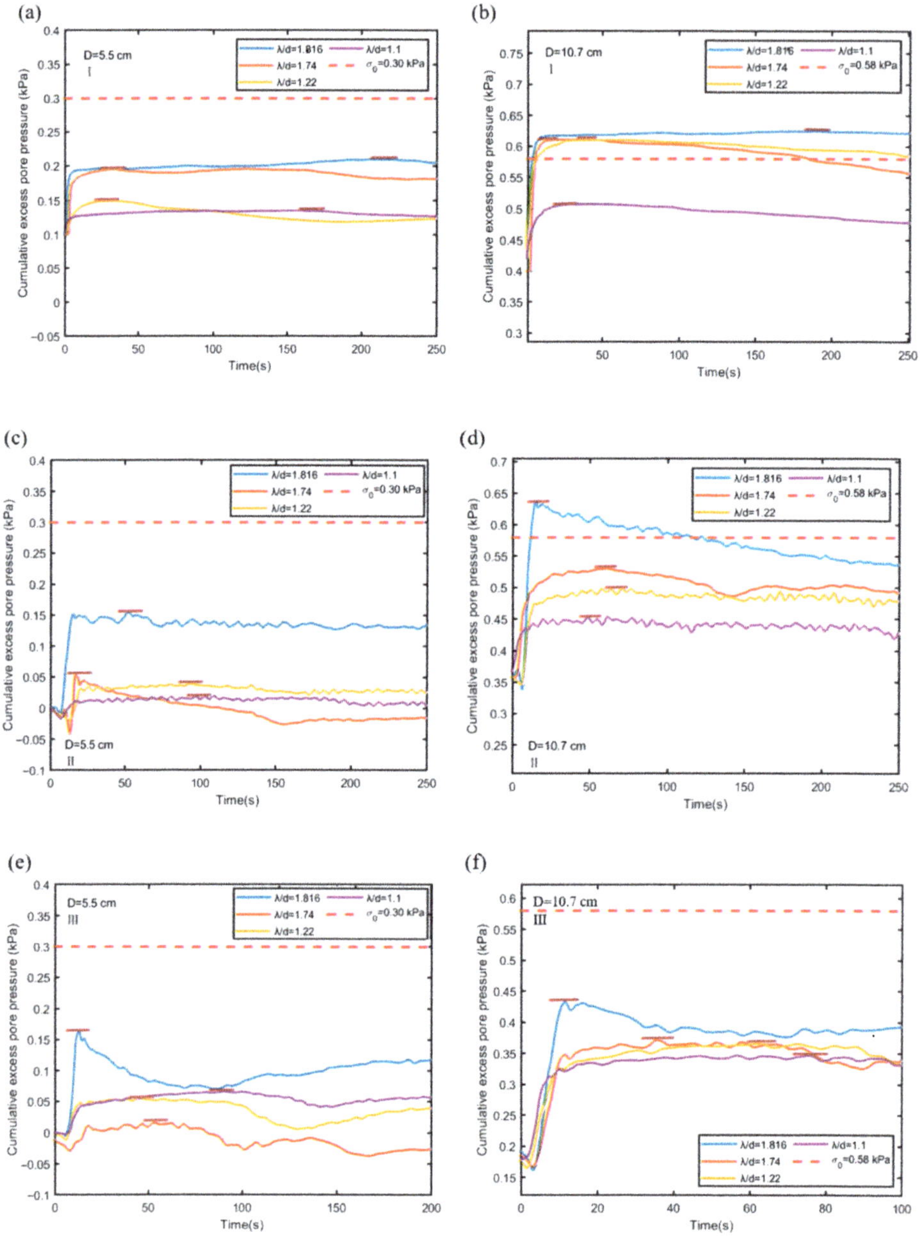

Figure 4. Time series of cumulative EPP at different depths in an initially liquefied soil under different relative wavelengths in I, II, and III rounds of experiments. (**a**,**c**,**e**) represent the cumulative EPP at D = 5.5 cm in rounds I, II, and III of the experiments. (**b**,**d**,**f**) represent the cumulative EPP at D = 10.7 cm in rounds I, II, and III of the experiments. Note: The red dashed line represents the initial average vertical effective stress of the seabed σ_0; the blue line corresponds to a relative wavelength (λ/d) = 1.816, the red line (λ/d) = 1.74, the yellow line (λ/d) = 1.22, and the purple line (λ/d) = 1.11, D represents the depth of the soil, and $f_d{}'$ is the initial liquefaction degree of the seabed. The short, horizontal, dark red line represents the cumulative magnitude value of EPP.

3.1. Cumulative EPP Response Under Different Wavelengths for Initially Liquefied Seabed

In this section, the cumulative EPP response to different relative wavelengths in initially liquefied seabed (rounds I, II, and III) is analyzed.

The measured time series of cumulative EPP in rounds I, II, and III of the experiments is shown in Figure 4 (for initially liquefied soil under different relative wavelengths). It can be seen that the initial seabed was in a liquified state. Specifically, due to the short consolidation time of the initial seabed, the initial EPP at 10.7 cm of the seabed was not 0 when wave loading started (the three panels on the right in Figure 4), and the soil moved in a circular manner under wave action. Therefore, the initial seabed was in a liquefied state in rounds I, II, and III of the experiments.

When the initial liquefaction degree is $f_d{}' = 0.68$ (Round I), under a wave action of $\lambda/d = 1.816$, 1.74, and 1.22 (Figure 4b), as well as when the initial liquefaction degree is $f_d{}' = 0.60$ (Round II), under a wave action of $\lambda/d = 1.816$ (Figure 4d), the cumulative EPP at 10.7 cm is greater than the initial effective stress of the seabed (Figure 4b,d). Therefore, at this time, the soil at the depth of the seabed is in a state of complete liquefaction. However, the EPP at $D = 5.5$ cm in the seabed is far from reaching the initial effective stress of the soil. The EPP in the deeper region of the seabed is relatively high before wave loading. This is because, during the natural sedimentation and consolidation process, the water content of the sediment is higher in the deeper regions of the seabed, which is consistent with the water content measurements obtained from the cylindrical samples collected in the Yellow River Delta. Under wave action, the difference in soil permeability leads to easier accumulation of pore water pressure at a certain depth within the seabed. In contrast, the pore water pressure in the shallow sediment is less likely to accumulate due to the relatively loose soil structure, and it dissipates more quickly.

When the initial liquefaction degree is $f_d{}' = 0.60$, especially for $\lambda/d = 1.816$, the EPP at $D = 10.7$ cm reaches the peak and then dissipates rapidly (Figure 4d), while the EPP of $D = 5.5$ cm reaches the peak and then dissipates slowly (Figure 4c), which may be due to the upward dissipation of EPP in the deeper part ($D = 10.7$ cm), which offsets the EPP of the shallow soil ($D = 5.5$ cm). This opinion is partly supported by the phenomenon that the fine-grained material moves up to the surface with the formation of seepage channels after the seabed is liquefied (Figure 5).

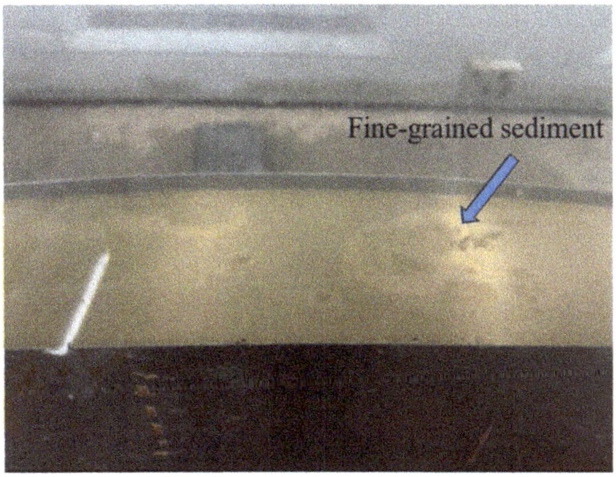

Figure 5. Experimental phenomenon on the seabed surface after soil liquefaction.

The peak value of the EPP was read from Figure 4 (see the short, horizontal, dark red lines in each sub-panel) and plotted in Figure 6. As shown in Figure 6b (data read from Figure 4), for the 10.7 cm depth, the EPP magnitude decreased with λ/d, indicating that EPP is most likely to be accumulated when the relative wavelength is the largest. At a depth of 5.5 cm, the EPP response in experiments I and II also conformed to the aforementioned law for D = 10.7 cm (Figure 6a). However, in three rounds of experiments, the response of the EPP did not conform to the aforementioned pattern (Figure 6a, red dashed frame), this is attributed to the phenomenon of "negative EPP".

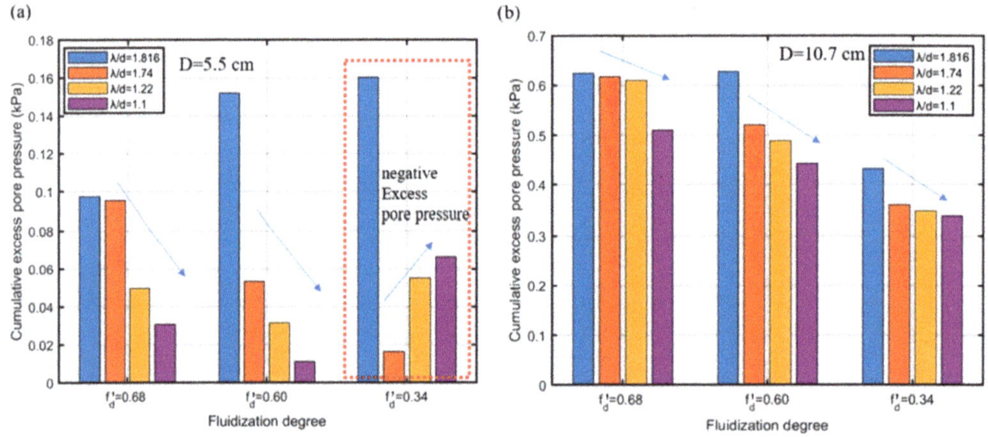

Figure 6. The bar charts of cumulative EPP magnitude in rounds I–III of the experiments (for initially liquefied seabed) under different relative wavelengths at different depths. (**a**) is the cumulative EPP magnitude at D = 5.5 cm in rounds I–III of the experiments (**b**) is the cumulative EPP magnitude at D = 10.7 cm in rounds I–III of the experiments Note: $f_d{}'$ is the initial liquefaction degree. The red dashed frame is affected by the "negative EPP" phenomenon.

In summary, the general trend in these three rounds of experiments (I, II, and III for initially liquefied seabed) is that when the seabed is in a liquified state, the cumulative magnitude of the EPP is proportional to the relative wavelength (λ/d) (Figure 6). This response law for liquefied seabed may be attributed to the fact that (1) The soil skeleton of a liquefied seabed is looser and has bigger pores for compression; (2) as the λ/d increases, the wave period increases, i.e., the time that the soil skeleton is compressed by waves increases (as a longer wave passes by the seabed), allowing EPP to continuously accumulate during the compression of the soil skeleton.

3.2. Cumulative EPP Response Under Different Wavelengths for Initially Non-Liquefied Seabed

In this section, the cumulative EPP response under different relative wavelengths when the seabed is in an initially non-liquefied state (rounds IV, V, and VI of the experiments) is analyzed as follows:

The peak value of EPP was read from Figure 7 (see the short, horizontal, dark red lines in each sub-panel) and plotted in Figure 8. As shown in Figure 8 (data read from Figure 7), non-liquefied silty seabed is not same as liquefied silty seabed. Instead, it adheres to a non-monotonic law, the cumulative EPP magnitude is related to the consolidation time, i.e., seabed in different consolidation times has its maximum (optimal) response wavelength (specifically to the relative wavelength (λ/d) at which the maximum cumulative EPP magnitude), and the optimal wavelength decreases with consolidation time (Figure 8).

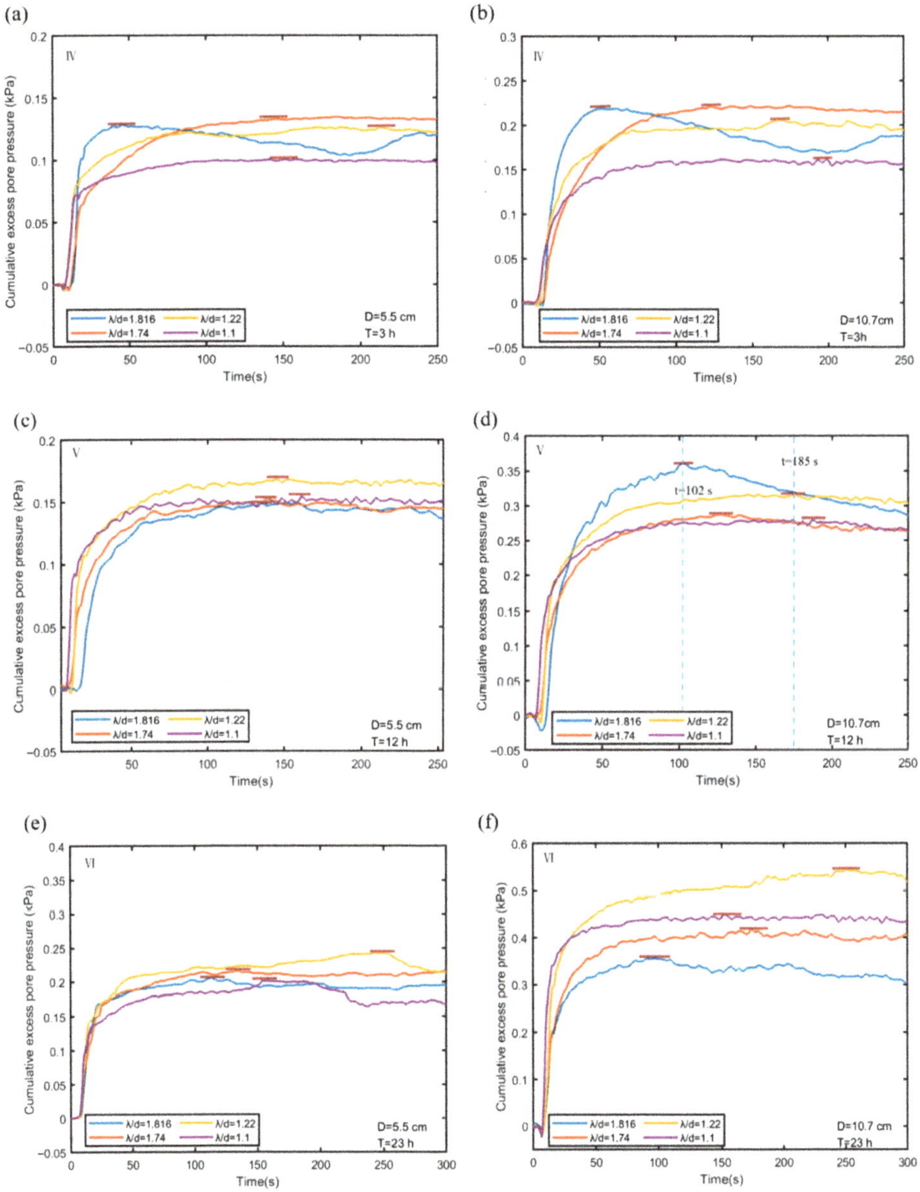

Figure 7. Time series of cumulative EPP at different depths of non-liquefied soil under different relative wavelengths in rounds IV, V, and VI of the experiments. (**a**,**c**,**e**) represent the cumulative EPP at D = 5.5 cm in rounds IV, V, and VI of the experiments. (**b**,**d**,**f**) represent the cumulative EPP at D = 10.7 cm in rounds IV, V, and VI of the experiments. Note: the blue line corresponds to the relative wavelength (λ/d) = 1.816, the red line (λ/d) = 1.74, the yellow line (λ/d) = 1.22, and the purple line (λ/d) = 1.11. D represent the depth of the soil, and T is the consolidation time. The short, horizontal, dark red line represents the cumulative magnitude value of EPP.

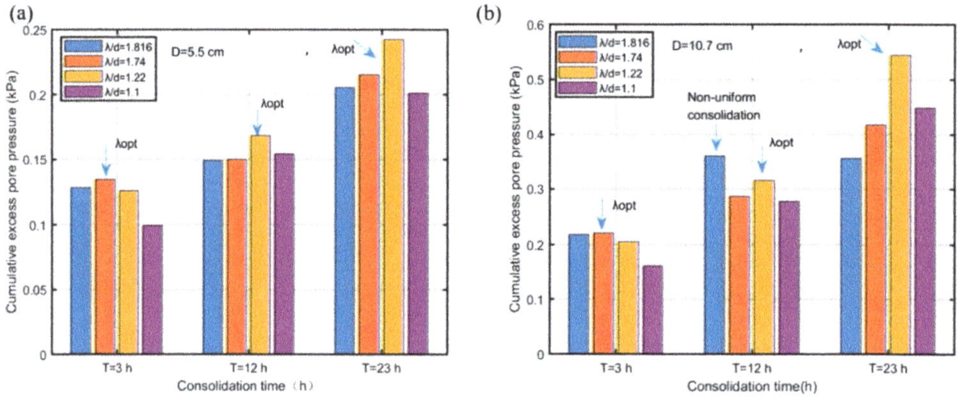

Figure 8. The bar charts of cumulative magnitude of EPP at different depths of IV–VI rounds of experiments under different relative wavelengths. (**a**) is the cumulative magnitude of EPP at D = 5.5 cm in rounds IV, V, and VI of the experiments. (**b**) is the cumulative magnitude of EPP at D = 10.7 cm in rounds IV, V, and VI of the experiments. Note: T is the consolidation time.

As can be seen from Figure 8, when the consolidation time is 3 h (round IV), the cumulative magnitude of EPP is largest under λ/d = 1.74 at both depths (this can also be found in Figure 7a,b).

When the consolidation time is 12 h (round V), the EPP magnitude peaks under λ/d = 1.22 at a depth of 5.5 cm (Figure 8a). It is noteworthy that at 10.7 cm, the cumulative magnitude of EPP peaks at λ/d = 1.816 (Figure 8b) when t $\leq$ 185 s and reaches a peak at t = 102 s (Figure 7d) and then decreases gradually thereafter. When t > 185 s, the cumulative magnitude of EPP under λ/d = 1.22 is still the largest (Figure 7d). This may be related to the non-uniform consolidation of the seabed in this round.

When the consolidation time is 23 h (round VI), under λ/d = 1.22, the EPP magnitude is the largest at both depths (Figure 8 or Figure 7e,f).

It can be concluded that the maximum response wavelength of the non-liquefied seabed is related to the consolidation time of the seabed. The difference in the cumulative magnitude of EPP at different relative wavelengths increases with the greater depth of the seabed (Figure 7, left 3 panels compared to the right 3 panels).

4. Discussion

This section further contrasts and compares the findings with the existing literature on the following aspects: the influence of relative wavelength on the cumulative EPP magnitude in liquefied seabed of different initial liquefaction degrees and in non-liquefied seabed with different consolidation times. Then, the influence of consolidation degrees on the cumulative EPP magnitude under the same relative wavelength is discussed.

4.1. Effect of Relative Wavelength on the Cumulative Magnitude of EPP

This section further compares the cumulative magnitude of EPP for liquefied seabed with different degrees of liquefaction and non-liquefied seabed with different consolidation times under the action of different relative wavelengths. The specific analysis results are shown in Tables 3 and 4.

Table 3. Cumulative magnitude of EPP in liquefied seabed.

f_d' \ λ/d	Depth (cm)	1.816	1.74	1.22	1.1
		Cumulative Magnitude of Excess Pore Pressure (kPa)			
f_d' = 0.68	5.5	0.1974	0.1953	0.1497	0.1308
	10.7	0.6240	0.6168	0.6102	0.5100
f_d' = 0.60	5.5	0.1520	0.0533	0.0314	0.0111
	10.7	0.6274	0.5205	0.4891	0.4430
f_d' = 0.34	5.5	0.1603	0.0163	0.0553	0.0664
	10.7	0.4324	0.3596	0.3468	0.3386

Note: f_d' is the initial liquefaction degree, λ/d is the relative wavelength, the vertical red frame indicates the maximum value of the cumulative magnitude of EPP, the horizontal red frame indicates outliers caused by the influence of "negative EPP".

Table 4. Cumulative magnitude of EPP in non-liquefied seabed.

T (h) \ λ/d	Depth (cm)	1.816	1.74	1.22	1.1
		Cumulative Magnitude of Excess Pore Pressure (kPa)			
T = 3	5.5	0.1284	0.1346	0.1259	0.0994
	10.7	0.2188	0.2209	0.2057	0.1614
T = 12	5.5	0.1492	0.1498	0.1684	0.1542
	10.7	0.3605	0.2870	0.3158	0.2793
T − 23	5.5	0.2056	0.2154	0.2424	0.2012
	10.7	0.3559	0.4177	0.5444	0.4488

Note: *T* is the seabed consolidation time, λ/d is the relative wavelength, the red frames indicates the maximum value of the cumulative magnitude of EPP, the blue frames indicates outliers caused by the influence of non-uniform consolidation.

As shown in Table 3, when the seabed is in a liquefied state, all the maximum EPP magnitudes occur under the maximum wavelength λ/d = 1.816, indicating that the optimal response wavelength is λ/d =1.816. In other words, when the seabed is in a liquefied state, consolidation time has no influence on the optimal wavelength. Additionally, the cumulative EPP magnitude is directly proportional to the relative wavelength (Figure 7). This is consistent with previous studies stating that long-period waves are more likely to cause pore water pressure accumulation [15–22].

As shown in Table 3, the influence of wavelength increases monotonically for all three liquefaction degrees, because the longer the wave is, the longer a wave action period is, which benefits the EPP accumulation.

It also can be seen from Table 3 that long waves are more likely to accumulate EPP in loose soils with a higher liquefaction degree (vertical red frame in Table 3), because when the seabed was in a higher liquefied state, the soil skeleton was looser due to the short consolidation time. The seabed had more space for compression, which is conducive to the accumulation of pore pressure.

It is worth noting that at a depth of 5.5 cm, when f_d' = 0.34, the response of EPP does not follow the rule above (horizontal red frame in Table 3), which is due to the influence of "negative EPP"

For non-liquefied seabed, the conclusion is different. The cumulative magnitude of EPP in non-liquefied seabed does not increase monotonically with relative wavelength but has a maximum response wavelength. This maximum response wavelength decreases with increasing consolidation time. Detailed analysis is as follows:

As shown in Table 4, when the consolidation time is 3 h, the cumulative magnitude of EPP reaches a maximum of 0.1346 kPa at D = 5.5 cm and a maximum of 0.2209 kPa at D = 10.7 cm under a wave action of λ/d = 1.74, indicating that the maximum response wavelength is λ/d = 1.74.

When the consolidation time is 12 h under a wave action of λ/d = 1.22, the cumulative magnitude of EPP reaches a maximum of 0.1684 kPa at D = 5.5 cm. However, it is noteworthy that at a depth of 10.7 cm, as shown in Figure 7d, when t $\leq$ 185 s, the cumulative magnitude of EPP reaches its peak under a wave action of λ/d = 1.816, peaking at t = 102 s and then gradually dissipating. This may be related to the non-uniform consolidation of the seabed (blue frames in Table 4). However, when t > 185 s, the cumulative magnitude of EPP is still highest under a wave action of λ/d = 1.22. Therefore, we consider this to be an occasional phenomenon, and when the consolidation time is 12 h, the maximum response wavelength is λ/d = 1.22.

When the seabed consolidation time is 23 h, the cumulative magnitude of EPP reaches a maximum of 0.2424 kPa at D = 5.5 cm and a maximum of 0.5444 kPa at D = 10.7 cm under a wave action with λ/d = 1.22, indicating that the maximum response wavelength is λ/d = 1.22.

The maximum response wavelengths for seabed consolidation times of 12 and 23 h are both λ/d = 1.22, which may be due to the limited consolidation times in the present experiments. It is uncertain whether the maximum response wavelength would continue to decrease with longer consolidation time.

In summary, when the seabed is in a non-liquefied state, there exists a maximum/optimal response wavelength for the EPP response in seabed of different consolidation times. The optimal wavelength decreases with increasing consolidation time (red frames in Table 4).

Okusa [15] suggested that the damping of long-period waves propagating through the seabed is smaller than that of short-period waves, and therefore long-period waves are more likely to contribute to the accumulation of pore water pressure. Xu et al. [22] also suggested that the short-wave component, although more energetic, dissipates energy more rapidly during its propagation through the seabed, and that the long-period component contributes more to the pore water pressure with the increase in seabed depth. In the present study, we found that the dissipation of wave energy is not only related to the wavelength, but also to the consolidation degree of the seabed. In this study, at the lowest consolidation degree of the seabed, greater wavelength lead to easier buildup of pore water pressure, which is consistent with the traditional conclusion. As wave pressure is transmitted to the seabed, more energy is consumed with the increase in the consolidation degree. Therefore, as consolidation degree increases, shorter relative wavelengths and higher frequency waves are more conducive to EPP accumulation.

4.2. *Effect of Consolidation Degree on the Cumulative Magnitude of EPP*

This section further analyzes the effect of consolidation degree on the cumulative magnitude of EPP under the action of the same relative wavelength.

As shown in Figure 9, as the consolidation time of the seabed increases, the degree of consolidation also increases accordingly. When the seabed is in a liquefied state, under the action of the same relative wavelength, the cumulative magnitude of EPP declines as the seabed consolidation degree increases. This may be attributed to the fact that when the consolidation degree is low, the liquefaction degree is high, resulting in a loose soil structure and a large pore ratio. As the consolidation degree increases and the liquefaction degree decreases, the soil gradually becomes more compact, the pore ratio decreases, and the pore water entering the soil structure is reduced accordingly.

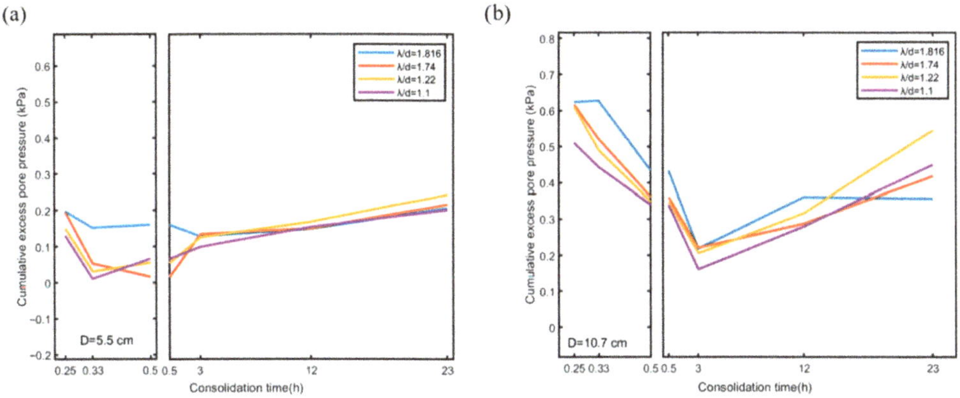

Figure 9. "$\surd$" shaped variation curves of cumulative magnitude of EPP with consolidation time derived from all the 24 groups of experiments. (**a**) represents the curve for EPP in liquefied seabed, and (**b**) represents the curve for EPP in non-liquefied seabed.

When the seabed is in a non-liquefied state, the cumulative magnitude of EPP at the same relative wavelength increases with consolidation time. This indicates that when the seabed is in a non-liquefied state, the higher the consolidation degree, the easier it is to accumulate EPP. This is due to the fact that the longer the consolidation time, the lower the permeability; thus, the pore water pressure can not be released in time, and it is easier to accumulate EPP, which is consistent with the traditional conclusion [25,34].

In summary, under the same relative wavelength, the relationship between the cumulative magnitude of the EPP and the consolidation time is a "$\surd$" curve which decreases first and then increases (Figure 9).

4.3. Limitation and Future Work

There are a few limitations in the present work. For example, the wave flume in our experiments is small, even though it makes multiple comparative experiments more convenient and controllable. Initial conditions may be more controllable, but this introduces many other negative aspects such as scale mismatching of wavelengths versus grain sizes (directly taken from the full-scale region), and constraining effects due to the narrowness of flume.

This study focuses on the Yellow River Delta area, where both waves and currents coexist in the actual marine environment. However, there are limitations in the functionality and dimensions of the experimental flume, which cannot simulate the complex wave–current coupling dynamics found in actual estuarine environments. Currents and wave–current interactions significantly influence wave conditions and the distribution of seabed surface pressures. Ye and Jeng [35] investigated soil response under combined wave–current action using numerical simulations based on Biot's poroelastic theory. Their results demonstrated that the difference in pore water pressure within the seabed between conditions with and without currents could reach up to 25%. Wen et al. [36] established a three-dimensional numerical model for pore-pressure response under combined short-crested waves and currents. The numerical results indicated that superimposing a following-current will result in larger pore pressure in the seabed. Qi et al. [37] investigated the response of excess pore water pressure in a sandy seabed under combined wave–current action through flume experiments. When the current flows in the same direction as the waves, the pore water pressure within the seabed increases; conversely, it decreases when the current opposes the wave direction. Experimental observations indicate that the presence of currents will alter wave conditions, such as wave height and wavelength. As the following current velocity increases, the wave height decreases while the wavelength significantly extends; conversely, as the opposing current velocity increases, the wave height increases and the wavelength shortens, indicating that opposing currents induce wave steepening [37]. Therefore, ignoring the effects of currents will lead to an underestimation of wave-induced seabed instability. Additionally, all simulated wave conditions fall within deep-water waves, lacking systematic investigation of intermediate-water waves and shallow-water waves. In subsequent research, we plan to conduct broader operational experiments in larger-scale flume to further validate and extend the principles derived from this study.

5. Conclusions

In the present study, the cumulative excess pore water pressure (EPP) response to various relative wavelengths and consolidation degrees in initially liquefied or non-liquefied seabed is investigated through 24 groups of controlled variable wave flume experiments. Main findings can be summarized as follows:

(1) The cumulative magnitude of EPP in the silty seabed of the Yellow River delta does not monotonically increase with relative wavelength. The effect of wavelength varies between liquefied and non-liquefied soils. In liquefied seabed, EPP monotonically increases with wavelength, while in non-liquefied seabed, EPP response has an optimal wavelength. The optimal wavelength is inversely related to the consolidation degree.

(2) The effect of the consolidation degree on the cumulative magnitude of EPP shows two different trends in liquefied and non-liquefied seabed. The relationship follows a "$\surd$" shaped curve, initially decreasing and finally increasing. Furthermore, when the seabed is in a non-liquefied state, a smaller relative wavelength facilitates greater accumulation of excess pore water pressure.

Author Contributions: Conceptualization, Y.Z. and S.Z.; methodology, Y.Z., M.W., and Z.Z.; formal analysis, H.L.; investigation, H.L. and A.M.; resources, S.Z.; writing—original draft preparation, H.L.; writing—review and editing, Y.Z., M.W., and S.Z.; project administration, S.Z. All authors have read and agreed to the published version of the manuscript.

Funding: This work was supported by the Natural Science Foundation of China [grant number 42276215], National Key Research and Development Program, Intergovernmental International Science and Technology Innovation Cooperation, China-Nigeria Estuary Delta Joint Laboratory [grant number 2024YFE0116400], and the Fundamental Research Funds for the Central Universities [grant number 202441015].

Data Availability Statement: The raw data supporting the conclusions of this article will be made available by the authors on request.

Conflicts of Interest: The authors declare no conflicts of interest.

References

1. Zen, K.; Umehara, Y.; Finn, W.D.L. A case study of the wave-induced liquefaction of sand layers under damaged breakwater. In Proceedings of the 3rd Canadian Conference on Marine Geotechnical Engineering, St. John's, NL, Canada, 19–21 June 1985; pp. 505–520.
2. Christian, J.; Taylor, P.; Yen, J.; Erali, D. Large diameter underwater pipeline for nuclear power plant designed against soil liquefaction. In Proceedings of the 6th Offshore Technology Conference, Houston, TX, USA, 6–8 May 1974; pp. 597–606.
3. Puzrin, A.M.; Alonso, E.E.; Pinyol, N.M. Caisson failure induced by liquefaction: Barcelona Harbour, Spain. In *Geomechanics of Failures*; Springer: Dordrecht, The Netherlands, 2010; pp. 85–148.

4. Xu, G.; Sun, Y.; Wang, X.; Hu, G.; Song, Y. Wave-induced shallow slides and their features on the subaqueous Yellow River delta. *Can. Geotech. J.* **2009**, *46*, 1406–1417. [CrossRef]
5. Li, M.; Chen, D.; Wu, H.; Tang, J.; Zhang, Y.; Luo, F.; Gou, F.; Gong, X.; Wang, Y. In-situ observations of wave-and current-supported fluid mud dynamics on a hyperturbid macrotidal mudflat. *Front. Mar. Sci.* **2024**, *11*, 1459899. [CrossRef]
6. Terzaghi, K. The theory of hydrodynamic stresses and its geotechnical applications. In Proceedings of the First International Congress for Applied Mechanics, Delft, The Netherlands, 22–26 April 1924; pp. 288–294.
7. Zen, K.; Yamazaki, H. Oscillatory pore pressure and liquefaction in seabed induced by ocean waves. *Soils Found* **1990**, *30*, 147–161. [CrossRef]
8. Sumer, B.M. *Liquefaction Around Marine Structures*; World Scientific: Hackensack, NJ, USA, 2014.
9. Alcérreca-Huerta, J.C.; Oumeraci, H. Wave-induced pressures in porous bonded revetments. Part I: Pressures on the revetment. *Coast. Eng.* **2016**, *110*, 87–101. [CrossRef]
10. Qi, W.G.; Gao, F.P. Wave induced instantaneously-liquefied soil depth in a non-cohesive seabed. *Ocean Eng.* **2018**, *153*, 412–423. [CrossRef]
11. Tzang, S.Y. Unfluidized soil responses of a silty seabed to monochromatic waves. *Coast. Eng.* **1998**, *35*, 283–301. [CrossRef]
12. Clukey, E.C.; Kulhawy, F.H.; Liu, L.F.; Tate, G.B. The impact of wave loads and pore-water pressure generation on initiation of sediment transport. *Geo-Mar. Lett.* **1985**, *5*, 177–183.
13. Feng, X.L.; Lin, L. The relationship between geotechnical parameters and sedimentary environment of soil layers since Holocene in modern Huanghe subaqueous delta. *Coast. Eng.* **1999**, *18*, 1–7.
14. Ippen, A.T. *Estuary and Coastline Hydrodynamics*; McGraw-Hill: New York, NY, USA, 1966; pp. 505–510.
15. Okusa, S.; Uchida, A. Pore-water pressure change in submarine sediments due to waves. *Mar. Georesour. Geotechnol.* **1980**, *4*, 145–161. [CrossRef]
16. Maeno, Y.; Hasegawa, T. In-situ measurements of wave-induced pore pressure for predicting properties of seabed deposits. *Coast. Eng. J.* **1987**, *30*, 99–115. [CrossRef]
17. Liu, Z.; Jeng, D.S.; Chan, A.H.C.; Luan, M. Wave-induced progressive liquefaction in a poro-elastoplastic seabed: A two-layered model. *Int. J. Numer. Anal. Methods Geomech.* **2010**, *33*, 591–610. [CrossRef]
18. Niu, J.; Xu, J.; Dong, P.; Li, G. Pore water pressure responses in silty sediment bed under random wave action. *Sci. Rep.* **2019**, *9*, 11685.
19. Zhang, Q.; Zhai, H.; Wang, P.; Wang, S.; Duan, L.; Chen, L.; Liu, Y.; Jeng, D.S. Experimental study on irregular wave-induced pore-water pressures in a porous seabed around a mono-pile. *Appl. Ocean. Res.* **2020**, *95*, 102041. [CrossRef]
20. Klammler, H.; Penko, A.M.; Staples, T.; Sheremet, A.; Calantoni, J. Observations and modeling of wave-induced burial and sediment entrainment: Likely importance of degree of liquefaction. *J. Geophys. Res. Ocean.* **2021**, *126*, e2021JC017378. [CrossRef]
21. Xu, X.; Xu, G.; Yang, J.; Xu, Z.; Ren, Y. Field observation of the wave-induced pore pressure response in a silty soil seabed. *Geo-Mar. Lett.* **2021**, *41*, 13. [CrossRef]
22. Xu, J.; Dong, J.; Zhang, S.; Sun, H.; Li, G.; Niu, J.; Dong, P. Pore-water pressure response of a silty seabed to random wave action: Importance of low-frequency waves. *Coast. Eng.* **2022**, *178*, 104214. [CrossRef]
23. Liu, X.; Jia, Y.; Zheng, J.; Yang, X.; Shan, H. Consolidation of sediments discharged from the Yellow River: Implications for sediment erodibility. *Ocean Dyn.* **2013**, *63*, 371–384. [CrossRef]
24. Ren, Y.; Xu, G.; Xu, X.; Zhao, T.; Wang, X. The initial wave induced failure of silty seabed: Liquefaction or shear failure. *Ocean. Eng.* **2020**, *200*, 106990. [CrossRef]
25. Chen, Z.; Ren, Y.; Xu, G.; Li, M. Effect of particle composition and consolidation degree on the wave-induced liquefaction of soil beds. *Acta Oceanol. Sin.* **2024**, *43*, 11–22. [CrossRef]
26. Sumer, B.M.; Hatipoglu, F.; Fredsøe, J.; Sumer, S.K. The sequence of sediment behaviour during wave-induced liquefaction. *Sedimentology* **2006**, *53*, 611–629. [CrossRef]
27. Okusa, S. Wave-induced stresses in unsaturated submarine sediments. *Geotechnique* **1985**, *32*, 517–532. [CrossRef]
28. Tsai, C.P. Wave-induced liquefaction potential in a porous seabed in front of a breakwater. *Ocean Eng.* **1995**, *22*, 1–18. [CrossRef]
29. Jeng, D.S.; Seymour, B.R. Wave-induced pore pressure and effective stresses in a porous seabed with variable permeability. Offshore Mech. *Arct. Eng.* **1997**, *119*, 226–233.
30. Jia, Y.; Zhang, L.; Zheng, J.; Liu, X.; Jeng, D.S.; Shan, H. Effects of wave-induced seabed liquefaction on sediment re-suspension in the yellow river delta. *Ocean Eng.* **2014**, *89*, 146–156. [CrossRef]
31. Wen, M.; Jia, Y.; Wang, Z.; Zhang, S.; Shan, H. Wave flume experiments on dynamics of the bottom boundary layer in silty seabed. *Acta Oceanol. Sin.* **2020**, *39*, 96–104. [CrossRef]
32. Dong, J.; Xu, J.; Li, G.; Li, A.; Zhang, S.; Niu, J.; Xu, X.; Wu, L. Experimental study on silty seabed liquefaction and its impact on sediment resuspension by random waves. *Mar. Sci. Eng.* **2022**, *10*, 437. [CrossRef]

33. Song, Y.P.; Sun, Y.F.; Du, X.; Dong, L.F.; Jiao, P.F. Comparative study on the difference of engineering geological char-acteristics between liquefied silt and non-liquefied silt in the Yellow River Delta. *Adv. Mar. Sci.* **2019**, *37*, 55–64.
34. Li, X.; Gao, F.; Yang, B.; Zang, J. Wave-induced Pore Pressure Responses and Soil Liquefaction Around Pile Foundation. *Int. J. Offshore Polar Eng.* **2011**, *21*, 233–239.
35. Ye, J.H.; Jeng, D.-S. Response of porous seabed to nature loadings: Waves and currents. *J. Eng. Mech. ASCE* **2012**, *138*, 601–613. [CrossRef]
36. Wen, F.; Wang, J.H.; Zhou, X.L. Response of saturated porous seabed under combined short-crested waves and current loading. *J. Coast Res.* **2016**, *32*, 286–300.
37. Qi, W.G.; Li, C.F.; Jeng, D.S.; Gao, F.P.; Liang, Z. Combined wave-current induced excess pore-pressure in a sandy seabed: Flume observations and comparisons with theoretical models. *Coast. Eng.* **2019**, *147*, 89–98. [CrossRef]

MDPI AG
Grosspeteranlage 5
4052 Basel
Switzerland
Tel.: +41 61 683 77 34

Water Editorial Office
E-mail: water@mdpi.com
www.mdpi.com/journal/water

www.ingramcontent.com/pod-product-compliance
Lightning Source LLC
Chambersburg PA
CBHW061546210926
51850CB00009B/2019